BRASSEY'S
NEW TITLES 1988-89

LIDDELL HART AND THE WEIGHT OF HISTORY

John J Mearsheimer

In the 1930s Liddell Hart acquired a reputation as a brilliant military strategist.... until now. Evidence is uncovered implying that he manipulated facts to create a false impression of his role in the interwar military debates.

0 08 036701 1 (H)   January 1989   **£15.95/US$28.75**

HOSTAGES TO FORTUNE
The Future of Western Interests in the Arabian Gulf

Michael Cunningham

The Arabian Gulf has sixty per cent of the world's known oil reserves and will remain the main source of oil for the next hundred years. How can the West guarantee access to that oil for as long as it is needed?

0 08 036259 1 (H)   November 1988   **£16.95/US$27.00**

SOVIET BREAKOUT: Strategies to Meet It

Joseph Churba

Offers a sobering and straightforward evaluation of Soviet intentions and outlines a feasible plan to thwart them – an important new book from a respected American strategist.

0 08 035981 7 (H)   October 1988   **£9.50/US$16.95**

DRIFTING APART? THE SUPERPOWERS AND THEIR EUROPEAN ALLIES

Edited by **Christopher Coker**

A collection of essays analysing the barriers of misunderstanding within the Alliance and attempting to understand similar dilemmas faced by the Soviet Bloc.

0 08 036711 9 (H)   March 1989   **£22.50/US$40.00** (approx)

THE TECHNOLOGY TRAP

Air Commodore Tim Garden

This book examines the success stories of sea, land and air weapon systems of the past hundred years. It reviews current advances in various scientific fields of research and suggests the most profitable applications of these advances with respect to future weapons systems.

0 08 036710 0 (H)   April 1989   **£18.95/US$34.00** (approx)

Available from bookshops, or by post from:

Brassey's, Headington Hill Hall, Oxford OX3 0BW, UK. Tel: (0865) 64881.
Pergamon-Brassey's, Maxwell House, Fairview Park, Elmsford, NY 10523, USA.
Tel: (914) 592-7700.

Members of Maxwell Pergamon Publishing Corporation plc

RUSI

and

BRASSEY's

Defence
Yearbook
1989

Edited by
The Royal United Services Institute for Defence Studies
London

99th Year of Publication

BRASSEY'S DEFENCE PUBLISHERS
a member of the Maxwell Pergamon Publishing Corporation plc
LONDON · OXFORD · WASHINGTON · NEW YORK
BEIJING · FRANKFURT · SÃO PAULO · SYDNEY · TOKYO · TORONTO

U.K. (Editorial)	Brassey's Defence Publishers Ltd., 24 Gray's Inn Road, London WC1X 8HR
(Orders)	Brassey's Defence Publishers Ltd., Headington Hill Hall, Oxford OX3 0BW, England
U.S.A. (Editorial)	Pergamon-Brassey's International Defense Publishers, Inc., 8000 Westpark Drive, Fourth Floor, McLean, Virginia 22102, U.S.A.
(Orders)	Pergamon Press, Inc., Maxwell House, Fairview Park, Elmsford, New York 10523, U.S.A.
PEOPLE'S REPUBLIC OF CHINA	Pergamon Press, Room 4037, Qianmen Hotel, Beijing, People's Republic of China
FEDERAL REPUBLIC OF GERMANY	Pergamon Press GmbH, Hammerweg 6, D-6242 Kronberg, Federal Republic of Germany
BRAZIL	Pergamon Editora Ltda, Rua Eça de Queiros, 346, CEP 04011, Paraiso, São Paulo, Brazil
AUSTRALIA	Pergamon-Brassey's Defence Publishers Pty Ltd., P.O. Box 544, Potts Point, N.S.W. 2011, Australia
JAPAN	Pergamon Press, 5th Floor, Matsuoka Central Building, 1–7–1 Nishishinjuku, Shinjuku-ku, Tokyo 160, Japan
CANADA	Pergamon Press Canada Ltd., Suite No. 271, 253 College Street, Toronto, Ontario, Canada M5T 1R5

This edition 1989

Library of Congress Catalog Card no. 75–641843

British Library Cataloguing in Publication Data
RUSI and Brassey's defence yearbook.—1989
1. Armed Forces—Periodicals
I. Royal United Services Institute for Defence Studies
355'.005   U1

ISBN 0–08–036698–8

Printed in Great Britain by A. Wheaton & Co. Ltd, Exeter

Contents

# Membership of the RUSI

Since 1831, the Royal United Services Institute for Defence Studies has pursued the aims of its Royal Charter, namely 'the Promotion and Advancement of the Military Sciences and Literature'. In its independent studies the Institute has sought to set the military sciences in their wider context of international security. It continues to do so.

As the professional association of the Armed Services, the RUSI provides a bridge between those interests and disciplines which have a proper role to play in defence and international security. It also aims to assist in creating a wider understanding of these issues and a better informed public debate.

Over the years, the RUSI has studied many aspects of defence and security with vigour and commonsense; these have ranged from strategic issues and operational military problems to applications of the latest defence technology. With its foundation in the military sciences, the RUSI brings a unique perspective and a strong practical viewpoint to current concerns and those of the future. This is evident from its studies programme and its range of publications.

In its work, the RUSI strives for objectivity, balance and clear judgement. In so doing it is fortunate in attracting and being able to draw upon a diverse membership with a wide range of responsibilities, experience and interests across the whole spectrum of defence, security and strategic studies.

Today, the contribution which the Institute makes to the study and discussion of matters affecting international security is of direct relevance and importance. Membership details can be obtained from the RUSI.

ROYAL UNITED SERVICES INSTITUTE FOR DEFENCE STUDIES
Whitehall, London SW1A 2ET
England
Tel: 01-930 5854

Preface

THE Royal United Services Institute is now in its 157th year, and is playing an ever-increasing role in the advancement of the science and literature of the Armed Forces, and in keeping the public informed on political–military affairs. The RUSI maintains a large research establishment, holds lectures and seminars on all aspects of national security, and produces a wide range of publications, including a quarterly Journal and a monthly News Brief. Brassey's Yearbook is now in its 99th year, and has an established reputation as one of the most authoritative sources in the defence field. As the *RUSI/Brassey's Defence Yearbook*, it carries on the tradition of high standards in the breadth and quality of its contents. Such is the reputation which the Yearbook has earned that the Editors can enlist the support of contributors of the highest prestige and qualifications as the contents list so clearly indicates.

Plus Ca Change . . . ?

DAVID BOLTON

Director, RUSI

PEACE IS breaking out almost everywhere! There is a ceasefire in the Iran–Iraq conflict with an end to the war in prospect; the Soviets are withdrawing from Afghanistan; South African forces have pulled back from Angola, and Cuban troops there could return home; even the Vietnamese are leaving Kampuchea. If that were not enough, President Gorbachev has proposed a restructuring of government in the USSR, with elections by secret ballot, and has also pushed the Soviet military into adopting a new defensive doctrine. Eastern Europe, however, is somewhat cautious in its approach to these reforms. In the United States, the Reagan era has ended and the new administration will be reviewing its policies and certainly those affecting its allies; and the arms control process will begin again. Even in the Middle East, the recently elected Israeli government might rethink its stance towards the Palestinians, and potential realignments in the Arab world could further unfold.

All these events suggest significant changes; but how radical, how deep-seated, are those changes likely to be? Are there not harsh underlying realities which could affect the outcome and, if a significant shift does occur, are the wider implications now foreseeable? If the Soviet Union is committed to *perestroika*, will the results bring tangible benefits to the population at large and be sufficient to overcome the resistance of those opposed to restructuring, swelled by those directly and adversely affected: where will the military fit in this equation? If Russia leads will others of the Eastern bloc follow? In the West, if, as seems likely, the US administration brings pressure to bear on its European allies to shoulder more of the common defence burden, will they respond positively to it, ignore it or react against it; yet again, could some NATO nations separately come to believe that arms control is the best means of curtailing defence expenditure, with improved security prospects relegated to second place?

What is certain is that 1989 will be an eventful period for international security affairs. The defence-security community will need to continue arguing its case at the formative level of policies which, in the long-run, could affect fundamentally both the responsibilities it is called upon to undertake and the wider context in which they fall, as well as the means of discharging those responsibilities.

THE SOVIET UNION AND EASTERN EUROPE

On 11 March 1989, Mr Gorbachev will have been in power for four years. Despite the policies of *glasnost* and *perestroika*, the harsh realities of Soviet life remain: food shortages, inadequate housing, substandard medical care, poor transportation, and the like. The problems stem from a system of subsidised prices, backward technology, an unproductive workforce and the absence of true incentives. If the economy of the Soviet Union is to be truly revitalised, Mr Gorbachev has come to realise that political reorganisation is a necessary prerequisite. To that end he has designed a new form of government seeking to balance party control with popular democracy. It is probably the most radical restructuring since the Bolshevik revolution. There will be a new super legislature, called the Congress of People's Deputies, comprising 1,500 elected delegates from districts across the country, along with 750 additional seats allocated to the Party, trade unions and other organisations. The Congress of People's Deputies should hold its first meeting in April 1989. The Congress would then choose a standing legislature of up to 450 members with its own president or chairman. It is expected that Mikhail Gorbachev will be that chairman. He sees the post as having broad policy-making powers at the head of a lively debating parliament, supported by a network of elected People's Councils or Soviets. Accountable to the public, he sees the Soviets as a means of cutting red tape and overcoming the inertia and graft often associated with entrenched party officials.

In attempting to overcome bureaucratic resistance by appealing directly to the public, as well as seeking to limit the terms in office of party leaders to ten years, Gorbachev has sought to generate activity in progressing his economic plans which have yet to produce tangible benefits for the man and woman in the street. Emphasis is, therefore, now being given to satisfying the consumer. There are to be improvements in the self-financing of enterprises, the introduction of new technology and the improvement of quality. There is to be more and better personal clothing, more cars and household appliances, as well as television sets and video recorders. Even the defence industry has been told to switch to the production of fridges and freezers, even if there is yet little evidence of it doing so. The military are also being asked to provide quality control in factories, based on the fact that they were the only "customers" in the past able to reject what the factories produced for them.

These are exciting times in the Soviet Union but many obstacles to change remain. At the very least there is a bureaucratic inertia which the Gorbachev reforms need to overcome, at worst this extends to actually resisting recon- struction. Many party officials, whilst recognising the need for change, believe that past performances should not be decried and that the rate of change need not, in fact should not, be so rapid. Despite the recent reshuffle in the Politburo, so far Mr Gorbachev has failed to fill vacant Central Committee seats with his own nominees and to establish a mandatory retirement age for party officials. He is thus unable to force through his policies: a major reshuffle of the Central Committee now seems unlikely before the next Party Congress in 1991. In seeking to override this impediment to more rapid change by appealing directly to the people, he must promote initiative from below after 70 years of orders from above. The workers, however, remain sceptical of dreams of "jam tomorrow" for they have heard that many times before; they need confidence that the new system will deliver what it promises and that it will not run the risk of being swept away with another reorganisa- tion. Then there is the resistance to be overcome from those who have been accustomed to giving the orders and receiving the perquisites that went with authority and which are now slowly being eroded. The publicising of grievances under *glasnost* has also brought nationality questions and aspirations into the open, not only in the Transcaucasus but in the Baltic states too. If law and order breaks down, as it did in the dispute over the Nagorno-Karabakh Region, and repression is again deemed necessary, will that bring with it the reassertion of central control over local democratisation, perhaps even extending to local enterprises? Mr Gorbachev may yet find that his own tenure in office is circumscribed and his powers constrained, possibly with *perestroika* remaining in little more than name only.

Mr Gorbachev has tackled so many internal problems and started so many initiatives that it is difficult to see how they can all be managed, let alone brought to completion. The same might also be said about his foreign policy, even by his own allies. If there are those who are cautious in the Soviet Union, then there are those who are more conservative in Eastern Europe. Some countries favouring change, such as Hungary, are seeking to discover the limits of what is permissible. Others, such as East Germany, officially do not recognise the need for reconstruction and regeneration, or they pay lip service to it, even though some young East Germans disagree with their older leaders. Poland, Romania and Yugoslavia are particular problem areas.

With his Eastern European allies, Mr Gorbachev appears to believe that stability and limited change is the best policy whilst he concentrates on his problems and difficulties at home. In Poland, after seven years of stagnation and economic failure, from the Soviet viewpoint General Jaruzelski seems to be the best leader available and someone who can be counted on to maintain order. Will 1989 see that view put to the test? 1989 may also see the end of the Ceauşescu regime in Romania but few in the Warsaw Pact, or elsewhere will lament his going. Ceauşescu's resolution to wipe out the nation's debt in one fell swoop has brought deprivation and suffering to his people. Minorities have also especially suffered under his hand. Bulgaria too could see the end of an era with the departure of Zhikov, who has been in power since 1954. The problems of minorities also extend to Yugoslavia, even if it is not formally part of the Eastern bloc. Oppression of Serbs by the Albanian majority in the province of Kosovo has brought calls for more direct intervention from Serbia. When added to the economic problems of 170 per cent inflation with an economy swamped by bureaucracy and the inability of the Republics to agree on implementing common policies, the situation becomes all the more serious. Radical and essential measures to reform the economy will bring further strains, producing a potentially explosive political situation which might need a military strongman to impose those policies on the whole of the country and thereby overcome the Tito legacy of separate republics rather than a unified nation. Despite assertions of respect for national independence within the Soviet bloc, should events in Eastern Europe take a dramatic turn it remains to be seen if the Brezhnev Doctrine of intervention to protect socialism remains in force.

In the sphere of armed forces, Marshal Akhromeev has acknowledged that the Soviet military might be perceived as largely offensive in its structure and doctrine. Whilst arguing that the Soviet forces were "counter-offensive" and designed to rebuff an attack from NATO, he wishes to correct the West's perception. He believes this can be achieved by adopting a "defensive strategy", greater openness in military matters, and through arms control measures. On defence spending, the Soviets themselves are dissatisfied with present methods. The Defence Ministry pays for military forces whilst research, development and procurement are financed by separate civilian ministries. This 'fudging' even confuses the Russians themselves. In future it is intended, therefore, to reorganise defence financing and ensure that it is clear and well understood, both within the Soviet Union and by countries in the West. However, this will take a year or more to achieve.

Apart from statements lauding the advantages for mutual security and stability of forces which adopt a defensive strategy, as yet there is little indication of what precisely this might mean. Claims that force levels should be reduced so that they are incapable of mounting a surprise attack yet able to defend against an aggressor might be a praiseworthy arms control goal, if difficult to comprehend in terms of force structure. Restricted deployment areas might help, but with today's mobile forces constant monitoring would be called for. As the military of both NATO and the Warsaw Pact readily accept, once battle is joined there is no such thing as a purely defensive force; air defence and anti-armour weapons can still have offensive connotations in support of an attacking force. Many arms control questions are also raised by such a statement. What baseline of numbers and weapons systems or balance of opposing forces should be agreed upon as a starting point for such reductions; what verification systems might be appropriate to prevent redeployments; and, in any phased reductions, what confidence-building measures should be agreed upon? In any event, it is doubtful if either side would eschew a counter-attack capability employing all available arms nor, again, deny themselves the right of modernisation. Joint

East–West study groups in the political-military field in the coming year might better clarify what is meant by "defensive defence", and arms control negotiations should test the validity of public political announcements. What is needed is a better mutual understanding of the principles and thoughts which might shape future defence policies: a formal and verifiable exchange of military data would facilitate this whole process.

On a practical level, and despite claims of reducing defence expenditure, there is as yet no evidence of a reduction in the numbers of tanks, artillery pieces, aircraft, and missiles coming into Warsaw Pact service. Whilst it may be argued that the outflow of production lines takes some time to reflect a change in policy, either by a reduction in numbers or by the closing down of particular lines, it would be reasonable to expect some indication of curtailed defence production in 1989, if that really is the professed intention of the Soviet leadership. A more likely outcome is one of reciprocal, if asymmetric, reductions.

The United States of America

A lack of continuity as American administrations change is a well known hazard in dealing with the United States. 1989 will be no exception. Yet many complicated and demanding issues, particularly in the national and international security areas, will confront the new President in 1989. The battle to reduce the budget and trade deficits will continue and this will impose its own constraints on defence spending. In its foreign policies, the administration will need to determine anew where its national interests best lie in Europe, the Middle East, Central America and the Pacific Basin. This, in turn, will determine its military force structures and procurement programmes in both the nuclear and conventional fields. Its goals and negotiating positions for arms control talks will also need to be decided upon. And all this with a new team! No wonder that few positive results can be expected in only a matter of months and on such complex issues.

What is certain is that any review of policy in the USA serves to promote vociferous public debate so that every shade of opinion is exposed to the administration. This year, in the political-military field, the debate will centre on the European allies making a greater contribution to NATO defence costs and, in terms of doctrine, that more emphasis is placed upon conventional forces and their improvement, as opposed to a reliance on the threatened early use of nuclear weapons. Recent legislation in Congress has established US troop ceilings in Europe and has called for the appointment of a special representative to conduct burden-sharing negotiations with NATO member nations. Americans have seen the economic growth of Europe and, when linked with its population of nearly 400 million, believe that there is not an equitable division of defence costs. It is argued that the USA might be a leading member of the Alliance but it is not, nor should it be, its permanent underwriter. As one possible remedy, it is being suggested that an increase of 1 per cent of GDP on the part of the European allies would be a significant additional contribution to defence; the $40 billion it would provide would exceed, for example, the defence budgets of the UK, Germany or France. How Europe responds to this challenge will significantly affect the future relationship of America with Europe, but the difficulties are not confined to Europe alone. Japan too is expected to do much more in securing its own sea lanes and extending support to other nations of the Pacific Basin. Whilst there may be disparities in the way defence costs are spelt out and, not least in the European view, the value which should be attributed to conscript forces and training areas located in or near densely populated European centres of population, the obverse of the argument should not be overlooked. If the Americans are seeking a greater sharing of the defence burden then there might well be a more equitable division of responsibilities within the Alliance as well, and this could serve to limit the leadership role which America has previously sought and accepted; always provided, of course, that the Europeans were prepared to accept such additional responsibilities.

In the field of nuclear weapons, a decision will be required on the next generation of strategic ICBMs. Should the small mobile Midgetman be cancelled and reliance placed on MX missiles based on trains at military installations, or should either programme be continued? Within the Strategic Arms Reduction Treaty (START) talks, the goals for sublimits within the 6,000 warhead ceiling would need to be determined afresh and particularly the correlation between those for land based and submarine launched missiles. It would also be necessary to decide on the staged reductions which would be required as well as agreeing on the necessary principles to govern the very complex area of verification. At the theatre level, the modernisation of nuclear weapons in Europe and how they might be deployed is a further area of concern and one fraught with some political difficulties; the sensitivities of the Germans would need to be met but without contravening the INF Treaty either in the letter or the spirit of that agreement.

It is entirely feasible that the next stage in any argument over burden-sharing with its European allies could result in the US withdrawing some of its troops from Europe. In any event, the extent of prepositioned stores for reinforcement forces will need to be reviewed and, should there not be a chemical weapons ban, then the need for a retaliatory chemical capability and where it might best be deployed, could once more be questioned. Then there remains the vexed question of the Strategic Defense Initiative. A sensible research programme will probably be continued, within strict financial constraints, and it might be that priorities for any future deployment could be to guard against the possibility of an accidental launch from a potential aggressor and then to develop a cost effective defence system to enhance the survivability of US retaliatory forces and command, control and communication systems.

Underlying these discussions will remain the view that, in Europe, emphasis should be placed upon developing improved conventional forces so that if deterrence fails, there is not an early recourse to nuclear weapons escalating to the possibility of a strategic exchange which would put the continent of the United States at risk. Force structures will also need to be reviewed for possible deployment to the Middle East, the Far East and Central America. Here the debate will centre on the value of light intervention forces as opposed to those with a heavy armour component and its consequent increased logistic lift and support demands. Inter-Service rivalry will also play a part in the competition for finite funds. A 600 ship navy with its 15 carrier task groups remains one area of contention, as does the need and requirement for further investment in the B2 or Stealth Bomber, the Advanced Tactical Fighter, and whether the Air Force's commitment to close air support can be sustained with a new generation of aircraft. Questions are also being asked concerning the restructuring at the highest levels of the defence establishment. By giving the Commanders-in-Chief a say in the budgetary and procurement processes, along with the necessary increases to their staffs, it is being suggested that the attention of the Commanders-in-Chief now, as a result, is being attenuated rather than focussed on their prime function and areas of responsibility. Moreover, to accommodate the inputs from Commanders-in-Chief, the Pentagon organisation itself has become even more diffuse. Despite the emphasis of Frank Carlucci on improved efficiency in defence procurement and budgeting, as well as facing up to hard choices in curtailing capabilities rather than "salami-slicing" when faced with defence cuts, the new administration will be faced with the need for an experienced defence secretary and staff which quickly comes together as a team and in whom both the military and industry will have confidence in order to tackle all these issues quickly and effectively. The Pentagon will also need to cooperate closely with the State Department to ensure that hard military realities feature in future arms control negotiating positions.

THE MIDDLE EAST

By 15 February 1989, the Soviet Union should have completed the withdrawal of its forces from Afghanistan under the agreement signed the previous April by both Pakistan and Afghanistan and guaranteed by the two Superpowers. In order to sustain the Kabul government for as long as possible it is expected that the bulk of the remaining Soviet forces, after the initial reductions in mid-1988, will remain until close to the final pull-out date. The last withdrawal could then very well be supported by airforce and artillery units based inside the Soviet Union but, in the flush of an agreement fulfilled, this contravention of the accord will probably be overlooked. Thereafter, it is likely that covert support for the Kabul administration by the USSR will be continued in a manner not dissimilar to that of US support for the guerilla groups through Pakistan. How long the Najibullah regime will survive is an open question. It is doubtful if he will be able to maintain his administration in Kabul and it is therefore more likely that he will seek to establish himself in the north of Afghanistan backing on to the Soviet Union. Whilst he remains, he provides a focus for the combined opposition of the guerilla groups. But as the Najibullah hold weakens strains between the guerilla groups can be expected to re-emerge as they seek to further their own interests and claims on power in the aftermath of the

Soviet withdrawal. There will also be increased tension between the guerilla leaders in the field and the seven political parties based in Peshawar, Pakistan, that make up the Afghan Guerilla Alliance. General Zia, the late Pakistani President, had favoured the fundamentalist guerrillas in the delivery of arms. With General Zia's death the Americans have cut back their arms supply to the fundamentalists to guard against the possibility that Afghanistan might become an Islamic republic similar to Iran. Moreover, by seeking to supply arms directly to the Mujahedin in Afghanistan, the Americans have sought to reduce the influence of the Guerilla Alliance in Pakistan, thereby allowing those who are doing the fighting in Afghanistan a greater say in the future of their country when the Soviet withdrawal is complete.

In the Iran–Iraq war, the protagonists were brought to the negotiating table by the Iranian acceptance of United Nations Resolution 598 which called for a ceasefire, a withdrawal to internationally recognised boundaries, the establishment of a monitoring team, an exchange of prisoners of war, and the setting up of an independent tribunal to examine the causes of the war. Iraq had accepted the UN resolution from its promulgation in July 1987. Iran came to accept what President Ayatollah Khomeyni regarded as a "poisoned chalice" because of a failing economy and an increasing war-weariness at home, diplomatic isolation internationally, and the problems of a transfer of power on the eventual demise of the Ayatollah himself. Arranging the succession could not be faced with an ongoing war, which was proving difficult to sustain, and international antipathy which could presage attempts to destabilise the country by supporting factional in-fighting within Iran itself. The right of succession is now being determined between the radicals and those who favour a more moderate approach to Iran's difficulties. Acceptance of UN Resolution 598 without the Ayatollah's blessing would have given rise to intolerable internal political strains: that Hashemi Rafsanjani, the speaker of the Majlis and acting Commander-in-Chief of the armed forces, was able to obtain the Ayatollah's reluctant endorsement seems to suggest that the so-called moderates are ahead in the internal struggle for power. The outstanding question now is whether the ceasefire reflects a genuine attempt to find a solution to the eight-year-old conflict, whilst at the same time wooing international opinion, or whether it is no more than a ruse to enable the economy to be rebuilt and war stocks accumulated so that, after a pause, the offensive might be resumed? A third possibility and the most likely, is that the Iranians have not yet decided where they perceive their best interests to lie and so they are keeping both options open with different factions favouring differing solutions.

If Iranian war aims have been modified, albeit temporarily, the same cannot be said for those of the Iraqis. At the time of the ceasefire, the Iraqi forces felt they were in the ascendency and their morale was high. They suspect that Iran's motives in agreeing to a ceasefire were tactical, Iran wishing only to replenish and regroup its forces. In consequence, the Iraqis will endeavour to prevent Iran gaining any military advantage from the ceasefire. One of the prime Iraqi war aims was to reassert full sovereignty over the Shatt al-Arab waterway. The Shatt is Iraq's only access to the Persian Gulf and it feels that the Algiers Treaty of 1975, in which the boundary with Iran passes down the waterway thereby providing Iran with a sanction over Iraq's maritime lifeline, was imposed upon Iraq when it was involved in a debilitating war with the Kurds whom the Shah was then supporting. Whilst the disputes of centuries continue to fester, President Saddam Hussein is determined that the Kurdish question should not influence the present negotiations with Iran: if the matter is not now "resolved", the Kurds' position will certainly be weakened by the Iraqi war machine being turned against them.

In the wider Middle Eastern field, Iraq will seek to develop an Arab "axis" in the region with countries that have supported it in the war, notably Saudi Arabia, Egypt and Jordan, and extending to certain smaller Gulf states such as Kuwait and Bahrain. In particular Saddam Hussein will look to the building of a sophisticated arms industry with Egypt. Iraq's objectives will be to ensure that Iran's regional influence is contained and its economic development matched, if not surpassed. Iraq will also remember the support which Syria afforded to Iran in the war, and President Asad will certainly be concerned with the prospect of a strong Iraq on one border and the fact that he can now place less reliance on a Soviet Union seeking to emphasise a more constructive and influential diplomatic role in the Middle East. Iraq will also support Jordan and the Palestinian people as a means of bringing pressure to bear upon Israel and Syria. Although Iraq's first priority will be its own domestic recovery, it is equipped with proven troops and the most advanced weapons of any Arab country, and this extends to chemical weapons. It thus poses a serious threat to both Syria and Israel, as well as having a destabilising potential elsewhere unless moderated by those who have supported it in the war. For its part, Saudia Arabia, remembering periods when both Iraq and Iran separately sought to impose their policies on the Gulf area, is determined that neither side will again become a regional power able to override Saudi national interests. This is evidenced by the recent purchases of weapons, from China and the United Kingdom, ranging from missiles through sophisticated aircraft and their bases, to the prospect of a large order of modern tanks and hopes for a submarine force. Saudi Arabia could have a moderating influence in the area as it matches political will and more diplomatic robustness to its economic strength and influence through OPEC, and with its burgeoning military capability enhancing the Gulf Cooperation Council GCC).

In the Arab-Israeli dispute, King Hussein's renunciation of Jordan's responsibility for the West Bank has forced both Israel and the Palestine Liberation Organisation to face up to hard choices. Is the PLO able to establish a provisional Palestinian Government-in-exile forging a consensus behind a new political programme and accepting the responsibilities which that incurs? Not least would be the unambiguous recognition of the state of Israel, along with determining policies with, and support of, the Palestinians within Gaza and the West Bank, as well as establishing more formal relations with both Jordan and Syria. The *intifadah*, or uprising, of the Palestinians in Gaza and the West Bank, borne out of a sense of desperation, could well give rise to new and extreme underground elements, even whilst still laying claim to the PLO cause, but impatient for results. Israel, for its part, needs to come to terms with the aspirations of the Palestinians within its present borders. The Israelis, in seeking to talk to the Palestinians, should accept that they will be representing the PLO, either overtly or covertly, no matter what face-saving devices the Israelis might use. Will a newly elected Israeli government feel it has a mandate to accept such realities at the same time as the PLO seeks to fashion new and realistic policies which attract strong Arab and international support? *Intifadah* and King Hussein have provided the catalyst for change, with President Mubarak a moderating influence: no doubt there will be further false starts but, just remotely, 1989 could see the beginnings of a move towards a solution of the Palestinian problem and the Arab-Israeli dispute.

WESTERN EUROPEAN SECURITY ISSUES

Even with domestic matters motivating many of the Soviet Union's policies and, in the United States, a new administration trying to determine how many of its earlier proclaimed policies might or might not be feasible, arms control issues will continue to underpin much of the discussion on European security in 1989. Whilst the Soviet Union claims that broad parity exists between NATO and the Warsaw Pact it also recognises that, within that overall balance, particular asymmetries exist. NATO for its part has declared that the conventional imbalance remains at the core of Western Europe's security concerns; the Warsaw Pact's preponderance of tanks and artillery, crucial to the ability to seize and hold ground, is seen as the means to launch a surprise attack or even to sustain a large-scale offensive. In response, the Soviet Union has claimed that it views NATO's air forces as having an offensive capability and particularly those able to carry both nuclear and conventional weapons, and it has suggested that there should be matching reductions in those aircraft to balance cuts in Warsaw Pact tanks and artillery tubes. More recently, Marshal Akhromeev, Chief of the General Staff of the Soviet Armed Forces, has claimed that if the West finds the preponderance of Soviet forces in Eastern Europe and in the Western Military Districts of the Soviet Union threatening, then the USSR finds NATO's maritime forces and its forward maritime strategy, particularly the US carrier tasks groups, as posing a real threat to the Soviet Union. In consequence, it is now calling for asymmetric reductions in the West's naval forces and an extension of confidence building measures (CBMs) to include a whole range of maritime capabilities.

This last demand by Marshal Akhromeev seems to suggest an expansion of the Conference on Security and Cooperation (CSCE) into conventional arms control and, thereby, an attempt to play down the human rights issues which the West consider to be a fundamental part of the CSCE process. The 35 CSCE participating nations are concerned with the full range of East-West contacts and their development extending, as they do, to confidence and security building measures. This process should not be masked by seeking to turn the CSCE into a European Security Conference as the Soviet Union has long wished. Conventional arms control should rather be confined to the 23 countries of NATO and the Warsaw Pact who are able to make the necessary reductions in their forces to ensure mutual security and stability at a lower level. The forum for this is what is now called the Conventional Stability Talks (CST). However, this does not preclude the possibility of confidence building measures being extended to maritime and air forces, always provided that this was mutually agreeable to NATO and the Warsaw Pact, and, for example, including pre-warning of major naval and air exercises.

What is significant in these various arms control claims and counter-claims is the apparent misunderstanding of each side's strategies and military doctrine, each by the other, and upon which their concepts of operations are based. The USSR with its long land frontier, internal lines of communication, and its historic dependence upon mass, fire-power and surprise, seeks to maintain its forces which have been developed in accordance with these precepts. At the same time, it sees an apparent advantage to the West in NATO's maritime capabilities which it seeks to limit by arms control measures. With NATO's dependence upon sea lines of communication for reinforcement in war and trade in peace, and its lack of strategic depth in continental Europe, its emphasis upon maritime forces should be understandable, even as it meets Alliance requirements for land and air forces whilst seeking to reduce the Warsaw Pact's predominance in these areas through arms control. If the stated objectives of improved security and stability at lower force levels really apply, then, as a matter of priority, emphasis must be given in arms control negotiations to reducing the perceived capability of either side to launch a surprise attack and sustain that action. A better mutual understanding of security thinking and military developments within both blocs can only facilitate this process.

Meanwhile, NATO, for its part, faces an increasing divergence of views over the possible employment of its forces in Europe. Increasingly, the Americans are giving emphasis to the need for improved conventional forces and less reliance upon nuclear weapons as a means of defence in Europe, thereby deterring any possible future European conflict. It is perfectly understandable that, should deterrence fail, the Americans would wish that there should be a protracted conventional phase in any hostilities before the prospect of the threatened use of nuclear weapons which might then put US cities at risk. The Europeans, for their part, have placed their reliance upon the threatened early use of nuclear weapons to prevent any war starting in Europe and, certainly, to avoid the prospect of any conventional conflict with modern weapons. However, if greater reliance is to be placed upon conventional forces as public opinion, even in Europe, finds the nuclear threat less credible then it carries with it the need to develop those forces both in numbers and capabilities. In short, the resources available for conventional defence must be increased; yet there is little evidence of either the political will or of public support for such a move.

This problem is further exacerbated for the Europeans by the decline in numbers of those approaching the recruitment age of 18–19 years old. Between 1985 and 1995 the fall will only be of 5 per cent in Spain but 26 per cent in the United Kingdom. In Western Germany it will be 43 per cent! This suggests that women and older people will need to be attracted to the armed forces by flexible terms of service, whilst those with relevant skills and experience may need to be retained beyond current retirement levels. This could well necessitate a review of the relationship between conscript, regular and reserve forces, as well as a reallocation of trades and duties within the military. Moreover, to offset this manpower shortfall there are increasing demands for reliance upon developing technologies but, again, these do not come cheaply and the pressures on resources are further exacerbated. Against this background, it is alarming to hear some leading politicians say that there is no cause for concern over demands for increased defence expenditure; budgets may be retained at their present level or even reduced, for arms control measures will ensure reductions commensurate with, if not exceeding, the demand for more money. This suggests diversified and conflicting aims in future arms control negotiations: level or reducing defence expenditure is unlikely to be compatible with maintaining stability and security at lower force levels, nor is it likely to enhance the West's negotiating position.

If national defence budgets are unlikely to rise by anything like the amounts required, then better use must be made of available resources, and the pressures are upon the European NATO nations to be seen to do so. Trade barriers and subsidies which hinder both competition and collaboration in defence procurement need to be removed. Future military doctrine and concepts should be responsive to new technologies, as well as taking account of what may be possible in arms control agreements. Large expenditures should be shared rather than avoided, either by multinational arrangements or NATO infrastructure funding techniques. The luxury of thinking nationally in military capabilities, defence procurement or even logistic support, is past.

PROSPECTS FOR CHANGE?

Re-reading previous Yearbooks shows continuing trends in many international security areas. The statements of earlier years can be as relevant today as when they were written. But new developments do occur, and fundamental change too, even if it is seldom rapid. It is therefore important to identify trends and to determine their progress as well as seeking evidence of any major variation. So, what are the prospects for 1989?

☐   Mr Gorbachev has brought about dramatic change in the Soviet Union. In seeking to restructure the economy he has been forced into a new and radical form of government to encourage individual initiative and popular democracy but with overall political control still vested in the party. The rate of progress has not been as fast as Mr Gorbachev might have wished but he has wrought fundamental changes which cannot be totally reversed; however they could be repressed, but probably not in 1989. Changes in the leadership of a number of Eastern European countries will reflect both the spread and limits of *perestroika*.

☐   In the USA, fundamental changes are very unlikely to be set in train but there will be a pause as the new administration takes up its responsibilities. The pace of change under the Reagan Administration had already slowed; growing interdependence in international and economic affairs will give rise to periodic stresses and strains. It will be important to discern whether these will increase the pressures for protectionism and neo-isolationism in the US.

☐   Fighting will continue in Afghanistan after the Soviet withdrawal. Once the Kabul regime is reduced to a

faction, the guerilla groups will probably not remain
united.

☐ Iran and Iraq have fought themselves to a standstill in a
war that neither can win, but the disputes of centuries are
unlikely to be resolved soon by negotiation. The best that
can be hoped for is that a period without armed conflict
can be extended for many years. More significant will be
the prospect of internal change within both Iran and Iraq.
In Iran the struggle for power will probably continue
between the various factions after Ayatollah Khomeini's
death. Whilst economic rebuilding takes place in Iraq,
indicators should be sought to determine possible internal
political developments and Iraq's role in the wider Middle
East.

☐ *Intifadah* and King Hussein's relinquishment of legal
administration responsibilities for the West Bank should
force Israel and the PLO to face up to the challenge of
beginning to find a solution to the Palestinian problem.
The Arab nations will also review their position and
alliances throughout the region.

Political-military issues in Europe have, in many instances, a
familiar ring. The exception is the call by the Soviet Union for
its armed forces to adopt a strategy of "war prevention"
through "defensive defence". What this might mean in terms
of force levels, deployments and military concepts is not yet
clear but the Soviet military themselves admit that if battle is
joined the aim is to win, no matter what doctrine has previously
been adopted. Nevertheless, until there is evidence of a
reduction in Soviet military funding, NATO needs to maintain
its guard and at the same time seek to develop a better
understanding with the Warsaw Pact at all levels, of why each
side might feel threatened and why it has developed current
force structures and military strategies. If arms control
negotiations are successfully to extend to conventional forces
and chemical weapons, as well as nuclear warheads and
delivery systems, then each side must feel secure. From the
viewpoint of the armed forces, this means taking military
capabilities fully into account and not placing too great a
reliance upon statements of intent. Within NATO, pressures
upon defence budgets have given rise, once again, to calls by
the US for a greater contribution by its European allies to their
common defence. Whilst a little more money may be found,

the solution will stem from making better use of the resources already available. If the "European Pillar" within NATO is to be strengthened then the onus lies with the European nations themselves, yet this should not run counter to but rather compliment transatlantic cooperation. The trend has long been established but it is slow in its development and a fresh impetus is needed.

In looking forward to 1989 it may be said that little has changed fundamentally but, at least, the prospects are better.

Part I — The Issues

Turkey in the Southern Flank

HIS EXCELLENCY TURGUT ÖZAL, PRIME MINISTER OF TURKEY

SECURITY IS a major concern for all states. NATO was born out of the necessity to preserve the security of its members and is the expression of the resolve of free peoples and sovereign nations to unite their efforts for collective defence and for the maintenance of peace and stability. After decades of intense efforts and through the solidarity of its members, today we all take pride in the fact that this collective endeavour has been a true success in its deterrent function.

There is no doubt that the security of our Alliance is indivisible and that there can be no regions with differing levels of security. We observe, however, a tendency among some defence analysts and planners to focus on the Central Region as the main area where the threat is faced and to try to evolve NATO's security policy and strategic calculations according to this ill-perceived priority.

Those who adopt such an approach fail to realise that the defence of the Central Region cannot be considered in isolation from that of the Southern or Northern Flanks. This requirement was in fact diagnosed in the early years of the Alliance and the rationale behind it was not challenged but proved to be valid through the past years.

From a purely military-strategic point, it has been established that the credibility of deterrence for the Central Region depends also on the ability to maintain an effective defensive posture in the Flanks. In other words, any weakness or instability in the Southern Flank may either induce a potential adversary to exploit this weakness or may provide it with a free hand to concentrate its full force in the Central Region. Thus, our collective efforts to enhance the defences of the Southern Flank would in turn benefit the whole Alliance. A strong Southern Flank, moreover, just by its perception as such by a potential aggressor, would be a factor in dissuading attempts to exploit destabilised hot-spots in adjacent areas.

A simple glance at the world map should be sufficient to demonstrate the geostrategic value of the Southern Flank, and notably Turkey, for the West. This region also borders the important waterways of the Mediterranean and the Aegean which are vital for the Alliance in military and economic terms. Mr. Gorbachev's statement to the press in Washington after the signing of the INF Treaty on 10 December 1987 maintaining that an imbalance existed in the Mediterranean in favour of NATO countries is certainly a calculated one giving an indication of their priority areas. This impression is reinforced by the proposals Mr. Gorbachev put forward on 16 March 1988 in Belgrade with respect to Mediterranean security.

Of course any evaluation of the Southern Flank outside the East-West context is bound to be misleading. This region, despite its crucial significance in the East-West equation, is the area where the military imbalance is qualitatively and quantitatively in the East's favour. Such a situation renders the region both a lucrative and vulnerable target. Indeed, in times of tension or conflict, the potential adversary, instead of risking a confrontation with the strong Central Front, may opt for an indirect approach by initiating a fait accompli in the Southern Flank. Adverse force ratios, quantitative and qualitative disadvantages in weapons systems, air defence deficiencies and destabilising factors in the Mediterranean basin accentuates the multi-dimensional defence problems of this Flank.

Turkey, being situated at the crossroads of three continents and important waterways, including the Turkish Straits has a vital role in, and makes a dedicated contribution to, the collective security of the Western world.

Notwithstanding its deficiencies in terms of modern equipment, the well-disciplined Turkish Army is the second largest in NATO after that of the US, rendering it a formidable asset for her Allies. Turkey, furthermore, shoulders much more than her fair share of the burden of the collective defence by spending about 4.5 per cent of her GNP for defence, which is well above the Alliance average.

Turkey, which maintains such a large standing army in peacetime due to obvious geostrategic reasons, is obliged to allocate substantial financial resources for defence in order to raise and maintain the combat effectiveness of Turkish Armed Forces. Moreover, achievement of desired modernisation in the medium term puts a heavy pressure on the national budget. For this reason, a continuous flow of security assistance from allied countries in order to realise modernisation is essential, at least in the short term.

On the other hand, the national defence industry, which has been significantly developed during the recent years, cannot be sufficient alone to meet the requirements for modern weapons systems and equipment of the Turkish Armed Forces. The objective of modernisation is to render all services of the Armed Forces fully capable in the fields of combat readiness, mobility, firepower, air defence, logistics, command, control and communications and intelligence.

Nevertheless, in the long run Turkey's preference is more foreign investment and joint ventures rather than security assistance. The emerging protectionist tendencies and trade barriers, despite their debatable advantages in the short run, would have disruptive and restrictive influences in the world economy in the years to come. Turkey favours liberalisation of trade and lifting of barriers. In fact, against this backdrop of unfavourable global economic trends, Turkish exports have risen from about $2 bn in 1986 to about $10 bn last year.

Turkey, with her rapid economic growth rate which averaged per cent in the 1983–87 period is on the way to being a valuable asset to her Allies by becoming an economic power, in addition to her military contribution to common defence.

Other economic indicators of Turkey, today a country of 55 million people—and a huge potential market—are promising for the future. Today, Turkey's GDP is approximate to that of Belgium and about 2.5 times more than that of Portugal or 1.5 times more than that of Greece. A marginal increase in Turkish per capita income would push this figure up significantly. Such a prospect provides economic opportunities for her Western allies which are her primary economic partners. To recapitulate, the efforts on the part of our Allies to invest in or support our economy would in return provide both security and economic benefits for them.

Joint ventures in defence industry and cooperative initiatives constitute an important area where such efforts can be concentrated. Turkey has a sufficient economic-technological base and resources to participate successfully in bilateral or multilateral enterprises. Certain economic considerations such as low-cost production inputs make Turkey's participation quite profitable.

The defence sector is only a dimension of a country's industry. It is both difficult and expensive to establish production facilities for Armed Forces before overall industrialisation attains a satisfactory level. On the other hand, it is widely acknowledged that there are close ties between the level of development and economic strength of a nation and its defence capability. There cannot be a sustained and balanced development in a country where security is imperiled; and a defence system which is not based on a solid economic strength is bound to be unstable.

The history of modern defence industry in Turkey goes back to the early years of the Republic when small armaments and ammunition manufacturing facilities were established by the government to Ankara and Kirikkale. In later years, establishment of production facilities serving the Turkish defence industry was maintained. Especially since the late 1970s efforts aimed at establishing and diversifying Turkish defence-related industries to meet military requirements have been greatly accelerated.

In 1987 the Turkish Defence Industry Support Fund was established in order to contribute to augmenting the combat capabilities of the Turkish Armed Forces through the development of our national defence industry. A considerable amount of funds including foreign currency has accumulated under the control of this Fund and has been put at the disposal of our defence procurement planners.

Turkey's special and crucial role for the Western world is not confined to geostrategic location, military contribution or economic prospects alone. It is my firm belief that in the long term, my country's most important contribution to the Western world would manifest itself in our determination to share and uphold the Western ideals which have bound us all and which we are all united and resolved to protect.

The Turkish people share the same aspirations and ideals as their allies and are determined to take part in the European integration process. The cultural richness of Europe would also be enhanced through this process. We believe that the success of Turkey's economic undertakings and her potential prestigious place in the Western family of nations as a staunch supporter of Western security and of Western ideals would constitute a remarkable precedent for the developing countries and especially for those with whom Turkey has traditional, cultural and historic ties. The spillover effect of Turkey's success in the path of taking full part in the European integration process should not be underestimated.

This brings me to the recent developments in East-West relations and to the question of cohesion in Europe and the Western world at large. We are living in an era where international relations have assumed a multi-dimensional character. Developments such as the INF Treaty and the "openness" and "restructuring" policies pursued in the East, are bound to have far reaching influences on the evolution of a new international environment. A comprehensive and coherent assessment of the opportunities and challenges which these changes offer us with respect to our common goal of providing a better and more secure life for our citizens, while defending their aspirations for a democratic way of life, is of prime importance.

In this connection, when we dwell upon issues such as the strategy of the Alliance, disarmament efforts for achieving both conventional and nuclear stability, transatlantic relations and recent trends for European defence arrangements, we should not overlook our basic notions of unity of the Western world and especially of Europe, and the indivisibility of defence.

The ideals and principles of the North Atlantic Treaty Organisation and the basic strategy of the Alliance which relies on the determination of the allies to perceive an attack on one as an attack on all maintain their validity. This is even more so in the post-INF Treaty security environment. Furthermore, the continued validity of NATO's strategy of Flexible Response and Forward Defence is of the utmost importance for the credibility of deterrence. It is, however, only natural that the dynamic relationship between the different components of NATO's strategy should be evaluated and updated, where necessary. Such an evaluation may take into account the fact that the INF Treaty completely eliminated a certain category of weapons which had a not too negligible place in NATO's strategy. The elimination of land-based INF's also made the conventional superiority of the Warsaw Pact over NATO even more pronounced and raised prospects for further arms control and disarmament efforts.

Hence, NATO, while preserving its basic military strategy, is faced with the task of maintaining its posture of credible deterrence as a matter of priority by improving its conventional capability, especially in the regions where the imbalance is most acute, and determining its priorities in the different but interrelated disarmament fields within an agreed framework with a view to enhancing security and improving stability at a lower level of armaments.

An improvement of NATO's conventional capabilities particularly in exposed areas such as the Southern Flank, where disparities are most acute, would induce the Warsaw Pact to be more forthcoming with regard to the conclusion of a conventional stability agreement.

Turkey, together with her allies, has always favoured a stable East-West relationship based on mutual respect. Global peace and security constitutes one of the most prominent aspirations of the Turkish people. On the other hand, as stated in the last NATO Summit Declaration the security in freedom and the prosperity of the European and North American allies are inextricably linked and the long-standing commitment of the North American democracies to the preservation of peace and security in Europe is vital.

In the changing and dynamic present international environment, the differences of opinion and interests that may arise between the Allies and the two coasts of the Atlantic are not something to be afraid of. In fact, in the past there were also diverging opinions within NATO on various subjects. In an alliance of free nations, this is natural. In the face of recent developments and concerns, while on the one hand the solidarity of the Alliance and the indivisibility of defence should be reaffirmed, on the other the emerging inclinations of regional approaches to defence must be avoided. The success of these initiatives depends on their contribution to common defence and to the strengthening of the European pillar. In connection with this, it should be stressed that the scope of the European pillar of the Alliance should be well defined to also include the flanks. In such a process the Western European Union can play an important role within the overall framework of Western defence. It was with this conviction that Turkey expressed its desire to join this institution.

I have so far tried to touch upon some of the issues pertaining to NATO, to their effect on the Southern Flank, to the importance and vulnerabilities of this region and to East-West relations in general, particularly from Turkey's vantage point. NATO will be viable only to the extent that it can respond, in a timely manner, to changes in the international environment and to the security requirements that arise from these changes. I am confident that we, as Allies, will meet such a challenge in unity, with vision and realism, in full recognition of our long term common interests.

How Well Will the Alliance Cope?

HIS EXCELLENCY SIR MICHAEL ALEXANDER, KCMG

The author is UK Permanent Representative to NATO. This contribution originated in a lecture given at the RUSI on 20 January 1988 and should be seen as a snapshot of the author's thinking at the time.

WHEN INVITED to speak in RUSI's annual series "The Year Ahead", my initial inclination was to pose the question "1988: Can the Alliance Cope?". Second thoughts suggested this question was too easy. The Alliance will cope: member governments know they have to cope and have long since acquired the habit of doing so. But there is a further question: will we cope coherently and actively or incoherently and reactively. The question cannot, in the nature of things, be answered definitively today. But it is worth asking because over the *next* 12 months it will be important that the Alliance does cope competently.

The Alliance is embarked, and has been for the last 12 months or so, on a critical phase in its history. Its foundations are shifting; intra-Alliance relations are increasingly complicated; the Warsaw Pact is in flux; public perceptions of the threat and of the future of East–West relations are changing.

More specifically 1988 will see the completion of the Reagan/Gorbachev dialogue, probably in Moscow just before a crucial CPSU party conference at the end of June; the first NATO Summit since 1982; major developments in arms control negotiations on strategic nuclear, on chemical and on conventional weapons; Presidential elections in France and the United States; the assumption of office by the first German Secretary General of NATO; and so on. In other words 1988 might well turn out to be one of those pivotal dates—like 1945, 1956 or 1968—around which historians like to organise their material.

BACKGROUND

The last 40 years of European history can be seen as the working out of President Truman's historic decisions to abandon US isolationism, to sustain a ruined Europe and to 'contain' the ambitions of a deeply unattractive regime in Moscow. If so, we are now arguably at the end of the post war era. The engagement of the US with the world may be reaching a point where the multiplicity and cost of her international responsibilities will cease to be regarded as sustainable in their totality. The ruin of Europe has long since passed into history. There is now a feeling that Western Europe is if anything the wealthier partner in a transatlantic relationship where competition sometimes seems to verge on confrontation. Fear of the Soviet Union is being replaced by a perception that the threat is diminishing, that the means used to contain the threat may no longer be appropriate and that the General Secretary in the Kremlin is one of the most attractive figures on the world stage. These changes in attitude have not occurred over night: some are better grounded than others. But all are coming sharply into focus. They present both opportunities and risks. It will probably become evident during 1988 whether the West (and for that matter the East) is going, on the one hand, to exploit the opportunities or, on the other, to stumble into the pitfalls.

PROSPECTS: THE BEST CASE

To make the point clearer we might imagine historians 20 years hence looking back at 1988. It is not entirely absurd to envisage them describing the 1990s as an era of unprecedented stability in superpower relations. They might date this era from the entry into force of the INF treaty in 1988; from the negotiation of a START agreement; and hence from the vindication of the Alliance approach to arms control. They might point to the elaboration by NATO in that year of a comprehensive philosophy firmly linking the requirements of security and those of arms control. They might acknowledge the sensible way in which NATO started the long negotiations on conventional forces. They might stress the remarkable advances made in verification and note how these facilitated the negotiation of a global agreement on chemical weapons —weapons which as they became cheaper, nastier and more difficult to monitor might if uncontrolled have turned out to be the biggest threat to international stability in the 1990s.

Our historians might comment that throughout the years of multilateral negotiation the Allies kept their nerve and ensured that their conventional and nuclear forces remained fully effective. As part of this pattern the European nations, including France, began in 1988 to accelerate the integration of their defence efforts and the harmonisation of their military doctrines, thus strengthening the Atlantic partnership at a crucial moment. The conventional negotiations eventually culminated in a security balance of markedly increased stability, with treaty based common ceilings for both sides, intrusive and mandatory verification, a managed reduction of Soviet force levels (and on a much smaller scale of US forces) in Europe, and the virtual elimination of the Soviet Union's capacity for surprise attack. In the resulting climate of increased confidence the post-war ice age in Europe began to dissolve and new relationships to develop between Eastern Europe and the Soviet Union as well as between the two halves of Europe.

PROSPECTS: THE WORST CASE

One could go on. But I am not particularly comfortable in the role of Dr Pangloss and in any case his views need to be balanced by those of Dr Gloom. For our historians might at least as easily find themselves looking back with regret at 1988s missed opportunities. This could be the year when the INF Treaty entered into force, but when its significance was misrepresented and alliance cohesion damaged as a result. It could be the year when a START agreement was not brought home, with much consequent recrimination; or when it was brought home and in the ensuing euphoria the Alliance governments chose to forget that successful negotiation from strength presupposes investment in strength. As a result the conventional stability talks, overtaken by unilateral and therefore unpoliced reductions on both sides, eventually followed the MBFR talks into stalemate and irrelevance.

1988 could also be the year when the Alliance disagreed about its policy priorities and failed to move towards the decisions necessary to maintain a credible mix of nuclear forces in Europe; when in consequence some started seriously to doubt the need for such forces and others to lose confidence in the strategy which underpinned them. "Beggar-my-neigh-bour" defence policies began to flourish among the European nations. Ignoring technological and military logic, they went their own way, picking and choosing from the Alliance menu, shedding tasks and turning out incompatible equipments from ever narrower national bases. The effort to develop a European defence identity within the Alliance fell apart and the burden sharing debate became increasingly acrimonious, precipitating a fundamental weakening of the transatlantic partnership. Finally, the historians might note how the Soviet government was faced with increasing domestic difficulties and major problems in Eastern Europe. Lacking both the reassurance of stability in the West and the challenge of firmness there, Moscow tried first to turn the clock back and then, rashly, to exploit the Alliance's disarray . . . and so on.

Neither of these caricatures is at all likely to prove accurate. I only want to highlight the point that we are in a period when the opportunities open to governments are considerable and the pitfalls are of similar dimensions. I shall use the rest of these remarks to describe in more detail the two sides of the coin and how the Alliance is trying to ensure it comes down the right way up. (Bar a couple of references to WEU I shall limit myself to the competencies of the North Atlantic Council. I shall not be dealing with economic issues, out of area problems or the third world in general though obviously these all affect the Alliance and may conspire to make 1988 an even more interesting year!)

THE OPPORTUNITIES

First, then, the opportunities—many of them linked with the INF Agreement. I should confess at the outset to some impatience with the Cassandras, on both sides of the Atlantic and of the Channel, who perceive in this agreement an own goal of epic dimensions. No doubt we can make it that if we so choose. Comment on the agreement has included the usual measure of self-fulfilling prophesy and overcritical kibbitzing. Of course the Russians gained something. If they had not done so there would have been no agreement. But a balanced view of the agreement must acknowledge that it contains major gains for the West. These, in my view, heavily outweight those made by the other side. Thus the agreement:

- [ ] confirms that strength and Alliance solidarity are the key to successful negotiation with the Russians;
- [ ] removes one direct and substantial threat to our population centres and critical military assets;
- [ ] establishes the principle of asymmetrical reduction as a basis for East–West arms control agreements;
- [ ] eliminates, on a mutual basis, a complete category of current nuclear weapons. This is unarguably without precedent in the history of nuclear armaments. I wonder how many genuine precedents there are even in the long history of conventional armaments;
- [ ] institutes a stringent and mandatory verification regime. Given that one of the signatories is the Soviet Union this aspect of the agreement (taken together with the Stockholm Agreement) is also a remarkable breakthrough;
- [ ] will, if implemented in good faith from the outset (as the Stockholm Agreement has been), markedly increase international confidence.

All these points, but notably those related to mandatory verification and asymmetry, are directly relevant to our future efforts in the arms control field. They will certainly be kept in mind in the next few weeks as NATO finalises its detailed proposals for the conventional stability talks. Since these proposals—and in due course the talks themselves—will have an important role in establishing the context for the implementation of the INF Agreement (and hence in the debate on ratification) it is important for NATO to reach a common position soon.

Of course the lessons and precedents of the INF negotiation must be consolidated and built upon. The opportunity—indeed the requirement—for the Alliance in 1988 is to agree arms control policies and proposals which are *positive*, in that they respond to the aspirations of our publics and wrest at least some of the initiative from Mr Gorbachev; *coherent*, in that they are complementary rather than contradictory; and *realistic*, in that they are sustainable, consistent with the true nature of East–West relations and with the requirements of our security. This is what the effort to develop the comprehensive concept—launched by Alliance Foreign Ministers in Reykjavik last summer—is all about. I hope the Council will make rapid progress in its work in the weeks immediately ahead.

The general improvement in the climate of East-West relations broadens the range of opportunities ahead. The West has always been ready for dialogue: Mr Gorbachev now seems ready to join in. This holds out the prospect of progress on a broad range of East-West issues: human rights, human contacts and the rest. The litmus test in Europe in 1988 will be the outcome of the CSCE review meeting. So far *glasnost* and *perestroika* have been more apparent in Moscow than in the contribution of the Soviet delegation in Vienna. The coordination of allied positions has not always been easy; given that the West cannot get everything it seeks, awkward questions of priority inevitably arise. But the overall bargaining position as between East and West is evenly balanced and the West is under no time pressure. Provided we maintain our basic solidarity there are good prospects for a satisfactory result in Vienna and for mutually beneficial progress thereafter.

So much for dialogue and detente. But there are also opportunities in regard to the other half of the doctrine elaborated by Pierre Harmel 20 years ago—deterrence, the maintenance of which is the prior condition for the pursuit of the rest of our agenda. Over the last few years the Alliance has in fact achieved a significant enhancement of its defence capability. (I am not referring to the deployment of Cruise and Pershing!) This effort can be continued, despite the restrictions on new money for defence currently apparent in most NATO countries.

Let me cite a few of the developments in train:

☐ NATO defence planning has become much more sophisticated. The introduction of the conceptual military framework encourages us to look further into the future than ever before. Opportunities for a collective approach to force planning are being actively examined, building on past successes such as NADGE, AWACS and the infrastructure programme;

☐ NATO's force planning is to be linked more directly to the work of the armaments community through an improved form of armaments planning which the North Atlantic Council (and therefore, significantly, France) has agreed should be the subject of an immediate trial in 1988/89;

☐ the Alliance now has the very real possibility of embarking on a much more comprehensive and structured approach to the effective employment of its total air assets through the ACCS programme;

☐ NATO planners have identified high priority force goals for each nation relating to the key deficiencies experienced by NATO's military commanders. Achieving these goals and thereby significantly strengthening NATO's overall posture, need not involve major additional expenditure;

☐ the Alliance should be able to agree in 1988 on the ways in which Spanish forces are to contribute to the common defence;

☐ there is increasing evidence of convergence in NATO and French thinking. This is apparent both at the working level and the political level, e.g. in the French Prime Minister's speech to France's National Defence College on 12 December 1987.

The arms control process is in my view going to have a direct effect on the defence side of the house in one slightly unexpected way. It will force the Alliance to think harder and more collectively than ever before about the equipment the other side has; why it has it; and how good that equipment really is. It will force us to think through the consequences of the answers for own forces. The incorporation of the outcomes in "net assessments" may in due course have considerable consequences for our perception of the nature of the East-West imbalance and for military planning.

There are therefore programmes and processes which, if pushed ahead with commitment this year and thereafter, will enable NATO to face the future with some confidence. That assumes, of course, an appropriate degree of commitment by member governments both internationally—to reach the necessary understandings—and domestically—to make the necessary funds available. HMG has made and is making such an effort. The same cannot be said of every member government. There have been problems in the last year. But there is a desire to overcome these.

1988 will also provide the opportunity to intensify defence cooperation at a European level. It has been evident for a good while that the European pillar of the Alliance will have to be strengthened, not least in organisational terms, if US expectations on burden sharing are to be met and if the whole structure is to remain stable. What has been encouraging in the last few weeks—literally—is the extent to which the need for such action has begun to become conventional wisdom. NATO's December Ministerial communiqué, for instance, contained for the first time an approving paragraph on the subject. Various senior US spokesmen, most notably the President himself in October, have voiced their support.

The evidence that something *is* happening is in the WEU security platform agreed in The Hague on 27 October 1987; in the naval deployments by five of the WEU nations to the Gulf; and in the active support for those deployments given by the other two members of the organisation. There is no shortage of other issues on which WEU can make progress in 1988, for example:

☐ the rationalisation of the organisation and in particular its colocation on one site—for which Brussels remains the obvious choice even if one partner still has doubts;

☐ the intensified use of the WEU as a forum for the exchange of views, and eventually the harmonisation of views, on matters of major security concern to member states. The joint meetings of Defence and Foreign Ministers, which do not take place in any other forum, have considerable potential;

☐ the expansion of the WEU's role as a ginger group at the political level for producing and refining ideas on defence collaboration, recognising that these will have to be followed up in NATO. For instance all the European contributors to NORTHAG are also members of WEU. As far as land forces are concerned NORTHAG is the Alliance's most successful test bed for the development of common military doctrine, common training and improvements in interoperability. The experience being gained in NORTHAG ought to be consolidated and extended into areas such as logistics and medical support (already under study in the Eurogroup) and eventually into the field of procurement (the responsibility of the IEPG).

None of this will be easy. It will be the job of the UK, which assumes the Presidency of the WEU at the beginning of July 1988, to ensure that the momentum gained in the last 15 months is sustained. The role of France in this context has perhaps received more attention from the media than that of the UK. Given the considerable evolution evident in recent

French speeches and writing on security matters this is certainly not a matter for complaint. But in fact by far the most detailed rationale and prospectus for European defence cooperation, including the part of the WEU, is contained in the speech given by Sir Geoffrey Howe in Brussels ten months ago—a speech which bears re-reading. Like the Hague platform it emphasises that the nations involved intend to develop a more cohesive identity in order to translate more effectively into practice their obligations of solidarity under both the Brussels and the North Atlantic Treaties. The Prime Minister, stressing that she is "all in favour of European countries increasing their cooperation in defence", noted last week that the clear and demonstrable effect of such cooperation must be to strengthen NATO. This is a criterion which can be met. Indeed it is not easy to envisage any other development of such potential value in strengthening NATO's foundations. In a period of change and uncertainty, increasing defence integration would have an evident stabilising effect. To set these developments firmly on the right track will therefore be one of 1988's more important opportunities.

PITFALLS

The pitfalls in the present situation are to some extent the obverse of the opportunities. The first is that of a general loss of perspective about the state of East-West relations. Most of us are susceptible to the excitement generated by Mr Gorbachev and most of us wish him well. He has taken up a formidable challenge: to resuscitate a derelict system and to change the character of the Russian people. On the evidence to date he offers a greater hope for positive change in East-West relations than any of his predecessors.

But there are any number of caveats to be entered. For instance:

☐ Mr Gorbachev is pursuing the national interests of the Soviet Union not those of the Alliance. As his references to interdependence make clear he recognises that some of these interests are mutual. But when our interests conflict we may expect to see less of Mr Gorbachev's "nice smile" and more of his "teeth of steel".

☐ Mr Gorbachev's success and hence his survival is far from certain. Whether or not he and his reforms prove durable, the facts of Euro–Asian geography; of Russian history; of ideology; and of Soviet military power will persist. The Soviet Union will remain both something very different from a Western democracy and a potential threat to Western Europe for as far ahead as we can reasonably look.

☐ The processes of *perestroika* and *glasnost* are bound to place stress on the existing structures in Eastern Europe. They will interact with the need for political change at the top compelled by age; with economic and ideological morbidity; with growing public expectations; and with uncertainty about Soviet intentions. The mixture looks distinctly unstable.

These uncertainties are inescapable. They impose on Western Governments a requirement for prudence and realism. But if 1988 sees a START agreement, rapid advances in other parts of the arms control spectrum and further triumphal progresses by Mr Gorbachev through our countries, the pressure on our governments to yield to popular enthusiasm, to relax their vigilance and to run down their defence effort will increase. This process will not come to a head in 1988. But it will be important for Alliance governments to show *now* that they can resist the temptation to use the improved international situation as an excuse for postponing difficult decisions. A period of detente of uncertain duration may not be the easiest moment to establish and communicate a consistent and convincing set of policies for the years ahead. But it may be the time when we need those policies most.

One such policy objective must be to reaffirm our present strategy. Flexible Response remains the only strategic concept capable of retaining general support within the Alliance. Of course it can be adapted and adjusted as circumstances change. But neither a move to significantly greater reliance on conventional forces—in an era of budgetary restrictions and adverse demographic trends; nor a reversion to the tripwire—in the post-Reykjavik era; nor the development of a nuclear war fighting capability—for use in Europe, seem to me plausible or consistent with Alliance solidarity. (I have already referred to the signs of convergence between attitudes in France and those of the Alliance. Obviously an understanding about the basics of our strategy will be a prerequisite for the full realisation of Europe's potential contribution to the Alliance.)

The continuing validity of Flexible Response does not mean that everyone accepts or understands the concept. It will have to be defended against criticism both from those who assert that the means of implementing the strategy are being removed, and from those who assert that the strategy is unnecessary or undesirable. To start with the first of these criticisms: obviously NATO is giving up two extremely capable and up-to-date delivery systems. In an ideal world we would have preferred to have begun elsewhere. But NATO has sacrificed neither its strategy nor its ability to implement that strategy. Flexible Response is a war deterring not a war fighting concept. It has never assumed a predetermined progress through a series of escalatory steps since this would be the equivalent of setting out with the intention of fighting a nuclear war. The infamous "ladder of escalation", in other words, is a myth. It is equally fanciful therefore to assert that the removal of the "rungs" labelled Cruise and Pershing will destroy the ladder or that, with over 300,000 US servicemen in Europe, it is a major step towards decoupling.

Flexible Response, however, *does* require the ability to respond appropriately to acts of aggression. This implies the availability of a mix of systems (both nuclear and conventional) sufficient to ensure that NATO's precise response could not be predicted by an aggressor. SACEUR's remaining 4,000 warheads must be deliverable by means that are varied, modern, survivable and widely deployed. They must hold at risk a broad selection of targets including those of the type which would have figured in the plans for Cruise and Pershing. The Soviet Union may have given up, and in larger numbers, systems which are at least as capable as ours. But it will not have removed from its target lists those NATO assets which were covered by the surrendered systems. These are uncomfortable truths. But to ignore them will be to stray into a particularly deep pit. The Alliance must therefore continue with the process which will enable it to retain an appropriate and up to date nuclear armoury. It must be prepared to insist that in doing so it is ignoring neither the letter nor the spirit of the INF treaty.

This will not be easy since for the second category of critic to which I have referred the principle attraction of the INF agreement is precisely the hope that it offers of further steps towards denuclearisation in Europe. Nuclear weapons, after all, frighten and outrage many of our citizens for precisely the same reasons that they deter generals and political leaders. NATO's deterrent strategy has worked; nuclear weapons are an essential part of such a strategy; conventional weapons, on their own, do not deter; the existence of more than one centre of nuclear decision making within the Alliance does enhance deterrence; a nuclear free world is a chimera. But such assertions are unpopular. Those not directly involved have a natural inclination to allow distaste to prevail over calculation. Paradoxically this inclination may become more of a problem as the nervous unilateralist fervour of the early 1980s ebbs and is replaced by complacency. The dilemma for the Alliance in 1988 will be that in arguing the case for nuclear deterrence in the new climate it may appear simply to be trying to keep itself in business while if it allows arguments for further nuclear disarmament to go unchallenged it may end up without a credible deterrence philosophy at all. The consequent need for vigilance in keeping arms control rhetoric in step with reality will be acute, for example, in considering the START/SDI relationship; in the handling of dual-capable systems in the conventional stability talks; and in forming Alliance policy on short range nuclear forces.

SNF will inevitably be the focus of much attention this year. There are evident pitfalls for the Alliance. SNF must not, for instance, be treated as an isolated problem: they have to be considered in the context of the requirements of NATO's defence posture as a whole. Nor must there be any implication that adjustments on the Warsaw Pact side of the balance—perhaps down to zero in some or all of the systems—will have to be matched proportionally on our side. A cut back in the 1,500 or so systems deployed by the Warsaw Pact would be very welcome. But even if this takes place NATO will require for some time yet an effective capability of the kind provided by its present deployment. Hence the language of the Reykjavik communiqué. Finally, the SNF imbalance must not be regarded as a source of concern to only one member of the Alliance. It concerns us all.

The SNF issue will I suppose, be presented by some commentators as a test of European solidarity within the Alliance. If so, it is important that they be answered with a show of Alliance solidarity. As European defence cooperation gathers momentum we must avoid any perception that there is, or could be, a European (or "continental") strategy pursued by some or all of the European members alongside that of the

Alliance itself. (The same goes for a "North American" strategy. A deterrent which discriminates is a deterrent liable to create more doubts than it resolves.) The security of NATO will be preserved if there is a common strategy accepted by both the European and North American members of the Alliance. The development of a European defence identity within the Alliance is a multilateral undertaking intended to enhance the security of the Alliance i.e. of Europe and of North America. It will do so if it is seen as such. It will not do so if it comes to be seen either as no more than an *ad hoc* network of bilateral relationships or as likely to result in the creation of a separate, and inevitably diversionary, pole of attraction.

This last point is directly relevant to my final pitfall for 1988: a mutual failure to pay due attention to the relationship between Europe and the United States. However the negotiations with Moscow develop, this could be an awkward year for the transatlantic partnership. The INF ratification debate is unlikely to be entirely straightforward. I have already mentioned "Discriminate deterrence". The approach of the Presidential election will increasingly make itself felt. The savings necessitated by the US budgetary cutback will have to be found somewhere—there will be hard choices for the Department of Defense in regard to both force levels and procurement. The US defence industry faces a period of involuntary retrenchment. All this may give a rougher edge to the two way street and to the burden sharing debate in general. So will the fact that in 1988 the US has had to take an unwelcome decision in regard to its base agreement with Spain, faces the continuation of difficult negotiations with Greece, and the possible opening of similar negotiations with Turkey and Portugal. A lot of careful Alliance management will be needed as well as a determined effort by everyone involved, on both sides of the Atlantic, to keep the big picture in mind.

CONCLUSION

It hope it is by now apparent why 1988 promises to be an interesting year. As I said at the outset no one can *know* now how effectively NATO is going to cope. But the Alliance is facing up to the issues, and this is at least half the battle. Much of the work fundamental to "coping well" is already in hand. The Alliance seems to me to be fully alive to the requirement

which I mentioned earlier for policies which are positive,
coherent, and realistic. It will evolve such policies from existing
and well tried approaches rather than *de novo*. The task will not
be easy and will, of course, never be definitively completed—-
certainly not in 1988. But the adaptation of its policies and
structures to changing situations is something the Alliance has
been doing with success ever since 1949. Given leadership in
capitals it can do so again. And I am, incidentally, in no doubt
that such leadership will be forthcoming in London.

The WEU and the European Dimension of Common Security

HIS EXCELLENCY AMBASSADOR ALFRED JEAN CAHEN

The author is Secretary-General of the Western European Union. This contribution originated in a lecture given at the RUSI.

THE CONTEXT in which Western European security problems were set during the 1970s and early 1980s has been static, both as regards the developments in the USSR and the evolution of East–West relations and as regards the transatlantic relationship.

In the Soviet Union, the last three years of General Secretary Brezhnev's reign were characterised by a rigid immobilism which—after Secretary General Andropov's short-lived leadership—was quietly and steadfastly pursued by General Secretary Chernenko. This certainly did not necessarily favour positive developments in the field of East–West relations. This was clearly demonstrated by the events between 1976 and 1985, which marked the end of the so-called "détente period" and by the progressive return to an atmosphere that was growing colder. But it did lend to this relationship between Moscow on the one hand, and Washington and its Atlantic allies on the other, a predictability which was doubtless not very constructive but nevertheless quite comfortable.

As far as the transatlantic relationship is concerned, it is a fact that it has lived through the late 1970s and early 1980s in a situation of status quo. Of course, there have been problems like those created on the old continent by some significant popular opposition to the deployment of the Euromissiles— cruise missiles and Pershing II—in accordance with the Allies' dual-track decision of December 1979. But such problems did not prevent things from remaining essentially the same in the Alliance. Its political doctrine, defined in the Harmel Report of 1967, stayed unchanged and was confirmed by the Foreign Ministers of the Alliance in Washington in 1984. The burden-sharing debate had lost much of its acuteness and had acquired

25

something of a ritualistic quality, with the possibility of United States troops stationed in Europe actually being recalled considered to be rather remote. Admittedly, status quo is not a very dynamic position to be in, but it is a rather comfortable one to which an Alliance easily becomes accustomed.

Today, the static context in which our Western European security had found itself for so many years has become a thing of the past. What is more, events have moved on at such a pace that it seems a long time since it disappeared. It has somewhat abruptly disintegrated and given way to entirely new and rapidly-evolving situations.

In Moscow, rigid immobilism has been replaced by a dynamic policy some aspects of which—particularly in the sphere of external relations and security—may appear to be tactical but which nevertheless have considerable importance and true substance. In addition, this new "thaw" in the Soviet Union has had an impact on the other Warsaw Pact countries where we see—at government level—a tendency to assert themselves nationally, at European level and even on the wider international plane. We also note the concern of the same governments at the awakening of public opinion to which their response is either cautious liberalisation or a sudden hardening of attitudes or an odd mixture of both. In any event, these changes have visible or even spectacular consequences for East–West relations, particularly in the sphere of arms control negotiations where there is a new surge of activity.

INF, START, reductions in conventional forces, the possibility of the progressive elimination of chemical weapons— progress is welcomed in all these areas. But there are as many questions as there are solutions. There is the question of the future of the nuclear weapons deployed in Europe, especially of the short-range variety, which is being vigorously debated in the Alliance. There is the problem of dual-capable weapons. There is the question of the future of the British and French nuclear arsenals. All these developments are, of course, part of the reason for the end of the status quo in which the Alliance had lived for many years and, in that respect, the Reykjavik Summit of October 1986 was undoubtedly something of a turning point.

But these factors are not the only ones which have led to the end of the status quo in the Alliance. Another no less important one is the growing debate on both sides of the Atlantic about the future of the transatlantic relationship. The transatlantic relationship has not always been an easy one. In the 1950s, Henry Kissinger spoke of a "troubled partnership" while years later, in September 1986, on a major BBC programme devoted to the future of the Alliance, his former aide and present friend and associate, Larry Eagleburger, referred to an "impossible relationship". When one looks at those two qualifications, one must agree that if some progress has been made, it has not always been in the right direction. As for a European view of the evolution of transatlantic relations, John Palmer, a British journalist, states in a recent book *Europe without America: The crisis in Atlantic relations*:

> The economic, military and political world of the Atlantic Alliance in which two generations of Americans and Europeans have grown to adulthood since 1945 is visibly crumbling. Even the most sober of observers now openly discuss the crisis in the Atlantic partnership and how long it can survive in anything like its present form.

Although this author may seem unduly pessimistic, there is a debate on the issue and more and more elements are being introduced into it. For instance, elements of reflection regarding the role of the US in the world and in the Atlantic zone, like in Professor Paul Kennedy's book *The Rise and Fall of the Great Powers* or in that of Professor Calleo *Beyond America Hegemony*. Then there are thoughts on US strategy both within and outside the Alliance, such as are contained in the Ikle/Wohlstetter report on "Discriminate deterrence". Added to these are US budgetary problems, and problems of burden-sharing which should be seen in a new light and in the context of the tour of European capitals undertaken last May by the US Deputy Secretary of Defense, William Taft. There are also elements which are not directly linked to the security field but which can have an impact on Western Europe's defence dialogue with the North Americans: for instance, growing difficulties between the US and the European Communities in the trade field. Furthermore, the Canadian Government has been rethinking some important respects of Canada's participation in common defence.

So the East–West relationship is in full evolution and the transatlantic relationship in a period of transition. It is therefore essential for the Western European allies to put their act together. If they do not succeed in doing so they will increasingly be bystanders rather than actors on an international scene that will be played out without them. There is then the risk that, confronted with this situation, they will adopt different, even diverging reactions and policies. That would be fatal both to the process of European construction, which would be diluted, and to the Alliance, which would progressively disintegrate. On the other hand, if the Europeans are able to act together they will strengthen the process of European construction, especially in the perspective of 1992. This means not only in the security field, but also in the economic area and in the area of political cooperation, the foreign policy field. They will also strengthen the Alliance, because if they get their act together they will be able to shoulder better their responsibilities within the Alliance and to give a better balance and more efficiency to it.

SECURITY COOPERATION IN EUROPE

Are the Europeans working in that direction? It is a fact that cooperation is growing among the states of Western Europe in the field of security. This movement is taking place both at the multi-lateral level and the bilateral level.

At the bi-lateral level the most spectacular manifestation is certainly the special relationship between the Federal Republic of Germany and France. But that special Franco-German relationship is by no means the only one. There are many others and they are growing in number and importance. There is, for example, very substantial defence cooperation between the United Kingdom and the Federal Republic of Germany. It is less visible than the Franco-German one which is generally called the "quiet alliance" but it is important enough for a book to have been dedicated to it by John Roper and Karl Kaiser entitled *British-German Defence Cooperation—Partners within the Alliance*.[1]

Multilateral cooperation finds its place in the framework of Western European Union (WEU), which has been re-activated for this purpose.

The question can of course be asked whether the fact that the European Allies inscribing their security problems on three levels: the European bilateral one, the European multilateral one, and the Alliance one is not a disruptive and divisive phenomenon for this Alliance, or whether it offers prospects for fruitful solidarity? I would think that it does offer the prospect for fruitful solidarity, if some conditions are met.

Those conditions have been formulated by Lord Carrington, and I could not agree more with this definition: in order to offer prospects of fruitful solidarity, the relationships between the bilateral fora and the multilateral European forum—WEU—on the one hand, and the Alliance on the other, must meet at least three criteria.

Firstly, they must be compatible and not contradictory. In other words, the actions of the WEU must be compatible with the objectives and the actions of the Alliance; and the actions within the bilateral fora must be compatible with the actions and objectives of the WEU and with the actions and objectives of the Alliance.

Secondly, they must be transparent. In other words, the allies who are not members of WEU and members of the bilateral fora must be fully informed of what is happening in the WEU and in the bilateral fora, and the members of the WEU who are not members of the bilateral fora must be equally well informed about what is happening in the bilateral fora so as not to feel excluded.

Thirdly, the activities in the bilateral fora and in the WEU must reinforce the commitments of the partners to the Alliance and certainly not diminish it.

As Secretary General of the WEU, I can state that we are certainly trying our best, and I think we are succeeding, in meeting those criteria with respect to the Alliance. I can also say regarding the bilateral fora that our member countries are respecting those conditions in relation to the WEU.

THE ROLE OF THE WEU

What is the precise role of the WEU in helping Europe put its act together within the framework of Atlantic solidarity? What is its role in shaping a European security dimension in the perspective of Atlantic solidarity and commitment? It is important to note that since the very beginning, since 1948, the WEU has been working in the double context of the process of European construction and of that of Atlantic solidarity. It has remained, since then, faithful to this double context. Its member states decided to reactivate it after a long sleep of 10 years in 1984–85. When they decided to reactivate it, they made sure that the commitment to this double context would be strengthened, that the reactivation would take place not outside but within the Alliance, and that the new WEU would be an element of the process of European construction. That is very clear.

Now what must WEU—as an element of the process of European construction, an element of the Atlantic solidarity, and possibly a beginning of the European pillar of the Alliance—do? The WEU rests on a treaty, the Treaty of Brussels of 17 March 1948 which was modified in 1954 by the Paris Agreements to allow the addition of the Federal Republic of Germany and Italy. The Treaty remains valid, and in it there are several dispositions, two of which are of great importance. There is Article V which creates a very compulsory alliance between the partners, and there is Article IV which formalises the links which exist between NATO and the WEU.

The Rome Declaration of October 1984—which was the "act of rebirth" of the Organisation—gave the WEU a new and important dimension by making it the European centre for common reflection and consultation on security problems among its member countries. Consequently, the WEU must assert itself as a forum for European political cooperation in the field of security. It must establish among its member States an ongoing dialogue with the aim of coming to converging or common positions and thereby developing a European security identity, and through that identity a public awareness about our problems of European security seen in the Atlantic perspective. The fact that the WEU has a parliamentary Assembly and that its Ministers have been meeting again since 1984 make it possible—in that perspective to develop a democratic, ie public, dialogue at European level on security problems and it is important for the emergence of such an awareness.

We know that growing sectors of public opinion have distanced themselves from the problems of our common security. This has been shown by the mass demonstrations which took place against the INF from 1979 to 1985. At the same time, the consensus on which any defence policy must be based has been eroded. This distancing could be reduced and this consensus restored, at least in part, by a democratic dialogue on security problems at the European level.

CAN THE WEU FILL THIS ROLE?

There were quite a few difficulties at the outset, but now the Organisation is doing what it was reactivated for. It is now established as a forum for European political cooperation in the field of security. It works out converging and common positions of its members in this field. It has begun shaping a European security identity, and it has done so in the best interests of the Alliance at a time when the Alliance is in transition.

European political cooperation is a success story because it brings together those representatives in its member States who have responsibility and authority in foreign affairs. This is essential because when those people—Ministers, Political Directors and experts—come to common conclusions and when they go back to their capitals and give their instructions, those common conclusions become converging policies or even common policies.

The same kind of cooperation now takes place in the WEU, and this is quite new. The WEU is also the only place in the West where Foreign Affairs and Defence Ministers meet together, and they do this twice a year. The Political Directors of the Foreign Ministries meet in theory at least four times a year with their Defence Ministry counterparts. In practice they meet considerably more often. Then there are meetings, two or three times a month, between the politico-military Directors of the Foreign Ministries and their Defence counterparts. In addition, expert groups from Foreign and Defence Ministries dealing with specific questions such as defence resources or security in the Mediterranean meet every one to three months. Coordinating all these activities is the Council of Permanent Representatives, made up of the Ambassadors of member States to the Court of St James's and a senior official from the United Kingdom Foreign and Commonwealth Office.

In these conditions, the first role assigned to the WEU by its member countries when they decided to reactivate it, to make of the Organisation a European political cooperation in the field of security, had been fulfilled.

What is more, thanks to these inter-governmental organs —as well as to the establishment of a special telex network between the seven capitals—an ongoing dialogue is now taking place among the member Governments allowing them to come to concerted or common positions regarding topical security problems. Achieving that was the second part entrusted to the Organisation when it was relaunched.

The need for the Europeans to get their act together had emerged very clearly following the Reykjavik Summit between

Mr Reagan and Mr Gorbachev in October 1986, which sent shock waves through Europe and the Alliance. It was not that there had been no consultations before Reykjavik in the Alliance—there had been. But those consultations had been placed in a step by step perspective, as was the practice for such negotiations. At the Summit, however, Mr Gorbachev changed the whole perspective by suggesting a global approach and the progressive, but ultimately complete, suppression of new nuclear weapons. Although this was not in fact agreed, the Europeans were left with the impression that it could have been and that all the strategies of the Alliance would have been changed overnight. That of course frightened them and after a number of high-level meetings at the level of politico-military Directors and Political Directors of the Ministries of Foreign Affairs with their Defence counterparts, the WEU Foreign and Defence Ministers had little difficulty in formulating a common position on these matters when they met on 13 November 1986 in Luxembourg. The conclusions of the WEU Ministers were immediately utilised by Mrs Thatcher in a meeting with President Reagan at Camp David the following day, 14 November.

Since then, the same phenomenon has occurred in relation to the varying course of East–West disarmament talks and more particularly those between the United States and the Soviet Union. This has not always been an easy matter. But the dialogue, now covering all essential security issues, has continued among our 14 (Foreign and Defence) Ministers and among their immediate collaborators, thus allowing views to be brought closer, harmonised and jointly expressed.

It was soon clear that such a convergence of views on topical problems was in line with the basic longer-term policies of the countries of Western Europe. This, in fact, is nothing new. It was recognised, for example, in the 1974 Ottawa Declaration by the Atlantic allies. This Declaration laid down the principle of the unity of the Alliance in the face of the threat to all its members. It also emphasised the specific vulnerability of Europe, which thus had a special position in the context of Atlantic solidarity. However, there had never been any attempt to clarify and define these fundamental joint policies, nor had their consequences been considered.

It is a task that WEU has undertaken at the instigation of its Member States. It met a need identified in November 1986 in the post-Reykjavik situation and the appeal made in December

1986 by the French Prime Minister, Jacques Chirac, in an address to the Parliamentary Assembly of the WEU in which he called for a definition of the "principles of Western European security". The task was a somewhat difficult one. There were of course differences in some national positions, but the political will to succeed was evident. Serious and in-depth work was begun and lasted nine months. It resulted in the seven Governments finding considerable common ground as to their security interests and options. The fruits of this work were embodied in a report which in turn led to a "Platform" adopted by the 14 Ministers on 27 October in The Hague.

In adopting this "Platform", the "Seven" had fulfilled the third essential role assigned to them—in the indispensable context of Atlantic solidarity—namely the creation of a European identity in the security area. This marked an important stage in the construction of Europe and in the strengthening of the European pillar of the Alliance. The "Platform" again underlines WEU's double vocation, one European and one Atlantic. As to the European vocation, the "Platform" states:

> We recall our commitment to build a European union in accordance with the Single European Act, which we all signed as members of the European Community. We are convinced that the construction of an integrated Europe will remain incomplete as long as it does not include security and defence.

On the Euro-Atlantic aspect, it states:

> We intend therefore to develop a more cohesive European defence identity which will translate more effectively into practice the obligations of solidarity to which we are committed through the modified Brussels and North Atlantic Treaties.

On the Atlantic vocation, it states:

> The security of the Western European Union countries can only be ensured in close association with our North American Allies. The security of the Alliance is indivisible. The partnership between the two sides of the Atlantic rests on the twin foundations of shared values and interests. Just as the commitment of the North American democracies is vital to Europe's security, a free, independent and increasingly more united Western Europe is vital to the Security of North America.

The "Platform" has three pivotal points: the criteria of European security, the conditions of European security and the responsibility of the Europeans regarding security in the defence field, in the arms control field and in the field of dialogue with the East. Let us quote some of its sections which are particularly significant.

Europe remains at the centre of East–West relations and, forty years after the end of the Second World War, a divided continent. The human consequences of this division remain unacceptable, although certain concrete improvements have been made on a bilateral level and on the basis of the Helsinki Final Act. We owe it to our people to overcome this situation and to exploit in the interest of all Europeans the opportunities for further improvements which may represent themselves.

It is our conviction that the balanced policy of the Harmel Report remains valid. Political solidarity and adequate military strength within the Atlantic Alliance, arms control, disarmament and the search for genuine détente continue to be integral parts of this policy. Military security and a policy of détente are not contradictory but complementary.

We recall the fundamental obligation of Article V of the modified Brussels Treaty to provide all the military and other aid and assistance in our power in the event of armed attack on any one of us. This pledge, which reflects our common destiny, reinforces our commitments under the Atlantic Alliance, to which we all belong, and which we are resolved to preserve.

We are determined to carry our share of the common defence in both the conventional and nuclear field, in accordance with the principles of risk and burden-sharing which are fundamental to allied cohesion.

—in the conventional field, all of us will continue to play our part in the ongoing efforts to improve our defences;
—in the nuclear field also, we shall continue to carry our share: some of us by pursuing appropriate cooperative arrangements with the US; the UK and France by continuing to maintain independent nuclear forces, the credibility of which they are determined to preserve".

. . . we shall ensure that our determination to defend any member country at its borders is made clearly manifest by means of appropriate arrangements.

The common responsibility of all Europeans is not only to preserve the peace but to shape it constructively. The Helsinki Final Act continues to serve as our guide to the fulfilment of the objective of gradually overcoming the division of Europe. We shall therefore continue to make full use of the CSCE process in order to promote comprehensive cooperation among all participating States.

Therefore, the "Platform" is an important document and it has been recognised as such on both sides of the Atlantic. All that has been done in a way that has not weakened the Alliance but strengthened it. Two examples can be given. Firstly, the "Platform" was widely used during the Alliance summit on 2–3 March 1988 in Brussels. When I was in Washington in February–March, I visited both the Pentagon and the State Department and was told how useful they found it for the preparation of the Summit and how much they hoped that a large part of it would influence the final Summit Communiqué. Indeed it did, and full sections of the Platform inspired the Communiqué. Secondly, WEU's action has proved to be useful to the West in what are generally called the "out-of-NATO", or "out-of-Europe" areas. It is true that the question of possible allied action outside the specific area of Atlantic solidarity has been a matter of controversy within the Alliance. Whenever the question has arisen, it has led to an academic discussion which has rarely—if ever—led to any solidarity of action. Doubtless, the reason for this lies in the ambiguity of Article VI of the . North Atlantic Treaty, but also in the diversity of political commitments on the part of the Alliance.

It is, however, evident that the WEU can take such action. This is stated in the modified Brussels Treaty, in the Rome Declaration and in the "Platform". So, when in August 1987 freedom of navigation in the Persian Gulf was threatened, the problem was taken up in the WEU. That very month, the Netherlands Presidency of the WEU Council called for a policy consultation, following which Italy, the Netherlands and Belgium decided to send a minesweeping task force to the Gulf to supplement the French and United Kingdom naval presence there. Although national missions, the units of the five naval forces were operating on the basis of WEU consultation. It was the first time that it had ever happened in Europe.

When the Red Sea was mined in 1984, there was a European naval presence there, which was completely uncoordinated. The British went with the Egyptians and the Americans, while the Italians and the French each acted separately. When in 1978 there was the attack from the direction of Angola on Shaba in Zaïre, two European countries, France and Belgium, had a very close interest in events. They tried to concert, but they not only failed in this, they succeeded in going in divergent directions.

In the Gulf, however, things were different. As it was the first time, we had to progress pragmatically. Later, two other WEU member countries which could not go to the Gulf, the Federal Republic of Germany because of the interpretation given to its Constitution, and Luxembourg because it has no naval vessels, showed solidarity with their five partners, the former by replacing units redeployed to the Gulf from the Mediterranean and Atlantic—the first time there have been FRG naval units in the Mediterranean—and the latter by making a financial contribution. So we have, in sequence, political planning by mutual agreement, solidarity and then coordination. The operations of the five WEU member countries have been coordinated throughout, under WEU aegis, both in the Gulf and in the capitals of Europe. All this happened alongside the US presence in the Gulf.

NEW MEMBERS FOR THE WEU

Since 1954, there have been seven members of the WEU. The success of its reactivation has, however, aroused the attention of other States. In October 1984, Portugal, and later Spain, showed an interest in joining the Organisation. Since then, Greece and Turkey have done likewise. The modified Brussels Treaty makes express provisions for the accession of new countries to the Organisation. Moreover, the "Seven" have always said that they did not want to form a "closed shop" and intended to open their Organisation, at the appropriate moment, to other European countries, provided they shared their democratic regime and politico-strategic options. But at the same time, they do not wish enlargement to damage the Organisation's cohesion. So, the member States decided to advance cautiously along this road, and in particular to wait until the WEU had been fully reactivated before issuing the first invitations.

It is in this perspective that the Foreign Affairs and Defence Ministers decided, on 19 April 1988, to invite the first two "candidates"—Portugal and Spain—to start discussions with the WEU with a view to their possible accession to the Organisation. These discussions have started on 26 May and, at the time of writing are still going on. The general aims of the discussions are:

- [ ] to elucidate the legal question and assess the implications thereof;
- [ ] to ensure that the candidate countries are aware of all the implications of accession as regards their political and military aspects;
- [ ] to make sure also of their determination fully to commit themselves in this connection;
- [ ] to identify, with the candidate States, the forms of their contribution to the objectives of the "Platform" and to their development.

CONCLUSIONS

It seems clear from all this that the WEU is now indeed effectively reactivated, that it is playing a full part in the process of the European construction and that it constitutes a strengthening element for the Alliance. Of course, difficulties remain as, for example, the problem of regrouping the WEU's administrative organs—now divided between London and Paris—into one place and restructuring them, which has not yet been solved. But these difficulties do not prevent the Organisation from fulfilling its role and from progressing. Of course, much remains to be done and many challenges lie ahead, but the political will of the member States to meet those challenges is evident. The Dutch Presidency (1 July 1987–31 June 1988) may be proud of WEU's achievements under its leadership (among others: adoption of the "Platform" and coordinated actions in the Gulf) and the British Presidency (1 July 1988–31 June 1989) has a programme that is at the same time realistic and ambitious. The Western European Union can look to the future with confidence.

NOTES

[1] John Roper and Karl Kaiser, *British–German Defence Cooperation—Partners within the Alliance*, Jane's in association with RIIA (London) and the Forschungs Institut der Deutschen Gesellschaft für Auswärtigen Politik (Bonn), 1988.

NATO after INF

MARC FIELDER

Marc Fielder is a researcher on the RUSI Western European Security Programme. The author wishes to express his debt to the participants of the Hanns Seidel-Stiftung C.V. Conference on the Problems of Security Policy and the German Question *(Wilbad Kreuth, January 1988) whose discussion contributed substantially to this work.*

THE FINAL months of the Reagan Administration were a poor contrast to the euphoria witnessed at the start of the President's term in office. The government's moral authority, carried on a tide of Christian fundamentalism, had been compromised during the tawdry "Irangate" affair in full view of the American public and serious economic problems existed over the nation's continued budget deficit. One of the few trophies which the administration could offer to domestic and world opinion was the success in securing an arms control initiative which resulted in the signing of the INF Treaty in Washington between the two Superpowers. Under the terms of the treaty both the Soviet and American arsenals of ground launched Intermediate Nuclear Force (INF) missiles (all those with a range between 500 and 5500km) were to be abolished, the first occasion on which missiles were scheduled for destruction rather than limitation below a specific number. But increasingly critics have argued that the INF agreement is less of a "Tyrian mantle" sitting on the shoulders of the West's arms control strategy than a loose thread which Gorbachev has tugged, threatening to unravel the whole skein of interconnected Alliance interests which were held together by the deployment of INF missiles in Western Europe. Yet the treaty is still overwhelmingly popular among the general public (and hence also among politicians) in the United States and Europe alike.

Failure to appreciate the full implications of the treaty and the dangers it poses for European security possibly arises from the fact that at a superficial level Gorbachev's acceptance of the treaty appears to represent a complete success for NATO's proscribed arms control goals. In the 1970s the West was confronted with a massive build up of Warsaw Pact military forces, one part of which involved the USSR starting to deploy the then new SS–20 missiles; with three independently targeted warheads integrated into a mobile missile system, this represented a massive increase in the capability of the Soviet's theatre nuclear forces beyond the old, less accurate, single warhead SS–4 and SS–5 systems previously deployed. These new weapons represented a massive increase in the threat levelled at Europe, particularly the airfields and ports etc. which would prove vital to achieve successful reinforcement from the United States. Soviet policy makers, who appeared to be woefully informed on the political realities in the West at the time, seemed to believe this would fail to elicit a serious response from NATO. An opinion which may at first have seemed justified as the US Administration vacillated awkwardly in the face of a European request for an adequate counter to the Soviet challenge. However, by December 1979 President Carter had been prevailed upon to reply to the Soviet threat by stationing American cruise and Pershing II missiles in Europe.

A NEGOTIATING ODYSSEY

Anticipating that the arrival of these new systems would provoke considerable concern among the general public of the Western democracies (who have always been uncomfortable with the arrival of "new" systems, regarding this as a dangerous escalatory rather than stabilising action) the systems were to be deployed as part of a "dual track" decision. This envisaged that while NATO planned the production of new INF, attempts should be made to reach an agreement with the Soviets which would negotiate limits on US deployment in exchange for reciprocal Soviet limitations. The USSR failed to respond effectively to this, and following a vote in the West German Bundestag in 1983 which agreed to accept deployment of INF the Russians responded by walking away from the arms negotiations talks in Geneva. They believed that Western resolve was not sufficiently strong to withstand the spectre of an end to detente and the start of a new Cold War. The Soviets were undoubtedly encouraged in taking this hard line by the appearance in the West of rapidly burgeoning unilateralist and pacifist movements which began to mount massive demonstrations in protest against the future arrival of the new weapons. The unilateralist movement gained greatest momentum in the five European states which were to host the cruise and Pershing missiles.

However, the USSR appeared to have miscalculated the overall effect of these movements and succumbed to the spectacle of the protests, confusing a vocal appeal to populism which movements such as the CND were making, with a general mobilisation of majority opinion behind the call for a unilateral withdrawal of nuclear arms.[1] Proof that the Soviets had incorrectly gauged the true feelings behind Western reactions were plainly shown by the re-election of right wing governments who were standing firmly on a deployment platform in both the United Kingdom and the Federal Republic of Germany at the height of the Euromissile debate in the early 1980s. By the time Mikhail Gorbachev had come to power in 1985 the Alliance was a long way towards its objectives and although the Western unilateralists were still a considerable force, the decline in numbers attending their rallies was the clearest indication that their movement had begun to flag. For the NATO authorities had succeeded in undermining their opponents' arguments by pointing out that the arrival of each new INF weapon had failed to be met with the apocalyptic repercussions which the peace movements had threatened would prove inevitable. This left the NATO establishment free to attack their critics with the most damaging of political taunts, that of being false prophets. The final triumph for supporters of the Alliance, and an apparent vindication of their stance, came with the return of the USSR to serious negotiations after Gorbachev's accession. The Soviet decision to return to the talks was promoted by at least two reasons which neatly dovetailed one into another.

First, left to its own dynamics the West was clearly not going to be deflected from its stated intention of eventually deploying upwards of 570 new INF missiles. Of these over 100 were Pershing IIs; highly accurate, with a range of approximately 1,800km, their flight time from launch to impact was something less than 15 minutes. Such characteristics posed a potentially lethal threat to the Soviet military command positions, marshalling areas and choke points etc. and so represented a substantial increase in the effectiveness of NATO's theatre nuclear arsenal. With such a short flight time the Soviets had no hope of finding a military counter to the threat, they must have reasoned therefore that a political settlement was likely to prove the only effective answer and this would only be achieved by a return to the negotiating forum. Second, the rise to power of Mr Gorbachev heralded his now famous "new thinking". The wisdom behind the previous policy which apparently sought to achieve a continual growth in the Soviet armed forces regardless of any serious evaluation of NATO capabilities was now refuted. It was replaced by Gorbachev's apparent intention to cut back on over investment in military expenditure in an attempt to release funds to support a growth in the Soviet Union's economic and commercial base. However, Gorbachev realised that a cut in arms production made as a result of negotiations prompted by economic exigencies alone would almost certainly prove unacceptable to the powerful Soviet political and military traditionalists. Such an apostasy had to be cloaked in arguments capable of placating "conservative" critics within the Party elite. The INF treaty had the advantage that it would secure the withdrawal of the dreaded Pershings, a coup for Gorbachev, and also provide him with leverage in future debates with the military. For he could reasonably claim that his policy had resulted in the withdrawal of the US weapons whereas it was the previous programme of continually new deployments which had provoked the American response in the first instance. Gorbachev's acceptance of the American proposals was something of a high risk policy as it is possible to see that there are many elements of the final treaty which might cause grave concern to Moscow's hawks. Soviet political conservatives may well have been alarmed as each of their provisos against what they believed were unacceptable elements of the so called "zero-zero" option were conceded in the face of American opposition. First the talks became global, rather than being confined to Europe as the Soviets wished, then they were extended to shorter as well as longer range INF, and finally, despite something of a negotiating odyssey, the Soviets relinquished their demand that 100 warheads should be retained. Although sceptics claim that Soviet negotiators created artificial impediments just to present the image that they were capable of compromise when these "objections" were later withdrawn. Nonetheless many NATO leaders took the opportunity to declare that the USSR had capitulated in the face of Alliance strength and unanimity. Furthermore surrendering the attempt to draw British and French warheads into the negotiations did appear to be a genuine concession, not the mere relinquishing of a bargaining chip. Even more significant, the treaty resulted in the abolition of nearly three times as many Soviet warheads as American, so creating an important precedent with the USSR accepting the validity of a call for asymmetrical reduction of weapons systems in the West's favour. These are not inconsiderable achievements, and appear to justify the Reagan Administration's position that a major political victory had been secured which enhances Western security. Despite this sceptics still claim that serious analysis reveals that, despite appearances to the contrary, it is in fact the Soviets who have excuted something of a diplomatic ambuscade by agreeing to the American proposals and gained a host of advantages largely unsuspected by the West.

FALLING FROM A TIGHTROPE

The *ancien regime* under Brezhnev would probably have refused to countenance any asymmetrical concessions, but now the Soviets are more flexible and appear to be prepared to make tactical withdrawals to gain strategic advantage. Through a more careful appreciation of the Western position Mr Gorbachev has identified serious flaws in NATO's negotiating position. First, he has understood that although the West's leaders successfully saw off the challenge of the unilateralists a high price was paid. With one eye always set on the snares that the unilateralists might lay for them in the battle to gain the support of the general public, NATO leaders may have overplayed their hand; as they consistently presented a negotiating package aimed as much at soothing the fears of their respective electorates over the issue of deployment, but achieved at the expense of fudging policy objectives.

The "dual track" was the first step in this process, whereby NATO leaders attempted to pre-empt domestic criticism of the call to build the cruise and Pershing missiles destined to be stationed in Europe. The second came with President Reagan's offer of the destruction of the weapons. It is worth considering that among the European governments it was felt that popular support could only be swung behind the arrival of the new missiles if due attention was paid to the sensibility of the general population. However much of a myth in the politician's mind, it was politically expedient to publically declare a clear policy to withdraw and destroy them at the first suitable opportunity. Such a position was a calculated risk because, as has already been stated, considerations of political solidarity aside there was a very real military requirement for such systems. The absence of such INF systems had first been realised in the late 1950s when the Soviets started to deploy their first SS–4 and SS–5 weapons. The NATO Council had responded to this in 1957 by agreeing to deploy the Thor and Jupiter medium-range ballistic missiles (MRBMs) in Europe. These arrived on the continent in 1959, where they served to increase both the credibility and the effectiveness of the American deterrent. They improved the Europeans' faith in the sincerity of America's extended nuclear guarantee, by providing a visible nuclear presence which served as a form of physical link between the two pillars of the Alliance. They also fulfilled a requirement outlined by the then Supreme Allied Commander Europe (SACEUR) General Norstad, that NATO needed to have secure land based weapons in Europe which were capable of providing what he saw as the ultimate deterrent to the Soviets, the ability to strike deep into the heart of Warsaw Pact territory and even into the Russian homeland itself. Yet by 1963 Thor and Jupiter had been outpaced by new technology, and replaced instead by land and sea based inter-continental ballistic missiles.[2] No equivalent land based system was adopted and so something of a gap appeared in NATO's forces. This was only filled by the arrival of cruise and Pershing (which again satisfied all the necessary criteria outlined by General Norstad) nearly two decades later. The problem for the Alliance was that this action was taken as a result of political calculations and not at the request from the military, and part of this political calculation was the offer of the "zero–zero" option. A clear gamble was taken by the execu-tives in which political considerations were underwritten by military requirements. Admittedly, in 1981, the odds seemed to rest heavily in favour of such a wager: the Soviets had remained so inflexible in the past that NATO's diplomats could make a reasonable assumption that Russian negotiators would never accept the abolition of a whole class of weapons. There was simply no precedent for Mr Gorbachev's sudden accep-tance of the offer which entirely wrongfooted his opponents. By accepting offers made in Geneva by the Americans he has exposed a lack of coordination between the military and the political wings of the Alliance. Just how deep this divergence of opinion runs only became obvious during the US Senate's

Hearings for the ratification of the treaty when the former
SACEUR General Bernard Rogers declared:

> I am convinced that under the aegis of the United States, the long
> term credibility of NATO's deterrent is being sacrificed by this
> treaty on the altar of short term expediency.[3]

In making this statement he possibly felt that no sooner had
a breach in NATO's defensive array been plugged by the
arrival of new missiles, than the politicians allowed them to slip
away and the hole to be blown open once more. Even the
present SACEUR, General John Galvin who supports the INF
treaty concedes that it represents the loss of several "bricks"
from his "defensive wall".[4] Beyond posing a complicated
challenge for the military, who need to sustain the credibility of
the forces in Europe after the loss of these highly potent
weapons, the treaty has posed political problems too. Serious
differences have started to become evident between the Allied
governments over the response they are to make to the new
political landscape which has developed since last December,
as well as the aims and objectives that need to be established
for future arms control negotiations.

EXTENDED GUARANTEES RUNNING OUT

The standard fear has centred on the European concern that
the United States remains practically invulnerable to any form
of Soviet conventional strike, and whose main security concern
centres on the threat of strategic nuclear weapons. Meanwhile
for the Europeans, as appalling as nuclear weapons may be, the
threat of a full scale conventional war is equally terrifying. This
long realised but often suppressed difference came to the fore
during the 1985 Reykjavik meeting between Gorbachev and
Reagan. Following earlier statements by Mr Reagan concerning the future development of the Strategic Defense Initiative,
it appeared that his personal view envisaged the abolition of
nuclear weapons as the ultimate goal of the arms control
process. Whether such an aim could seriously be achieved is a
matter for debate, but the mere suspicion that American
thinking is gravitating towards the scenario of a world without
strategic missiles, seriously undermines the credibility of the
extended guarantee to Europe. There is the very real fear
therefore that the abolition of the INF systems is the first break
in the coupling between the two pillars of the Alliance

Alliance, and that it is the first step towards the eventual denuclearisation of Europe, which would leave the continent at an even greater disadvantage in the face of the Soviet conventional superiority.

It is when these broader issues are investigated that it becomes easier to speculate that the genuine political triumph belongs to Mr Gorbachev for accepting the agreement, rather than the West for proposing it. The Soviets may well have lost their SS-20s, but any sacrifice they have incurred in their military capability (and this point is open to debate) is entirely offset in the light of the equivalent Western loss coupled with the possible political discord that has been provoked.

Apologists for the treaty have consistently refuted the critics' claim that an entire "rung" has been lost from the ladder of escalation, but this claim has been seriously vitiated by the sudden urgency, even among some treaty supporters, to now to focus attention on the need to overhaul the existing stocks of short range nuclear forces (SNF) which will remain in Europe after the INF are withdrawn.

These SNF tactical weapons were first introduced into Europe in 1954. Then it was intended that they should compensate for Soviet conventional superiority, providing an additional deterrent to the Soviet forces beyond the threat of an all out strategic strike and provide a further link in the nuclear chain between the US strategic weapons and Europe. A major problem concerning this stock of SNF today is that they have evolved as a monument to a pattern of uncoordinated development and deployment over a period of several years. The greater the threat believed to be presented by the Soviet strategic nuclear and conventional forces, the greater the reliance placed on SNF to deter any potential Soviet aggression in Europe. This led to a massive build-up in the number of weapons, which had reached a total deployment approaching 7,000 warheads by 1966. This stockpile remained in existence until the mid 1970s, by which time the arsenal ranged across every imaginable system from the Lance surface-to-surface missile to nuclear artillery shells. Many of these weapons were ageing rapidly by the 1970s, and some, such as the atomic mines, proved a liability to store safely even in peacetime. Others, with their extremely short range would, in the event of a conflict need to be deployed well forward, yet at the same time protected in a secure environment. Such complications almost caused more problems than the weapons

merited. Such proliferation of weapons, supposedly performing so many tasks, led to an increasingly confused understanding of the precise role they were to play in military planning.
This rather blurred picture coupled with a failure to establish a
comprehensive understanding of how they were to be used,
and exactly what they were supposed to achieve seriously
undermined the deterrent effect they were meant to present.
In an attempt to rationalise this situation in the 1970s the
United States Defence Secretary James Schlesinger attempted
to introduce weapons with "lower yield, increased accuracy
and tailored effects".[5] Yet attempts to introduce these "neutron bombs" floundered as protests mounted in Europe
centring on the argument that the discriminatory nature of
these weapons would make their use more acceptable politically, and hence by association more likely. In the face of the
fierce opposition in Europe, Schlesinger's plans were set
aside.[6] It was not until the 1979 decision to deploy cruise and
Pershing that the problem of SNF was once again seriously
addressed, for at this time it was announced that the new INF
deployments would be offset in Europe by a reduction of
1,000 warheads from the short range stockpile.

Members of the NATO establishment pointed to this
unilateral withdrawal of weapons as an indication that the West
was committed to preserving a nuclear balance through the
maintenance of credible forces, but determined that this should
be achieved by deploying the minimum number of weapons that
were felt to be realistically capable of dissuading aggression
from an enemy. Sceptics argued that this action was in fact a
highly astute political manoeuvre that offered the general
public a distraction from the forthcoming arrival of new
missiles, but was in reality no more than a political sop which
would make a major increase in capability more palatable to
Western public opinion. At the same time it would giving an
opportunity to overhaul the existing arsenal and "weed out"
those weapons which were virtually obsolete, or so vulnerable to
premptive strikes as to be militarily worthless. Some claimed
that the only reason more publicity was not given to this decision
lay in the fact that the NATO leaders felt some trepidation in
announcing a unilateral withdrawal too loudly at the time of the
Soviet invasion of Afghanistan, or to risk the decision being
attacked by opponents as a publicity stunt.

The process of rationalising the weapons was further
developed at a meeting of the NATO Nuclear Planning Group

at Montebello in 1983 where it was decided to dismantle a
further 1,400 SNF warheads, a decision which would ultimately bring the number of warheads to its lowest level for
two decades. But unlike the simple unilateral withdrawal
announced in 1979, Ministers declared that the remaining
systems must provide effective deterrence and consequently
"both the delivery systems and the warheads must be survivable and responsive".[7] To achieve this a range of improvements were identified. However, although weapons were
withdrawn no action was taken on the call for modernisation
and this aspect of the decision fell into abeyance. The key
point is that regardless of the interpretation one places on the
motives of NATO's leaders in making these decisions the
emphasis was obviously placed on reducing the overall
number of tactical nuclear weapons available in Europe, while
the issue of modernisation of the remaining systems was
relegated to the political backburner. It is worthwhile to
speculate that if the INF Agreement was not felt to represent
a diminution of the West's deterrent, why it was that the issue
of modernising the remaining weapons suddenly developed
into a key topic at Alliance meetings after "zero-zero"
became imminent. The answer, which many in the NATO
hierarchy would vigorously deny, may be that the SNF debate,
and the Montebello decision in particular, are now no longer
being seen as a mere supplement to a broader discussion of
nuclear strategy, as was the case during the Cruise and
Pershing debate. Instead it has become clear that Montebello
is now being viewed by many on the political right as a useful
vehicle for offsetting some of the deficiences that they believe
will be created in the panoply of European ground based
weapons. For those who remain suspicious of the Gorbachev
"charm offensive" the issue is simple. Failure to modernise
the current stock of SNF would result within a few years in
"structural disarmament", as the existing systems could no
longer pose any genuinely viable threat to Soviet forces.
Proponents of modernisation will claim that this has now
been given greater emphasis, although is not a precedent
aimed at circumventing the treaty, but part of a long held
NATO policy. To this end NATO's military leaders have
suggested a follow on to the Lance missile, which with its
current range of only 120km remains unsatisfactory. Instead
it could be replaced some time after its proposed withdrawal
in the mid 1990s with a weapon similar to the Joint Tactical

Missile System. Currently designed as a conventional weapon, to be launched from NATO's Multi Launch Rocket System (MLRS), reports suggest that the US Defence Secretary Mr Frank Carlucci will seek to gain Congressional approval to integrate the same missiles with a nuclear head. Whichever system is finally adopted it is almost certain that it will be capable of a range more closely approaching the 500km limit which delineates SNF from Intermediate range systems. Similarly a new generation of artillery shell, with improved range and accuracy is likely to be produced.[8] The debate has intensified with a growing realisation, especially among conservatives in the United States that the INF treaty had perhaps succeeded in losing the most effective theatre nuclear weapons in Europe, while retaining the least useful. The momentum has gone so far that even General Galvin has declared (specifically in reference to Lance) that if he was not provided with a follow on surface-to-surface theatre missile he would be forced to task aircraft with new roles, and "as each element is removed, there is less flexibility and fewer options. This could mean that we might have to reassess our strategy".[9]

For the British government too, the issue is clear. The disparity in conventional forces needs to be countered, and existing weapons are apparently thought inadequate. Yet to accept the stationing of flights of B–52s armed with Advanced Cruise Missiles as was first suggested, presented too high a risk of provoking accusations that the spirit of the treaty was being broken; accusations which would prove difficult to refute. Implementation of the Montebello decisions gives the government scope for manoeuvre around the contentious arguments which would surround the acceptance of an entirely "new" weapon, yet still achieve the effect of increasing capability to meet any gaps in defence.

GERMAN QUESTIONS UNANSWERED

What appeared an obvious policy to both the US and the British administration has not been so readily accepted within the rest of Europe. Denmark, Spain and Greece have long been hostile to many aspects of American nuclear policy but the US has traditionally been able during times of crisis (such as the Euromissile saga) to forge a consensus with key members of the Alliance—Britain, Belgium, Holland, Italy and the Federal Republic of Germany. During the deployment debate even President Mitterrand, during his famous address to the Bundestag, was prepared to throw his weight behind the decision to deploy. But now that consensus appears to be broken. This became very clear in the wake of Secretary of State George Schultz's visit to the FRG in January 1988, when he emerged from a meeting with the West German Foreign Minister, Hanns-Dietrich Genscher, and each subsequently gave embarassingly divergent accounts of the conclusions they had reached after discussing future arms control policies. Mr Shultz had stated the importance of pursuing negotiations over strategic, conventional, and chemical weapon disarmament. In contrast Herr Genscher stated that the highest priority had been placed on securing negotiations over reductions in short range weapons. These contradictions stemmed from the fact that the United States (and Britain too) had failed to realise the full effect of the repercussions resulting from the Washington treaty. The most significant of these was that a cardinal principle of Alliance security had been broken, namely that there must be an equally shared level of security among Alliance partners, coupled with an equally shared level of risk. While a full panoply of nuclear forces were deployed throughout Europe this axiom was satisfied, but the elimination of a whole class of weapons appeared to upset this careful balancing act. The Germans suddenly felt that the whole pattern of nuclear deployment had shifted unfavourably. Short range weapons now held much greater significance in the nuclear equation than before, but the burden of fielding them fell almost exclusively on the FRG.

It was Helmut Schmidt who had realised the dangers of INF being deployed in the FRG alone, which resulted in the decision to spread the bases among as many of the Alliance partners as possible, although from an operational point of view it would have been entirely possible to base them all in the FRG. But had this been carried out Schmidt would have been left open to the challenge that he was being turned into some kind of cat's paw, and that the FRG was being assigned the role of America's client state in Europe. Having been careful to side-step this political gin in 1980, the Allies promptly stepped straight into it in 1988. In addition, as onerous as the Germans might find the feeling that their nation is being requisitioned as the peace time Quartermasters store for Europe's short range nuclear warheads, this is but a small consideration in comparison to the fear that these weapons if used, would fall exclusively on German territory. As Volker Ruhe, Chancellor Kohl's special envoy has stated so succinctly, "the shorter the range the deader the Germans". Thus the Germans display little interest in modernising weapons, for even the most favourable scenario suggests that the SNF that possessed sufficient range to land beyond the borders of the FRG would in all probability fall on the German Democratic Republic, and although this possibility may cause the majority of NATO nations little concern, the West Germans, with the sense of shared community between themselves and the GDR are bound to find such a prospect abhorrent.

German temporising over the issue of modernisation and complaints that Germany has been "singularised" by the threat has led to a welter of criticism being inveighed against them. It is claimed by exasperated Americans that the Soviets have retargeted some of their land based SS–24 and SS–25 missiles to perform a role similar to that of the SS–20, as well as redeploying ballistic missile submarines to cover the Western European theatre.[10] Even the American Ambassador in Bonn, Richard Burt, normally most sympathetic to German "angst", has pointed out his belief that the Soviets place such a premium on the destruction of nuclear weapons and their platforms there is no reason to suggest that Germany is markedly less secure than the United Kingdom which hosts squadrons of US F–111s.[11]

However, such arguments only carry any direct relevance to the defence specialists concerned with such arcanum. The general public in the FRG remains unconvinced, and adheres instead to the sentiment that several different zones of security are starting to develop within the Alliance. It may well be the case that the argument proffered by Richard Burt and others is correct, but as always in a democracy, if a misconception has been seized upon as the truth it proves almost impossible to refute; illusion once accepted as reality becomes more important than a disputed fact. Furthermore, this dissension in NATO ranks is examined in the full glare of the Western media, and is in no way subdued by the platitudes which the allies are always capable of promoting after various Alliance meetings. Gorbachev and the Warsaw Pact have no doubt been surprised that the solidarity the European governments were capable of displaying in the face of Soviet intransigence, should collapse when trying to counter the protean flexibility of current Soviet strategy. Whether Gorbachev expected to garner quite such as rich harvest as this falling out among the Allies, prompted by their inability to achieve an effective response to his actions, is a matter for speculation. But he has undoubtedly responded quickly to the political fruits which have been offered. Perceiving the precarious state of the internal debate in the FRG, he no doubt supported (if not inspired) the East German leader Erich Honecker in his offer of an inner-German compromise over the issue of non-modernisation of short-range weapons. It was suggested by Herr Honecker that the GDR would undertake to have no modernised weapons on its soil if the West Germans would undertake to do the same.[12] This follows an offer made by the GDR in October 1986 which suggested the creation of a 300km nuclear free corridor running through the centre of the two Germanys. There has also been an increasing barrage of "expert" Soviet military opinion both admonishing and regretting that NATO is engaged in an attempt to adopt compensatory weapons to replace the INF scheduled for destruction.[13] Of course it can be argued that such verbal broadsides are little more than the customary political sparring that it usually carried out between the opposing Alliances. Yet this seems more sinister due to the fact that a coordinated pattern is steadily emerging from Moscow's initiatives. At the height of the modernisation debate in January 1988, the Soviet Foreign Minister, Eduard Shevardnadze made visits to both West Germany and Spain (the latter was at the time engaged in an acrimonius dispute with the United States over the future of American bases on Spanish soil). Later in March, Mr Gorbachev himself, while on a visit to Yugoslavia, stated that the Soviet Union was prepared to accept a proposal (first made public by Bulgaria) to turn the Balkan peninsula into a nuclear free zone, and rid the area of any foreign military presence. Gorbachev singled out Greece for "initiatives . . . directed towards the reduction of military activities in the area". The Greeks were at that time engaged in discussions aimed at closing US military bases in their country.[14] The Soviets were clearly not shooting from the hip, but placing carefully aimed shots between the chinks in the NATO armour.

In the final analysis the NATO establishment will probably prove capable of pulling the disparate threads of Alliance interest into some form of coherent policy. Despite the virulent opposition of the German government to modernising weapons which was cited above, there was already evidence of the possibility for an eventual compromise solution by the time the NATO summit was convened in Brussels in March 1988. The final communiqué from the summit managed to reduce the bitter dispute to the customary anodyne pronouncements which mentioned the need to modernise "wherever this was necessary". Ultimately a solution will be found because although the Allies may disagree over the future policies and development of the organisation, a basic belief in its value and function remains across the 16 nations, but consensus will only be achieved through concession. This raises the issue of the nature of the price to pay attendant on such a compromise. The German's can always be pressed into line behind the other Alliance partners. This was the case over the demand that they should surrender the Pershing 1A missiles which the Soviets stated were to be destroyed under the terms of the INF Agreement, as they were armed with US nuclear warheads. The Germans initially stated that this was impossible, but when the issue threatened to block the entire INF Agreement which the Reagan Administration was determined to secure they finally relinquished the missiles. However, the sensibilities of many Germans were affronted, and these were not the anti-NATO Greens, but a large body of older conservatives who felt the Alliance was prepared to squander their interests for short term political gains (Pershing 1A, although old, had a range of 740km and therefore the important psychological advantage of being technically capable of striking deep into Warsaw Pact territory). Similarly the population of the smaller nations such as Denmark who are opposed to the response supported by the larger nations such as Britain may also feel slighted that their interests are not taken fully into consideration. No individual incident is likely to provoke a critical problem for NATO but the cumulative effect of continual minor dissension could prove fatal. It is this more insidious challenge that is being posed in the wake of the INF treaty. The full scope of the danger may not yet be realised, but still held in incubation, only to be hatched by future Soviet interjections.

The Swedish politician, Trygve Lie is claimed to have stated that a real diplomat is one "who can cut his neighbour's throat, without having his neighbour notice". Mr Gorbachev may yet prove the consumate diplomat.

NOTES

[1] J. Joffe, "Why the European Anti-nuclear Movement Failed", *International Security,* Spring 1987, pp. 3–40.

[2] North Atlantic Assembly Papers, *Nuclear Weapons In Europe,* Chapter 1, "The Evolution of NATO's INF Doctrine and Capabilities", North Atlantic Assembly, Brussels, November 1984.

[3] "Senate Hearings Assess INF Treaty's Impact on NATO", *Army Magazine,* March 1988, p. 22.

[4] *The Financial Times,* 19 January 1988.

[5] North Atlantic Assembly, *Nuclear Weapons In Europe,* op cit, pp. 5–6.

[6] Committee On Foreign Relations, *Nuclear Weapons and Foreign Policy:* US Congress, Senate, 93rd Congress, 1974, p. 209.

[7] NATO Final Communiqués 1981–1985, *The NATO Nuclear Planning Group Autumn Ministerial Meeting, Montebello Quebec, Canada (27–28 October 1983),* Annex To The Final Communiqué, Brussels, NATO Information Service, pp. 106–107.

[8] SIPRI, *World Armaments and Disarmament Yearbook 1986,* Part II, Chapter 3, "Nuclear Weapons", Oxford, OUP, 1986, pp. 37–80.

[9] *Army Magazine,* op cit, p. 26.

[10] *The Times,* London, 15.3.88.

[11] Speaking at the annual Wehrkunde Defence Conference in Munich, 7 February 1988, *The International Herald Tribune,* 8 February 1988.

[12] *Neues Deutschland,* 6 January 1988.

[13] Col. General Chervov in *Trud,* (Moscow), 9 January 1988. See also, *Joint Publication Research Service,* JPRS–UMA–88–004, 29 February 1988, p. 87, referring to reports in *Krasnaya Zvezda* on the 10 January 1988.

[14] Address to Federal Assembly, 16 March 1988, *BBC Summary of World Broadcasts,* Soviet TV 1530 gmt 16.3.88, EE 0103 C/3, 18 March 1988.

The Effects of Demographic Development on the Armed Services of the Federal Republic of Germany

DIETHELM SPRANGER

Colonel (General Staff) Diethelm Spranger is the Head of the Personnel Planning Branch on the Armed Forces Staff of the Federal Ministry of Defence. He is a member of the ad hoc Group on Manpower Practices and Problems (AGHM) of the NATO Eurogroup.

THE DECLINING BIRTH RATE

SINCE THE last century, birth rates in the Federal Republic of Germany, as in most industrialised countries, have been falling continuously. While the average family of the last century still had six liveborn children, the figure had decreased to as few as four by the first decade of this century and to barely more than two children per family by the middle of this century. Nevertheless, the population has been growing constantly because the death rate, and in particular infant mortality, have been falling noticeably at the same time, while the expectation of life at birth has been rising continuously. The only setbacks in that development were caused by the high casualty rates of both world wars, although the birth deficits which resulted were made up relatively quickly in the following decades.

The demographic development of the Federal Republic of Germany has been characterised since its foundation after the Second World War, not only by the development of the birth and mortality rates, but also by a considerable augmentation of the population by immigrants, at first refugees from the former German territories in the East and the GDR, and later by immigrant workers from all over Europe. From 1950 until 1973, the population increased from 50 to 62 million inhabitants.

In the long term, a population can only remain stable or grow if there is a birth rate of at least 210 babies for every 100 women. This has not been the case in the Federal Republic of Germany since 1970. In a significant development, the birth rate dropped in the period between 1964 and 1974 from 254 to 151 children for every 100 women. By 1985, it had declined further to 128 children. It is only in the last two years that a slight increase has been noted. In absolute figures, this means that in the course of the past 16 years, 3.5 million fewer children were born than would have been necessary to maintain the population. When taking a look at the population pyramid, one can see that the long-term population losses caused by that development are far greater than those caused by the two World Wars.

Internationally, the development described above is in line with the trend noted in all industrialised countries. But nowhere is the recession in the birth rate as serious as in the Federal Republic of Germany where it has been the lowest worldwide since 1973.

There are many reasons for such an extreme decline in the birth rate. It would certainly be wrong to only blame the invention of modern contraceptives. In addition to the improved possibilities of birth control and family planning, the same influence ought to be exercised by other factors, such as a change of moral and ethical values as well as the secularisation process, emancipation and women becoming economically active, together with alternative forms of life and anxieties of the future.

In a young democracy such as the Federal Republic of Germany, the Government is reluctant to exercise an influence affecting the sovereign right of life planning of its adult citizens by taking radical measures favouring a rate of population growth. It is true, however, that a whole series of measures have been initiated to benefit families. All in all, however, it is not expected that these measures will result in any fundamental change of the population's regeneration behaviour.

EFFECTS ON STATE AND SOCIETY

The effects of the declining birth rates on the various domains of state and society are manifold and deep-seated. Some domains are directly affected, while others will only be concerned on a medium or long-term basis.

While the population as a whole has been decreasing only

very slowly, the age composition has been changing relatively fast. In the year 2000 the overall population, with a total of approximately 61 million inhabitants, will only fall short of the 1973 maximum peak by about 1 million. However, the number of people under 20 years of age will drop from 18 to 13 million, while the age group 20 to 60 will increase from 31 to 34 million and that of people older than 60 from 12 to more than 13 million. In the time frame from the year 2000 to 2030 a decline of the population figure by more than 10 million is projected. In the course of that development, the number of people over 60 years of age will continue to rise. Simultaneously, the group of economically active people will be continually decreasing.

Directly affected by the aforementioned development is the entire sector of culture and education. The growing unemployment of teachers is a serious problem these days in a difficult process of adaptation for the educational sector. Trade and industry are confronted by the problem that in the next few years there will continue to be an excessive supply of older labour forces and a growing lack of people just entering a professional career.

All in all, it is worth noting that there is almost no sector of state, trade and industry, or our society at large which would not be affected in one way or another by the decline in the birth rate and the decrease of our population. The challenge for our politicians is great indeed and rules out any thinking in terms of legislative periods. Prospective, long-term planning will be of decisive significance.

EFFECTS ON THE ARMED FORCES

After a time-lag of 20 years, the Federal Armed Forces are particularly badly affected by shrinking birth rates. The armed services of the Bundeswehr are a conscript army with a great proportional share of volunteers and a strong element of reservists needed for mobilisation in a state of defence. With a peacetime strength of 495,000 servicemen, the armed forces so far used to require some 225,000 conscripts and volunteers aged between 19 and 28 to make up for the remainder each year. That strength can be safeguarded until 1988. From 1989 to 1993, however, the potential number of 19-year-old men fit for military service will drop to about 135,000. Without any remedial action being initiated, the peacetime strength of the armed forces would therefore have to be reduced to approximately 300,000 men over a period of 10 years. This theoretical computation is based on the fact that, now as in the past, some 35 per cent of our conscripts would be exempted from their 15-month military service on medical or other grounds, that more than 10 per cent would be needed in other public functions, and that the recruitment of volunteers would be affected by the shrinking age groups in the same way as the manpower resources in terms of conscripts.

The Federal Government has faced up to the problems in a timely fashion and set up a long-term commission to investigate, as early as in 1981, all possible means of maintaining the personnel readiness posture of the armed forces. The investigations were conducted with a view to the possibility of reducing the peacetime strength of the armed services and quickly revealed that any curtailment would be fiercely contradictory to the political goals of the NATO Alliance, these being, to modify the order of significance of nuclear weapons in the Alliance and to strengthen the conventional defence capability. A curtailment of the defence efforts of the Federal Republic, which provides for more than 50 per cent of NATO forces in Central Europe in peacetime, would have serious security-political repercussions and would affect the policy of the Alliance. Also and in particular, it had to be taken into account that disarmament talks aimed at eliminating the Warsaw Pact's conventional superiority would hardly be successful when negotiating from a position of inadequate conventional strength.

At first, the investigations were conducted with a view to the possibility of maintaining the armed forces' peacetime and wartime strengths covering a wide spectrum and including all possible alternatives for the expansion of potential and the reduction of requirements. It was soon revealed, however, that simple solutions, such as military service for women and foreign nationals had to be rejected on constitutional grounds; also abandoned for structural reasons was the idea of extending basic military service from 15 to 24 months. In the ultimate analysis, a whole series of measures were submitted to the Federal Government for decision-making which were suitable to be implemented and—perhaps equally important—were affordable.

On 17 October 1984, the Federal Government decided that the armed services' peacetime strength of 495,000 men and their wartime strength of 1.34 million men should be maintained and that these establishments be safeguarded through a whole series of measures. In the spring of 1985, the Federal Minister of Defence translated that decision into a "Personnel Implementation Plan" coordinating these measures with one another and preparing to implement them. That implementation plan is continuously adapted to updated resource figure projections and to the results of the structural analyses of the three services.

Under current planning, the composition of the armed forces' peacetime strength in the 1990s is to be as follows:

☐ 206,000 conscripts in basic military service,
☐ 250,000 career and temporary-career servicemen,
☐   39,000 reservists.

The rapid mobilisation capability of the armed forces in establishing their wartime strength of 1.34 million men in a state of defence is to be fundamentally improved at the same time. To that effect a new reservist concept has been elaborated.

CONSCRIPTS IN BASIC MILITARY SERVICE

General conscription is—as one of our Federal Presidents once put it—a "legitimate child of democracy". It is a clear indication of the responsibility shared by all citizens for national defence. In the foreseeable future, general conscription will be indispensable with a view to the personnel readiness posture of the armed forces of the Federal Republic of Germany. While it might be possible, by making extreme financial sacrifices, to recruit an adequate number of volunteers to provide for our peacetime strength, it would be altogether impossible to grow to a rapidly mobilisable wartime strength of 1.34 million men from a voluntary army which would be in need of a reservist potential of more than one million men.

The drop in the birth rate initially has a very positive effect with a view to the full exploitation of our conscript resources. For the first time ever, it will be possible to call up to basic military service all those who are fit and available for duty in the armed forces, and to subsequently earmark them for wartime assignments as members of the reserve force. The problem of equity in conscription used to be a major burden weighing on its general acceptance. During the past few years, an effort has been made, through generous exemptions from service and strict fitness criteria, to restrict our too large potential of conscript personnel. The criticism of that procedure was perfectly justifiable, however. Too many young men, who were indeed fit for duty in the armed forces, were not called up and were thus able—while others were serving in the forces—to obtain considerable advantages in their training or professional careers.

The Federal Government has been using the possibility of improving equity in conscription and gave the directive that prior to the extension of basic military service the fitness criteria were to be changed, the possibilities of exempting conscripts from military service were to be reduced and members of the stronger age groups not yet inducted were to be called up for military service as far as possible. Experience shows that it is indeed possible to call up for military service, without impairing the armed forces' operational readiness significantly, about one third of the young men so far exempted from duty on health or other grounds.

Despite that measure it was indispensable to extend the duration of basic military service from 15 to 18 months with effect from 1989. The Federal Government passed the necessary legislation to that effect as early as in 1986. Resistance against this indispensable, yet unpopular measure, was surprisingly moderate, both in public and the political opposition, and shows a high degree of understanding of the needs of national defence among the population and the responsible politicians. It became also clearly discernible that this measure was understood by our NATO allies and in the Warsaw Pact as an unmistakable token of German preparedness for defence and NATO solidarity.

The extension of the duration of basic military service is of significance, not only in relation to quantity, but also from the viewpoint of quality. Through such an extension, the number of recruits not yet fully trained will be reduced, and the possibilities of an improved training, in particular with a view to preparation for wartime assignments, will be considerably enhanced. In the ultimate analysis, the extension of the duration of basic military service constitutes a significant measure towards improving our conventional defence capability, without changing our peacetime strength.

Through the measures aimed at improving equity in conscription, the induction of members of the stronger age groups not called up for military service so far, and through the extension of the term of military service, it will be possible to maintain the necessary proportional share of conscripts in basic military service beyond the time frame of the current Armed Forces Plan, i.e. beyond the year 2000.

CAREER AND TEMPORARY-CAREER PERSONNEL

Not all the categories of personnel from which volunteers are recruited are affected by shrinking age groups. For officers and career NCOs, the situation with a view to the number of applicants has always been so good that even when reducing the annual strength figures by 50 per cent, a shortage of applications would not be expected. This is where an augmentation of our establishment and at the same time a considerable improvement in our officer/NCO cadres could be made. It goes without saying that such an increase is limited for structural reasons. With too great a proportional share of career personnel serving between 30 to 40 years in the armed forces, there would be the consequent problem of over-age, and career expectations would deteriorate as a matter of necessity. Structural analyses of the armed services have revealed that the share of career personnel should be augmented in line with structural requirements by a total of around 15 per cent in the course of the next 20 years.

Another group not affected by a lack of applicants is that of short-service volunteers enlisting for a term of two years. This group is mainly recruited from secondary school graduates who want to enter a university after their two-year service. It is a group of great importance in the recruitment of junior NCOs and officers of the reserve, i.e. personnel not suited for a regular officer or NCO career in staff or special assignments. Their training requirements are very great when related to their length of duty. That is why an increase in the scope of that category of personnel must continue to be confined to a few thousand.

Fully affected by shrinking age groups is the category of NCOs and men with terms of duty between four and 15 years. Even in the past, that category could never attain its full authorised strength, and therefore a serious collapse will have to be expected unless remedial action is taken. This is a category which is of central significance in relation to the personnel readiness of the armed forces. A number of this group are junior-grade NCOs (OR–5/6) who, as commanders of tank vehicles, team or squad leaders and as specialists in technically exacting assignments, are the very backbone of leadership and training and are the superiors of their voluntary or conscripted crews.

In the past, barely 10 per cent of the young men available for military service could be recruited as short-service volunteers, with the majority of them enlisting only for four years. That number is not sufficient to retain the 130,000 men of that category as needed in the 1990s as a minimum requirement. That goal can be attained only if we succeed in recruiting 12 per cent of the men fit for military service and have the majority of them enlist for a term of eight to 15 years. As part of the Personnel Implementation Plan a whole series of measures to improve the attractiveness of a temporary service career have been initiated and will be implemented. The plan covers measures aimed at improving the attractiveness of duty and of the individual careers; the provision of greater financial incentives and an improvement in retirement benefits and vocational training. The latter, in particular, is of decisive importance as the temporary-career serviceman must be convinced that he will find safe and qualified employment in civilian life after his separation from the armed forces. In facing that difficult challenge, the Bundeswehr can reply on its well-proven Vocational Advancement Service which cooperates closely with trade and industry and which can provide for an improved re-integration into a professional civilian career as a result of its wide spectrum of advancement options.

For voluntary personnel, a trend prognosis of the potential number of applicants is always rather vague since the manpower resources, as a function of the self-regulating law of supply and demand, are influenced by competition between trade and industry and other employers. Experience in the past, especially experience of the periods of full employment in the 1960s and early 1970s, and the assessment of the effects taken by measures initiated since 1983, have convinced our responsible Bundeswehr leaders that it will be possible, in the same way as for conscripts in basic military service, to maintain the necessary number of short-service volunteers in the 1990s.

The Reservist Concept

Under the military legislation of the Federal Republic of

Germany, general conscription does not terminate at the end of active duty of a conscript or volunteer, but continues to be valid in the form of a reserve duty obligation. Depending on the rank attained in the course of active duty, conscripts may be earmarked for wartime assignments and indeed called up for reserve duty training at least until the age of 32 and up to the age of 60.

The potential of trained and available reservists—far more than two million—has been so great that only one out of two reservists could be earmarked for a wartime assignment, and only one out of four could indeed be called back to reserve duty training. The result has been a considerable inequity, similar to that caused in connection with basic military service, so that the acceptance of general conscription at large was burdened.

Due to the shrinking age groups, the reservist potential, has also been decreasing considerably. In October 1984, the Federal Government took the decision to better exploit—in addition to the measures aimed at extending our active personnel strength—our reservist potential as well, both on a peacetime and wartime footing, and by doing so, increase our mobilisation capability and the sustainability of our armed forces. That decision is in line with our changing Alliance strategy, with greater emphasis being placed on conventional defence capabilities.

The declining reservist potential, accompanied by an increase in the quantitative and qualitative reservist requirements, has necessitated the preparation of a new reservist concept over the past two years. The reservist concept is focused on a better training of reservists, in keeping with their wartime assignments, as early as possible during their active duty; other accents are a more even distribution of wartime assignments and burden-sharing in terms of reserve duty training as well as an improvement of our mobilisation capability. To that effect, the number of posts to be used for reserve duty training in the peacetime establishment will be almost doubled; at the same time, a stand-by readiness component consisting of command/control and other cadre personnel will be set up. That component can be called up at short notice, even prior to general mobilisation, and is to be used to pave the way for the in-processing of reservist personnel. The future aim is that every reservist, if possible, will be earmarked as a member of the alert or personnel reserve and can be called up for reserve duty training on a regular rotational basis.

The new reservist concept presupposes a major intellectual reappraisal, both in the armed services and our society at large, and imposes substantial sacrifices, in particular on trade and industry. However, as a result of the closer integration of reservists into active units, this concept also promises a closer relationship between the population and the Bundeswehr and a keener interest of the public in the problems of national defence.

CONCLUSION

The sharp decline in the birth rate since the mid-1960s has confronted the Federal Republic of Germany with difficult decisions. Almost all sectors of state and society are affected on a short, medium or long-term basis by this fall in the birth rate which is not a temporary phenomenon, but one which will continue to exist. With a time-lag of 20 years following the first decline in the birth rate, the Federal Armed Forces are beginning to experience a recession in the number of men available and fit for military service. In the course of the next few years, the manpower potential is projected to decrease by almost 50 per cent.

The Federal Government has faced up to these problems at an early stage and has examined all possible solutions to solve the problem. In its decision-making, our Government did not chose the easiest way, namely to reduce the strength of our armed forces, but decided, for urgent defence and security-political reasons and in line with our NATO commitments, to maintain the peacetime and wartime strengths of our armed forces and to safeguard the manpower levels needed to that effect through a whole series of measures. In the course of the last legislative term, a bill was passed to improve equity in conscription and to extend the duration of basic military service, so that an adequate number of conscript personnel can be provided beyond the year 2000. In the current legislative term, the prerequisities are being created to also safeguard the necessary manning levels of career and temporary-career servicemen. Through a new reservist concept it will be possible, at the same time, to enhance the armed forces' mobilisation capability and sustainability and to improve the status of operational readiness of reserve units.

Despite a considerable reduction in manpower resources, the Federal Armed Forces Plans for the 1990s preserve the armed forces' capability of rapid reaction against surprise attacks, while improving at the same time their mobilisation capability and sustainability in war. They thus strengthen our allied forces' forward defence capability in Central Europe. Last but not least, they are making a contribution, through an improvement of conventional defences, towards the continuation of the process of nuclear and conventional arms control negotiations, without any one-sided advance concessions impairing that development.

In conclusion, a critical comment may be permitted: that the drop in the birth rate, below a level as would be needed to preserve our population, should be a genuine cause for concern, rather than some detailed issues such as old-age pensions or the strength of the armed forces. It is to be hoped that the peoples of the Old World will not be indifferent enough to silently and gradually die out, but see to it that a young generation grow up which—beyond the fostering of their self-image and the pursuit of their own happiness—will more clearly perceive their own share of responsibility for future life, not only of their own nation, but of our cultural community at large.

Chemical Weapons Proliferation in the Developing World

ELISA D. HARRIS

The author is an SSRC–MacArthur Foundation fellow in international peace and security studies, and a doctoral candidate in international relations at the University of Oxford. This chapter was written while she was a visiting research fellow at the Royal United Services Institute for Defence Studies.

THROUGHOUT THE 1960s and 1970s, concerns about nuclear proliferation and about limiting Superpower nuclear arsenals overshadowed the subject of chemical weapons (CW) and their control. This situation changed abruptly in 1984, when the United Nations confirmed that chemical weapons were being used in the Gulf War.[1] More recent reports of biological weapons proliferation,[2] and of the proliferation of ballistic missiles in unstable regions such as the Middle East,[3] have only served to heighten concerns about the spread of non-nuclear weapons of mass destruction in the developing world.

Many believe that the threat to international security posed by chemical weapons proliferation and use now rivals that of nuclear weapons. As Kenneth Adelman, the director of the US Arms Control and Disarmament Agency (ACDA), told a Senate hearing in 1984,

> when I look to the remainder of this century and what kind of threats there are to security around the world, I personally put the threat of a nuclear war low, very low. I personally put the increasing use of chemical weapons around the world high.[4]

Others, however, point to differing estimates of the number of CW possessor states, and to conflicting reports concerning the identities and capabilities of such states, and question whether the CW proliferation problem is as serious as government officials suggest.[5]

This article explores the nature of the CW proliferation problem in the developing world by examining the following questions. Why might Third World countries decide to acquire chemical weapons, and what factors may constrain such acqusition? How extensive is the CW proliferation problem?

What have been the national and international responses to the threat of CW proliferation? Finally, what are the likely implications if the CW proliferation problem is as serious as some suggest, and is not brought under control?

INCENTIVES FOR AND CONSTRAINTS ON CW PROLIFERATION

It is difficult to know precisely why Third World countries might decide to acquire chemical weapons. In each individual case, policymakers will undoubtedly be influenced by a variety of factors unique to their particular situation. In what follows, an attempt will be made to identify in more general terms possible incentives for, as well as constraints on, CW proliferation in the developing world.[6] The incentives for acquiring chemical weapons may be both political and military in nature. The constraints on CW acquisition are likely to be of a moral, legal, and technical character.

One reason why Third World countries may develop an interest in chemical weapons is because they may believe that such weapons have political value. In the 1950s, the United Kingdom and France acquired nuclear weapons in part because membership of the "nuclear club" ensured these countries great power status. As Lord Cherwell argued at the time,

> [i]f we are unable to make bombs ourselves and have to rely entirely on the United States army for this vital weapon we shall sink to the rank of a second-class nation, only permitted to supply auxiliary troops, like the native levies who were allowed small arms but not artillery.[7]

Although chemical weapons may not have the same prestige value as nuclear weapons, for developing countries unable to acquire a nuclear option CW may be an attractive alternative. Chemical weapons have long been characterised as "the poor man's weapon of mass destruction". Third World interest in acquiring CW may in part reflect this perception.

A second reason why Third World countries may develop an interest in chemical weapons is because they may believe that such weapons have military value. In the West, chemical weapons have generally been viewed as having relatively low military utility. On the battlefield, sudden changes in weather conditions can render the effects of chemical weapons use unpredictable. In addition, most Western military forces are equipped with protective gear such as gas masks and suits which are able to counter the effects of traditional chemical agents. Finally, Western forces have available to them conventional as well as nuclear weapons with equal or better military characteristics.

The military utility of chemical weapons may, however, be judged differently in the developing world, particularly in situations where predictability is less important, and where both chemical protection and sophisticated conventional and nuclear weapons are simply unavailable. As Ricardo Fraile, a Chemical and Biological Warfare (CBW) consultant, explains:

> The chemical weapons used by the Third World do not have to be sophisticated since the people they are used against do not have any protection. A simple gas like mustard gas can be very effective against men who do not have protective clothing.[8]

Every confirmed use of chemical weapons since the First World War has, in fact, involved a Third World country lacking chemical protection, and has shown CW to be effective both in military terms and in terms of its terror-inducing effects. This was true of the use of chemical weapons by Spain in Morocco in 1925, by the Soviet Union in China in 1934, by Italy in Ethiopia between 1935–36, by Japan in China between 1937–45, by Egypt in Yemen between 1963–67[9] and, more recently, by Iraq in the Gulf War.

Third World countries may therefore see CW as an effective means of accomplishing a number of different military missions. They can be used in counter insurgency operations to flush guerrillas out of hard to reach areas and to terrorise their domestic supporters. Chemical weapons can also be used to counter an adverary's superior conventional forces. For a country facing large concentrations of enemy troops, CW can be a vital force multiplier. They can also be used to paralyse enemy air power on the ground, rendering it vulnerable to follow on attacks with conventional high explosives. Chemical weapons can also be acquired for the traditional purpose of deterring enemy chemical use through the threat of retaliation in-kind. Finally, because of their potential for mass destruction, CW can serve as a counter to an adversary's nuclear capability.

In deciding whether to acquire chemical weapons, developing countries are likely to consider these types of political and military incentives.The CW policy decisions of Third World countries are also likely to be influenced by various moral,

DY—D

legal, and technical constraints. These constraints on the acquisition of CW may now, however, be breaking down.

In the West, chemical weapons have traditionally been viewed as particularly immoral and inhumane. The Romans, for example, are said to have decreed "'Armis bella non veneris geri,' war is waged with weapons, not with poisons".[10] In 1855, Lord Dundonald's plan to use poison sulphur in the Crimean War was rejected on the grounds that the "effect of the chemical weapon is so horrible that no honourable combatant can use it".[11] But it was the First World War, in which some 113,000 tons of chemical agents resulted in an estimated 1.3 million gas casualties,[12] that really crystallised Western revulsion to poison gas. Although Third World countries did not share this particular experience, a number of them did, as noted earlier, fall victim to chemical attacks in subsequent years. Third World interest in chemical weapons, like that in much of the West, might therefore be dampened by the moral revulsion arising from these earlier experiences.

It is possible, however, that the example set by the United States and the Soviet Union in the CW area might be eroding this longstanding moral constraint. Interestingly enough, the US Arms Control and Disarmament Agency opposed new US chemical weapons production in 1974 partly out of concern about the effect of such a step on CW proliferation. As Fred Ikle, the Director of ACDA, explained at the time,

> If we start on a new type of production program . . . such an effort, particularly if it became large-scale, might stimulate third countries to make new efforts in the production of chemical weapons, a type of weapon which could be adversely used against our allies, our friends, or even us, and where the technology is not all that difficult to obtain for a large number of countries.[13]

In December 1987, the United States began to produce chemical weapons again for the first time since 1969. Although General Secretary Mikhail Gorbachev announced the cessation of Soviet chemical weapons production in April 1987, the Soviet Union is believed to have maintained an active CW production programme throughout the 1970s and 1980s.[14] This programme may have influenced Third World thinking about the morality of chemical weapons. The new US CW production programme may have a similar effect.

International law may provide a second possible constraint on CW proliferation in the developing world. The 1925

Geneva Protocol bans the wartime use of chemical as well as biological weapons, but not the development, production, possession, or transfer of such weapons. Over 100 countries, many of whom are from the developing world, are parties to this agreement.[15] A number of Third World countries are also participating in negotiations in Geneva aimed at banning chemical weapons.

Both the Geneva Protocol and the longstanding effort to ban chemical weapons may have helped hinder Third World CW proliferation by underscoring that the use of such weapons was immoral and inhumane, by inspiring confidence that others were not acquiring chemical weapons, and by raising the political costs of pursuing a CW capability. But the international response to Iraq's repeated use of chemical weapons may have weakened the constraints on CW use embodied in international law and, by implication, on CW proliferation, as well. As was noted earlier, the UN confirmed in 1984 that chemical weapons were being used in the Gulf War. Not until 1986, however, did the UN officially identify Iraq as the guilty party. Baghdad's war effort, moreover, continued to receive support from the Soviet Bloc, France, China, and Egypt, all of whom provided substantial quantities of conventional arms to Iraq.[16]

In the past, technological barriers have served as a third constraint on the ability of Third World countries to develop chemical weapons. But these technological barriers have been gradually lowered as the industrialisation/modernisation process has brought petrochemical, pesticide, fertilizer, and pharmaceutical industries to the developing world. Any country with such industries has the potential, in terms of equipment, raw materials, and expertise, to produce some chemical warfare agents. Moreover, because much of the equipment and many of the chemicals that can be used to make chemical warfare agents have legitimate commercial uses, countries without such facilities can readily acquire the necessary components on the international market. Of course, technological barriers can be overcome most easily by the direct transfer of chemical weapons from a possessor state.

Third World countries may thus have both political and military incentives for acquiring chemical weapons, and may also face fewer moral, legal, and technical constraints on such acquisition than was previously the case. This does not mean, however, that all developing countries will necessarily acquire chemical weapons. For a majority of these states, the incentives to acquire CW either do not exist or are outweighed by existing constraints, however weakened the latter may have become. Only in a relatively limited number of Third World countries, therefore, has the combination of strong incentives and diminishing constraints appeared to produce a decision to acquire chemical weapons. These countries are considered below.

THE EXTENT OF THE CW PROLIFERATION PROBLEM

Any effort to explore the nature and extent of the CW proliferation problem in the develping world is hampered by the reluctance of most governments to identify specific chemical weapons states. Instead, government officials speaking on the record have tended to characterise the problem in terms of the *number* of such states. However, even these numerical estimates are imprecise and somewhat contradictory.

In June 1987, the head of the Soviet delegation to the Conference on Disarmament suggested that there were between nine and 15 countries with chemical weapons. One year earlier, the official Soviet news agency, TASS, reported that 13 to 15 countries had chemical weapons.[17] In the United Kingdom, on the other hand, government ministers stated in 1986 that possibly more than 20 countries either had chemical weapons or were considering acquiring them. In 1985, a government minister told the House of Lords that more than 12 states were believed to possess chemical weapons, compared to about six a decade earlier.[18] In the United States, the Director of the Arms Control and Disarmament Agency stated in 1987 that at least 15 states possessed chemical weapons, with others attempting to acquire them.[19] One year earlier, the US Army Chief of Staff and an official in the Office of the Secretary of Defense offered differing estimates: the former referred to 14 chemical weapons states with two others possible; the latter stated that 16 countries had chemical weapons, and six more were probable.[20] All of these estimates are believed to include both Third World countries and members of NATO and the Warsaw Pact.[21]

In addition to being imprecise and somewhat contradictory, official government statements generally do not explain how they *define* a chemical weapons state.[22] Do they distinguish, for example, between countries with chemical weapons research programmes and those with a militarily useful chemical weapons capability? If so, what *is* a military useful CW capability? Does it involve access to a certain quantity of chemical weapons, or of certain chemical agents? Finally, do official statements distinguish between those countries with chemical weapons under national control, and those with foreign chemical weapons deployed on their territory?

One possible reason why most governments do not identify CW possessor states and are not more specific in discussing capabilities is because they may be uncertain about precisely who does in fact have what capability. As was noted earlier, any country with a commercial chemical industry has the ability to produce some chemical warfare agents. Moreover, without direct access to such facilities, it is probably impossible to know whether activities being undertaken are of a commercial or military nature. Equipment and chemical feedstock purchases on the international market may not reveal much either, as they could either be for commercial or military purposes. Finally, even chemical delivery systems don't have distinctive features. An artillery shell, bomb, or missile warhead carrying a chemical warfare agent looks essentially the same as one carrying conventional high explosives.

Despite these difficulties, it is possible to explore the nature and extent of the Third World CW proliferation problem using publicly available information. In the discussion which follows, this information is organised according to specific criteria. It should be emphasised, however, that these criteria are merely a tool for organising the available information on CW proliferation, and that conclusions derived from them about the CW status of particular countries are based largely on this information. The criteria used in this discussion are as follows:

☐ *Known* CW states are those who have either declared that they possess chemical weapons or whose use of such weapons has been definitively confirmed.

☐ *Probable* CW states are those reported by NATO or Western government officials, on or off the record, as producing chemical weapons, possessing chemical weapons under national control, or using chemical weapons.

☐ *Seeking to possess* CW states are those reported by NATO or Western government officials, on or off the record, as trying to acquire chemical weapons.

☐ *Doubtful* CW states are those reported, generally by

domestic or foreign adversaries, as possessing CW under national control or of using chemical weapons, with no confirmation by NATO or Western government officials.

On the basis of these criteria, Iraq is the only Third World country known to possess chemical weapons.[23] Eleven other Third World countries are probable CW states. The CW capabilities of seven of these countries have been mentioned in press reports citing US government officials speaking on the record. They include Burma,[24] China,[25] Iran,[26] North Korea,[27] Syria,[28] Taiwan,[29] and Vietnam.[30] The four remaining countries, Egypt,[31] Ethiopia,[32] Israel,[33] and Libya,[34] have been discussed in press reports citing unnamed US government officials and classified documents. South Korea has been identified by unnamed US government officials as a country seeking to possess chemical weapons.[35]

At least 18 other countries have been accused of either possessing or using chemical weapons, but there has been no confirmation of these claims, either on or off the record, by NATO or Western government officials. These doubtful chemical weapons states include Afghanistan,[36] Angola,[37] Argentina,[38] Chad,[39] Chile,[40] Cuba,[41] El Salvador,[42] Guatemala,[43] India,[44] Indonesia,[45] Laos,[46] Mozambique,[47] Nicaragua,[48] Pakistan,[49] Peru,[50] the Philippines,[51] South Africa,[52] and Thailand.[53]

It is difficult to know with any certainty why Iraq, the 11 probable CW states, and South Korea might have decided to acquire chemical weapons. Nevertheless, it is possible to draw some conclusions, however tentative, about the role of political and military incentives in the CW policy decisions of these countries based on publicly available information.

It would appear that none of these Third World countries acquired chemical weapons for reasons of political prestige. In order for chemical weapons to have *political* value as the "poor man's weapon of mass destruction", possessor status would have to be known *and* acknowledged by the country concerned. Yet even Iraq did not openly acknowledge its CW activities until July 1988.[54] Indeed, many Third World countries have emphasised just the opposite—that they do not possess chemical weapons. Among the probable CW states that have denied possession, are Burma, China, Egypt, Ethiopia, and Israel. Doubtful CW states that have denied possession include India, Indonesia, Pakistan, Peru, and the Philippines. This suggests that chemical weapons do not have the same sort of prestige value as nuclear weapons, and that decisions in the Third World to acquire chemical weapons are based largely on military rather than political incentives.

As noted earlier, Third World countries may view the military utility of chemical weapons differently than Western states. In at least five countries, chemical weapons may have been acquired for use in counter insurgency operations in their own or neighbouring states. Iraq's initial decision to obtain chemical weapons may have been related to its long history of problems with its Kurdish population. Burma is said to have acquired chemical weapons to quell domestic insurgents.[55] Egypt used chemical weapons during its intervention in the civil war in Yemen in the 1960s.[56] Ethiopia has had a longstanding problem with Eritrean secessionists. Finally, Vietnam has been accused by the US of using the so-called yellow rain mycotoxins and other unidentified chemical agents against the Hmong in neighbouring Laos and against resistance forces in Kampuchea.[57]

Third World countries facing superior conventional forces also have a powerful incentive for acquiring chemical weapons. Iraq's more recent chemical weapons programme is clearly a response to the Iranian human wave attacks in the Gulf War. The Egyptian and Syrian CW efforts may be related to Israel's conventional superiority, particularly its highly effective air force. Ethiopia may value chemical weapons for use in its longstanding conflict with Somalia over the Ogaden. Libya is reported to have acquired and used chemical weapons in its war with neighbouring Chad.[58] North Korea may have obtained chemical weapons for use against South Korean and US conventional forces. Taiwan's chemical weapons are reportedly designed to thwart an invasion from mainland China.[59] Finally, Vietnam is reported to have used chemical weapons in 1979 against Chinese intervention forces.[60]

At least four countries in the Third World may have acquired chemical weapons in order to deter enemy chemical use through the threat of retaliation in kind. Reports that China suffered chemical attacks during border skirmishes with the Soviet Union in 1969 and during its intervention in Vietnam in 1979 suggest a strong motive for a Chinese CW programme.[61] Iran has almost certainly acquired chemical weapons in response to Iraqi CW use in the Gulf War. Israel is reported to have obtained chemical weapons in the 1970s in response to the CW threat from its Arab neighbours, particularly Egypt.[62] South Korea's efforts to acquire chemical weapons are probably linked to the CW programme of its chief rival, North Korea.

Finally, at least some Third World countries may have acquired chemical weapons in order to counter an adversary's nuclear capability. Israel's alleged nuclear programme, for example, may have stimulated the CW efforts of a number of Arab countries, including Iraq, Egypt, and Syria.[63]

Iraq, South Korea, and the 11 probable CW states may thus have had a variety of military incentives for acquiring chemical weapons. These military incentives probably outweighed existing moral and legal constraints on CW proliferation which, as was suggested earlier, may have weakened in recent years. Third World CW armament programmes probably have been helped even more by the erosion of technological barriers to CW acquisition.

As was noted earlier, Third World countries interested in acquiring chemical weapons can either bypass the technological problems altogether by procuring such weapons from other CW states, or can seek to develop their own chemical weapons production capability by purchasing the necessary equipment and chemicals on the international market. US intelligence officials have acknowledged the role of commercial firms in Third World CW programmes.[64] But the United States has often emphasised the former route to proliferation, identifying Soviet training, technical assistance, and chemical weapons transfers to allies in the developing world as a key factor in the proliferation problem. In 1984, for example, columnist Jack Anderson reported that the CIA had concluded that "Soviet military assistance has been a common source and major stimulus to the momentum of chemical weapons proliferation in the Third World".[65] General Secretary Gorbachev has denied the US charges, stating that the Soviet Union has not transferred chemical weapons to anyone, has not deployed them in the territory of other states, and "has always strictly abided by those principles in its practical policies".[66]

Information attributed to Western government officials suggests that Iraq, South Korea, and the 11 probable CW states may have benefited from both Soviet and commercial assistance. The Soviet role has almost certainly been overstated, however, as indigenous production programmes, relying largely on equipment and chemicals purchased from

Western firms, have been reported for a majority of these states.

Five Third World countries may have previously received chemical weapons directly from the Soviet Union. These include Egypt,[67] Ethiopia,[68] Libya,[69] Syria[70] and Vietnam[71]. Chemical weapons and technical assistance may also have been provided by one Third World country to another. Syria is reported to have helped the Iranian CW effort,[72] Iran is said to have supplied chemical weapons to Libya in return for Soviet-made mines,[73] while Israel is reported to have provided chemical agents to Taiwan, and advice on CW matters to the Chinese.[74] A majority of these countries, however, are now said to possess indigenous chemical weapons production capabilities, often developed with equipment and chemicals purchased from Western companies. These include Iraq,[75] Burma,[76] China,[77] Iran,[78] Israel,[79] Libya,[80] North Korea,[81] Syria,[82] Taiwan,[83] and Vietnam[84]. Egypt may also have its own CW production capability.[85]

A number of conclusions emerge from this discussion of publicly available information on CW proliferation in the developing world. First, Iraq is the only known CW state, although 11 other countries are probable CW states, and South Korea might be seeking to possess chemical weapons. The CW status of 18 other countries is doubtful. Second, Third World countries that may have acquired chemical weapons have done so largely for military rather than political reasons. Finally, a majority of the countries that are reported to have acquired chemical weapons appear to have done so through indigenous production programmes based on equipment and chemicals purchased from Western companies. The Soviet role in the Third World proliferation problem has thus almost surely been overstated.

These conclusions, it must be acknowledged, are derived from criteria that are far from ideal. Laos, for example, has been accused by the US government of using yellow rain and other unidentified chemical agents. However, because the reports of yellow rain warfare have been shown to be unreliable,[86] Laos has been classified here as a doubtful, rather than a probable, CW state. In addition, particular emphasis has been given to reports involving NATO and Western government officials, speaking on or off the record, and to reports said to be based on classified information. If these reports are incorrect, the conclusions, derived from them will be similarly flawed.

The problem of credible sources is not necessarily helped by
multiple press reports containing the same basic information.
Two of the most detailed accounts of the proliferation
problem, by Oberdorfer and Ember, draw heavily from an
earlier report, by the columnist Jack Anderson, purportedly
based on intelligence sources and classified documents. Other
reports may also be derived from the same initial source or
report, so the possibility of erroneous information being
unintentionally perpetuated cannot be overlooked.

It is also possible that false or exaggerated reports of CW
proliferation may be propagated intentionally.[87] CW possession or use charges can be useful for discrediting a domestic or
foreign adversary. Many of the claims concerning doubtful CW
states probably fall in this category. CW proliferation fears can
also help increase demand for chemical protective equipment.
Finally, concern about CW proliferation can be used to
increase support for national CW armament programmes.

Without access to intelligence information, it is impossible to
know exactly how extensive the CW proliferation problem in
the developing world really is. Moreover, judging from the
official government statements noted earlier, even with access
to intelligence information, it may still be impossible to know
precisely who has what capability. In some respects, however,
this may no longer matter. It is now widely believed that
chemical weapons are spreading to more and more countries.
This perception of a growing CW proliferation problem has
been driving national as well as international policy.

NATIONAL AND INTERNATIONAL RESPONSES

Evidence in 1984 that Iraq's chemical weapons had been
produced with chemicals and equipment purchased from
Western chemical companies stimulated both national and
international efforts to control the sale of the relevant
chemicals and technology. In March 1984, the United States
banned the sale to both Iraq and Iran of five chemicals that
could be used to manufacture mustard and nerve gases.[88] Two
years later, the United States extended its list of controlled
chemicals and added Syria to the ban, following reports that
Syria had begun to produce chemical weapons and had
discussed cooperation in the field with Iran.[89] In 1987, the
United States added an additional eight chemicals to its export
control list for Iraq, Iran and Syria, in response to information

that these countries had altered their purchasing policies to evade existing controls. At the same time, the US extended its 1984 export control list to all but 18 Western industrialised countries, out of concern about the wider proliferation problem.[90]

Other governments, including those of Australia, Canada, Japan, the Netherlands, Denmark, the Republic of Ireland and the United Kingdom also imposed export controls on a variety of different chemicals in the spring and summer of 1984.[91] The Federal Republic of Germany also adopted chemical export controls in the spring of 1984 and, a few months later, placed controls on a wide range of equipment that could be used to produce chemical weapons, following reports linking a West German built pesticide plant to Iraqi efforts to produce nerve agents.[92] Chemical export controls were also approved by the European Community in 1984, thus providing the first multilateral response to the problem of CW proliferation.[93] In January 1986, the Soviet Union imposed export controls on nine chemicals and one chemical group that could be used to make chemical warfare agents.[94] Relevant members of the Council for Mutual Economic Assistance (CMEA) are reported to have met in June 1987 to discuss similar steps.[95]

One glaring weakness in these early national non-proliferation efforts was that countries tended to include different chemicals in their export control lists. In March 1985, the US Secretary of State, George Shultz, called for multilateral as well as bilateral efforts to curb exports of chemicals that could be used to make chemical weapons.[96] A few months later, Australia took the lead in coordinating national export control policies, hosting the first in a series of meetings of Western industrialised states.[97]

At its first meeting in June 1985, the so-called Australian Group included the 10 member states of the European Community, plus Australia, Canada, Japan, New Zealand, and the United States.[98] By September 1987, its membership had grown to 20, the newcomers being Portugal, Spain, Norway, Switzerland, and the European Commission itself. Perhaps more importantly, the Group had succeeded in developing a uniform list of chemicals to be controlled by all member states. Formal export licences are now required for eight specific chemicals, and a warning list of over 30 other chemicals has been circulated to private chemical companies in the hope that they will cooperate with government agencies in controlling the sale of these substances.[99]

In the autumn of 1985, the Soviet Union began to express an interest in working with the United States to halt the spread of chemical weapons.[100] This culminated in a promise, at the Reagan–Gorbachev summit in November 1985, that the two sides would "initiate a dialogue on preventing the proliferation of chemical weapons".[101] US officials hoped that this dialogue would lead to a cooperative relationship similar to the longstanding US–Soviet effort to prevent nuclear proliferation. In early 1986, Kenneth Adelman, the Director of the US Arms Control and Disarmament Agency, suggested that the two sides should harmonise export controls on chemicals and equipment to problem countries in the Third World, as well as share intelligence information on such countries.[102]

At least initially, the Soviet Union was interested in a much more specific bilateral effort. Shortly before the November 1985 Reagan–Gorbachev summit, the General Secretary advanced the "thought" that the two sides could develop a treaty to curb the spread of chemical weapons, just as they had concluded a treaty to halt the spread of nuclear weapons.[103] A few months later, in a major foreign policy address, Gorbachev proposed the idea of a chemical non-proliferation treaty as an interim step prior to the conclusion of a chemical weapons ban.[104] The United States rejected the Soviet proposal, arguing that the most effective means of halting CW proliferation was by concluding a ban on such weapons. As the US State Department spokesman, Bernard Kalb, explained,

> We believe that an effective and verifiable global ban on all chemical weapons is the way to solve the triple problems of existing chemical weapons capabilities, their use, and their further spread.[105]

A chemical non-proliferation treaty probably was not, therefore, on the agenda of the US–Soviet CW non-proliferation discussions held in 1986 and 1987. Whether the two sides achieved progress in the areas suggested earlier by the United States is not publicly known.[106]

Neither the chemical export controls promoted largely by Western industrialised states nor the chemical non-proliferation treaty suggested by the Soviet Union provide a solution to the problem of chemical weapons proliferation in the developing world. Export control policies will not prevent countries that already possess civil chemical industries from developing at least some chemical warfare agents. Moreover, it is impossible to control all of the chemicals that could be used to produce chemical warfare agents, as many have legitimate commercial uses. Finally, a determined state may be able to evade export controls entirely, either by setting up front companies, or by dealing with unscrupulous chemical traders or manufacturers.[107] Chemical export controls will not, therefore, prevent a country that really wants chemical weapons from acquiring them. At best, export controls will make it more difficult and costly for the potential proliferator to acquire chemical weapons, and may provide supplier countries with early warning of such proliferation. Or as a US government official acknowledged in 1984, chemical export controls

> can achieve valuable objectives, such as the disruption of a given state's plans to produce CW quickly for immediate use in battle and the imposition of higher economic costs on such a state. But no export control policy can erect an insurmountable barrier against acquisition or at-home production of CW.[108]

Verification and, more importantly, political obstacles rule out a chemical non-proliferation treaty as a solution to the problem of CW proliferation. Developing countries oppose the creation of another discriminatory, Nuclear Non-proliferation Treaty type of arrangement, which would allow existing members of the chemical club to retain their chemical weapons, but would prohibit the "have-nots" from acquiring them. Western countries view efforts to conclude such an agreement as a distraction from the negotiations in Geneva aimed at banning chemical weapons. As the US Ambassador to the Geneva negotiations, Donald Lowitz, explained in 1986: "the focus of our efforts is and must remain a comprehensive agreement that eliminates forever the scourge of these terrible weapons".[109] A West German official agreed, noting that "[p]roliferation will be halted when countries can verify and control the global production of chemical weapons".[110]

Third World countries may be reluctant, however, to abandon their chemical weapons capabilities and participate in a CW disarmament agreement if security corners have prompted them to acquire chemical weapons. As the British Foreign Secretary, Sir Geoffrey Howe, told the third UN special session on disarmament in June 1988, disarmament cannot proceed in a vacuum . . . It is inextricably tied up with security. No one is going to discuss disarmament seriously if he already feels insecure.[111]

Third World countries that have acquired CW must therefore be convinced that their security interests can be met by means other than chemical weapons.

Conclusions/Implications

In the absence of a chemical weapons ban encompassing all CW states, the proliferation of chemical weapons, particularly in the developing world, is likely to have profound implications. As chemical weapons become more fashionable, various sub-national groups, including terrorist organisations, may become more interested in acquiring a CW capability.[112] With a basic knowledge of chemistry and a small amount of money, such groups can easily produce enough chemical agent to threaten an average-size city. Research quantities of ready made agents such as mustard gas can also be purchased directly from some chemical manufacturers.[113] Opportunities to steal chemical weapons may also increase as more and more countries acquire them. Finally, sub-national groups might also obtain chemical weapons directly from states such as Libya and Iran who are sympathetic to their political aspirations.

At the national level, the spread of chemical weapons may lead to further proliferation in three ways. First, non-CW states may move to acquire chemical weapons in response to the CW threat posed by neighbouring states. In addition, chemical weapons may increasingly be seen as a legitimate means of countering conventional military threats, either domestic or foreign. Finally, proliferation may lead to chemical weapons being transferred more freely between states, and hence to further proliferation.

CW proliferation could also make conflict itself more likely, particularly in unstable regions. Following confirmation of Iraq's use of chemical weapons in the Gulf War, there were reports that the United States had examined the feasibility of air strikes against Iraqi CW sites.[114] More recently, Israel was reported to be considering a pre-emptive strike against a Syrian installation said to be developing nerve agent warheads for SS–21 and SCUD missiles.[115] Either of these strikes could have led to a military confrontation between the countries involved.

CW proliferation could also make conflict more destructive, particularly in areas such as the Middle East and Asia, where ballistic missiles are also proliferating. During the Gulf War, Iraq and Iran used conventionally armed missiles in their attacks on each other's cities. In the final months of the war, however, Iraq reportedly threatened to launch chemical strikes against large Iranian cities.[116] At least seven other probable CW states also possess ballistic missiles which could be modified to carry chemical warheads, including China, Egypt, Israel, Libya, North Korea, Syria, and Taiwan. South Korea also has a conventional ballistic missile capability.[117] The use of chemically armed ballistic missiles by any of these countries would lead to unprecedented destruction. Moreover, if chemically armed ballistic missiles were used against Israel, it would almost certainly trigger a nuclear response.

The proliferation of chemical weapons, and of sophisticated delivery means such as ballistic missiles, also has implications for the Superpowers themselves. With such military power in the hands of Third World countries, both the United States and the Soviet Union will find it increasingly difficult to pursue their regional interests. Moreover, the use of chemically armed ballistic missiles in a region like the Middle East could draw the Superpowers in on the side of their respective allies, culminating in conflict between East and West as well.[118]

The West, thus, has a clear interest in halting the spread of chemical weapons in the developing world. Although the identities and capabilities of CW possessor states may still be unclear, Iraq's acquisition and repeated use of chemical weapons provides firm evidence of a Third World CW proliferation problem. Chemical export controls are of limited value in preventing developing countries from acquiring indigenous CW production capabilities. A ban on chemical weapons could go a long way toward solving the proliferation problem, but only if all existing CW states can be convinced that their security interests can be met by means other than chemical weapons. If countries like Iraq refuse to participate in the chemical weapons ban, the treaty's non-proliferation objectives will not be realised, and other countries may not agree to ratification. CW proliferation in the developing world may thus make the elimination of chemical weapons both more urgent and, at the same time, more difficult to achieve.

NOTES

I am grateful to Ivo H. Daalder, Julian Perry Robinson, and Nicholas A. Sims for their helpful comments on an earlier draft. This chapter is based on information available through July 1988.

[1] For the various UN Security Council reports investigating the use of chemical weapons in the Gulf War, see S/16433 of 26 March 1984, S/17911 of 12 March 1986, and S/18852 of 8 May 1987, S/19823 of 25 April 1988, S/20060 of 20 July 1988 and S/20063 of 25 July 1988.

[2] John H. Cushman, Jr., "US Suspects Toxic Arms Development," *International Herald Tribune*, 5 May 1988.

[3] See, for example, Geoffrey Kemp, "Mideast Missile Madness: A Bazaar for Doomsday," *Washington Post*, 27 March 1988; and David Ottaway, "In Mideast, Warfare Takes a New Turn," *Washington Post*, 5 April 1988.

[4] US Senate, Committee on Foreign Relations and Subcommittee on Energy, Nuclear Proliferation and Government Processes of the Committee on Governmental Affairs, *Joint Hearing on Chemical Warfare: Arms Control and Nonproliferation*, 98th Cong., 2nd Sess. Washington, DC, US Government Printing Office, 1984, p. 34.

[5] See, for example, J.P. Perry Robinson, "Some Developments Over the Past Year in the Field of Chemical Warfare Armament," Appendix to Background Paper No. 7, 13th Workshop of the Pugwash Study Group on Chemical Warfare, Monitoring a Chemical Weapons Treaty, Geneva, Switzerland, 23–24 January 1988.

[6] Many of the ideas reflected here are drawn from US House of Representatives, Subcommittee on International Security and Scientific Affairs of the Committee on Foreign Affairs, *Report on Binary Weapons: Implications of the US Chemical Stockpile Modernisation Program for Chemical Weapons Proliferation*, 98th Cong., 2nd Sess. Washington, DC: US Government Printing Office, 1984, pp. 18–27.

[7] As cited in Margaret Gowing, *Independence and Deterrence: Britain and Atomic Energy, 1945–52 Volume I Policy Making*, London, Macmillan, 1974, p. 407.

[8] As cited in William E. Smith, "Clouds of Desperation," *Time*, 19 March 1984, p. 6.

[9] Stockholm International Peace Research Institute, *World Armaments and Disarmament, SIPRI Yearbook 1982* London: Taylor & Francis Ltd., 1982, p. 336 (hereafter cited as *SIPRI Yearbook*, with appropriate year). See also, Stockholm International Peace Research Institute, *The Problem of Chemical and Biological Warfare, Vol. I, The Rise of CB Weapons*, Stockholm, Almqvist & Wiksell, 1971, pp. 142–52, 159–61.

[10] As cited in Dr. A.J.J. Ooms, "What makes chemical weapons so peculiar?" in *The Holmenkollen Report on the Chemical Weapons Convention*, Oslo, Royal Norwegian Ministry of Foreign Affairs, 1987, p. 32.

[11] As cited in A. Jack Ooms, "Chemical Weapons: Is Revulsion a Safeguard?" *Atlantic Community Quarterly*, Vol. 24, No. 2, Summer 1986, p. 157.

[12] These figures are drawn from SIPRI, *The Problem of Chemical and Biological Warfare, Vol. I, The Rise of CB Weapons*, pp.128–29.

[13] US House of Representatives, Special Subcommittee on Arms Control and Disarmament of the Committee on Armed Services, *Hearings on the Review of Arms Control and Disarmament Activities*, 93rd Cong., 2nd Sess. Washington, DC, US Government Printing Office, 1974, p. 9.

[14] For the Gorbachev statement, see "Speech by Mikhail Gorbachev at Friendship Rally," Soviet TV, 10 April 1987 as reprinted in SWB–EE/8541, 13 April 1987. For a US view on the Soviet programme, see US Defense Intelligence Agency, Directorate for Scientific and Technical Intelligence, *Soviet Chemical Weapons Threat*, Washington, DC, Defense Intelligence Agency, 1985 DST–1620F–051–85, p. 9.

[15] The text and list of Parties to the Geneva Protocol may be found in US Arms Control and Disarmament Agency, *Arms Control and Disarmament Agreements*, Washington, DC, Arms Control and Disarmament Agency, 1982, pp. 14–18.

[16] See, Anthony H. Cordesman, *The Impact of Arms Transfers on the Iran/Iraq War*, Whitehall Paper London, Royal United Services Institute for Defence Studies, 1987, pp. 6, 12, 17.

[17] "Nazarkin Views Chemical Weapons Convention Prospects," *Komsomolskaya Pravda*, (in Russian) 16 June 1987 as reprinted in FBIS–SU, 22 June 1987; and "US Role in Chemical Weapons Ban Reviewed," *Tass*, (in English) 9 April 1986 as reprinted in FBIS–SU, April 1986.

[18] "Speech by Mr Tim Renton MP, Minister of State for Foreign and Commonwealth Affairs of the United Kingdom of Great Britain and Northern Ireland to the Conference on Disarmament," 15 July 1986; Hansard (Commons), 16 December 1986, col 500; and Hansard (Lords) 24 April 1985, col 1178.

[19] Kenneth L. Adelman, "USA's Need for a Chemical Deterrent," ACDA Occasional Papers, Number 2, 11 February 1987, p. 1.

[20] US House of Representatives, Subcommittee on Defense of the Committee on Appropriations, *Hearings on Department of Defense Appropriations for 1987*, 99th Cong., 2nd Sess., Washington, DC, US Government Printing Office, 1986, Part 1, p. 114; and US Senate, Subcommittee on Strategic and Theater Nuclear Forces of the Committee on Armed Forces, *Hearings on the Department of Defense Authorization for Appropriations for Fiscal Year 1987*, 99th Cong., 2nd Sess., Washington, DC, US Government Printing Office, 1986, Part 4, p.1730.

[21] Within the two military alliances, the United States, the Soviet Union, and France are known CW states. Czechoslovakia, the German Democratic Republic, Romania, and Poland can be classified as probable CW states, on the basis of press reports citing unnamed NATO and Western government officials. See, Don Oberdorfer, "Chemical Arms Curbs Are Sought," *Washington Post*, 9 September 1985; "Welt Am Sonntag Says Warsaw Pact States Producing Poison Gas," *Radio Free Europe*, 2 March 1986; and Robert C. Toth, "Germ, Chemical Arms Reported Proliferating," *Los Angeles Times*, 27 May 1986. See also Karl Feldmeyer, "Der Warschauer Pakt übt mit scharfer chemischer Munition," *Frankfurter Allgemeine Zeitung*, 19 August 1985, and " CDU/CSU Spokesman on USSR Chemical Weapons," Hamburg *DPA*, (in German) 12 August 1985 as reprinted in FBIS–WE, August 1985.

[22] One notable exception may be found in a 1986 interview with Kenneth Adelman, the Director of the US Arms Control and Disarmament Agency. Adelman said that stockpile size is "sufficient so that it could give a military utility to the possessing country, and sufficient to cause a great deal of damage to the other side. We are not talking about experimental possession, research possession." He also said that in the main, Third World countries have purchased rather than produced chemical weapons, and that the most widely possessed agent was mustard gas, although nerve gas and incapacitants were also stockpiled. See, Lois Ember, "Worldwide Spread of Chemical Arms Receiving Increased Attention," *Chemical and Engineering News*, 14 April 1986, p. 13.

[23] Robert J. McCartney, "Iraqi Official Acknowledges Chemical–Arms Use in War," *International Herald Tribune*, 2–3 July 1988; and "US Says Iraq Uses Chemical Weapons in War with Iran," *International Herald Tribune*, 6 March 1984.

[24] "Statement by Rear Admiral William O. Studeman, US Navy, Director of Naval Intelligence, before the Seapower and Strategic and Critical Materials Subcommittee of the House Armed Services Committee, on Intelligence Issues," 1 March 1988, p. 48. For other reports based on information from US government sources, see Oberdorfer, "Chemical Arms Curbs Are Sought"; and Ember, "Worldwide Spread of Chemical Arms Receiving Increased Attention," p. 9.

[25] Studeman Statement, 1 March 1988, p. 48. For other reports based on information from US government sources, see "China and Israel," *Foreign Report*, 12 July 1984; and Jack Anderson, "The Growing Chemical Club," *Washington Post*, 26 August 1984.

[26] "US Says Iran May Use Chemical Arms," *International Herald Tribune*, 26 April 1985; Tom Diaz, "Syria said to have offered chemical weapons to Iran," *Washington Times*, 9 December 1985; and William Beecher, "US seeking action to halt chemical arms spread," *Boston Globe*, 4 February 1986.

[27] Studeman Statement, 1 March 1988, p. 48. See also the comments by US Secretary of Defense Caspar Weinberger in Rick Atkinson, "Weinberger Maintains Opposition to Dutch Plan for Cruise Missiles," the *Washington Post*, 13 May 1984, and the testimony of the US Army Chief of Staff General John A. Wickham, Jr., in US House of Representatives, Subcommittee on Defense of the Committee on Appropriations, *Hearings on Department of Defense Appropriations for 1986*, 99th Cong., 1st Sess., Washington, DC, US Government Office, 1985, Part 2, p. 85. For a later report on what was said to be secret testimony to the Congress by General Wickham, see Chuck Vinch, "US: N. Korea has chemical arms," *Pacific StArs & Stripes*, 30 June 1986.

[28] Diaz, "Syria said to have offered chemical weapons to Iran." For other reports based on information from US government sources, see Oberdorfer, "Chemical Arms Curbs Are Sought"; and Gaylord Shaw, "Syria Reported to be Making Chemical Arms," *Los Angeles Times*, 26 March 1986.

[29] Studeman Statement, 1 March 1988, p. 48. For other reports based on information from US

government sources, see "China and Israel"; and Oberdorfer, "Chemical Arms Curbs Are Sought."

[30] US Department of State, "Chemical Warfare in Southeast Asia and Afghanistan," Report to the Congress from Secretary of State Alexander M. Haig, Jr., 22 March 1982, Special Report No. 98 (hereafter cited as *The Haig Report*); and US Department of State, "Chemical Warfare in Southeast Asia and Afghanistan: An Update," Report from Secretary of State George P. Shultz, November 1982, Special Report No. 104 (hereafter cited as *The Shultz Report*).

[31] Richard Halloran, "US Finds 14 Nations Now have Chemical Arms," *New York Times*, 20 May 1984; and Anderson, "The Growing Chemical Club".

[32] Anderson, "The Growing Chemical Club"; and Ember, "Worldwide Spread of Chemical Arms Receiving Increased Attention," p. 9.

[33] "China and Israel"; and Anderson, "The Growing Chemical Club".

[34] Michael Gordon, "US and Soviet to Meet Again on Curbing Chemical Weapons," *New York Times*, 26 August 1986; and Elaine Sciolino, "US Sends 2,000 Gas Masks to the Chadians," *New York Times*, 25 September 1987.

[35] Oberdorfer, "Chemical Arms Curbs Are Sought"; and Ember, "Worldwide Spread of Chemical Weapons Receiving Increased Attention," p. 9.

[36] The Haig Report refers to *"some evidence* that Afghan Government forces *may* have used Soviet supplied chemical weapons against the *mujahidin* even before the Soviet invasion." Emphasis added. *The Haig Report*, p. 6.

[37] Paul Moorcroft, "A new heart for the UNITA army," *Jane's Defence Weekly*, 13 September 1986; and James Morrison, "Angola again tied to use of nerve gas," *Washington Times*, 11 March 1988.

[38] "Argentine 'Gas Shells' Found in Falklands," *Daily Telegraph*, 18 August 1982.

[39] *TASS*, 21 April 1986, as reported in FBIS—SU, 25 April 1986, cited in *Arms Control Reporter, 1986*, p. 704.B.172.

[40] D. Shribman, "FBI learns Chilean plot to kill Letelier in '76 involved nerve gas," *New York Times*, 14 December 1981 and *Krasnaya Zveda*, 18 August 1984, both cited in *SIPRI Yearbook 1987*, p. 111 and note 32.

[41] David Wood, "Nicaragua has gear for waging chemical war, Pentagon says," *New Orleans Times–Picayune*, 5 December 1984.

[42] "Rights group charges massacre by El Salvador," *New York Times*, 13 August 1981.

[43] Havanna international radio, 6 July 1982, cited in *SIPRI Yearbook 1987*, p. 111 and note 36.

[44] Pierre Simonitsch, "Chemical Weapons: everyone has stocks, despite treaty," *Frankfurter Rundschau*, 29 August 1982 as reprinted in *The German Tribune*, 12 September 1982.

[45] "Chemicals claim," *Jane's Defence Weekly*, 24 August 1985.

[46] *The Haig Report;* and *The Shultz Report*.

[47] James Morrison, "Mozambique accused of chemical warfare," *Washington Times*, 31 December 1986.

[48] Wood, "Nicaragua has gear for waging chemical war, Pentagon says"; and *ACAN*, 16 February 1986 as reported in FBIS–LA, 18 February 1986, cited in *Arms Control Reporter, 1986*, p. 704.B.167.

[49] Jack Anderson, "Powderkeg fuse on our planet burning shorter," *Washington Post*, 3 December 1981, and "Phnom Penh says US gives Asian allies chemical weapons," *New York Times*, 14 April 1982, cited in *SIPRI Yearbook 1983*, p. 407 and note 164.

[50] H.G. Brauch, "Chemical Weapons: arsenals and recent developments," a paper presented at the Conference on Non Nuclear War in Europe, Groningen, 28 Nov–1 Dec 1984, cited in *SIPRI Yearbook 1987*, p. 111 and note 33.

[51] K. Dalton, "Manilla investigation into napalm bombing claim," *Times*, 26 September 1984, cited in *SIPRI Yearbook 1987*, p. 113 and note 39.

[52] UNHCR/WHO investigatory report, 1 June 1978 in UN Security Council document *S/13473*, 27 July 1979, cited in *SIPRI Yearbook 1987*, p. 111 and note 41; and Johannesburg *SAPA*, 3 September 1986 as reported in FBIS–ME, 4 September 1986, cited in *Arms Control Reporter, 1986*, p. 704.B.200.

[53] "Phnom Penh says US gives Asian allies chemical weapons," cited in *SIPRI Yearbook 1983*, p. 407 and note 164; and Kampuchean news agency statement as reported in FBIS–AP, 25 February 1985, cited in *Arms Control Reporter, 1985*, pp. 704.B.117–18.

[54] McCartney, "Iraqi Official Acknowledges Chemical–Arms Use in War." Two months earlier, the

Iraqi Foreign Minister had warned publicly that Iraq had to repel aggression by "all means, including the use of chemical weapons, against those who seek to occupy its territory." Patrick Tyler, "Iraq Seems to Abandon Diplomacy for Missiles," *International Herald Tribune*, 12 May 1988.

[55] Oberdorfer, "Chemical Arms Curbs Are Sought".

[56] M. Meselson and D.E. Viney, "The Yemen," in Steven Rose, ed., *CBW: Chemical and Biological Warfare*, Boston, Beacon Press, 1969, pp. 99–102; and "China and Israel."

[57] *The Haig Report;* and *The Shultz Report.*

[58] Sciolino, "US Sends 2,000 Gas Masks to the Chadians"; and Michael Gordon, "US Thinks Libya May Plan to Make Chemical Weapons," *New York Times,* 24 December 1987.

[59] "China and Israel"; and Anderson, "The Growing Chemical Club."

[60] "China and Israel"; and Anderson, "The Growing Chemical Club."

[61] "China and Israel"; and Anderson, "The Growing Chemical Club".

[62] "China and Israel"; and Anderson, "The Growing Chemical Club".

[63] In this regard, see the comments by the former head of the Egyptian chemical warfare programme in "Arabs 'need chemical weapons,' " *Independent,* 28 July 1988.

[64] Tom Diaz, "Chemical weapons proliferation spurred by business, panel says," *Washington Times,* 10 May 1985; and Jack Anderson and Dale Van Atta, "Iran May Turn Chemical Tables on Iraq," *Washington Post,* 2 October 1985.

[65] Anderson, "The Growing Chemical Club".

[66] Soviet Embassy, Information Department, Press release, "Statement by Mikhail Gorbachev," Washington, 16 January 1986, p. 7.

[67] Anderson, "The Growing Chemical Club"; and "China and Israel".

[68] Anderson, "The Growing Chemical Club".

[69] Simon O'Dwyer-Russell, "Gaddafi arms Syria with gas warheads," *Sunday Telegraph,* 23 November 1986.

[70] Oberdorfer, "Chemical Arms Curbs Are Sought".

[71] *The Haig Report;* and *The Shultz Report.*

[72] Anderson and Van Atta, "Iran May Turn Chemical Tables on Iraq"; and Beecher, "US seeking action to halt chemical arms spread."

[73] Sciolino, "US Sends 2,000 Gas Masks to the Chadians."

[74] "China and Israel."

[75] William Greer, "US Stops Chemicals Shipment to Iraq," *International Herald Tribune,* 2 April 1984; and Robert Harris, "The poor man's atom bomb," *The Listener,* 30 October 1986.

[76] Studeman Statement, 1 March 1988, p. 48; Oberdorfer, "Chemical Arms Curbs Are Sought"; and E.A. Wayne, "Tracking chemical weapons in the Gulf War," *Christian Science Monitor,* 13 April 1988.

[77] Studeman Statement, 1 March 1988, p. 48; and Anderson, "The Growing Chemical Club".

[78] Diaz, "Syria said to have offered chemical weapons to Iran"; and Harvey Morris, "Iran military wants to use chemical arms," *Independent,* 27 December 1987.

[79] Oberdorfer, "Chemical Arms Curbs Are Sought"; and Harris, "The poor man's atom bomb".

[80] Gordon, "US and Soviet to Meet Again on Curbing Chemical Weapons." See also, Gordon, "US Thinks Libya May Plan to Make Chemical Weapons".

[81] Studeman Statement, 1 March 1988, p. 48; and Atkinson, "Weinberger Maintains Opposition to Dutch Plan for Cruise Missiles."

[82] Diaz, "Syria said to have offered chemical weapons to Iran"; and Shaw, "Syria Reported to Be Making Chemical Arms".

[83] Studeman Statement, 1 March 1988, p. 48; "China and Israel"; and Oberdorfer, "Chemical Arms Curbs Are Sought".

[84] Studeman Statement, 1 March 1988, p. 48.

[85] Israeli, rather than US, sources, have stated this. See, for example, "Analyst Warns of Chemical Weapons," *Near East Report,* 25 August 1986.

[86] See, Elisa D. Harris, "Sverdlovsk and Yellow Rain: Two Cases of Soviet Non-compliance?" *International Security,* Vol. II, No. 4, 1987, pp 41–95.

[87] These ideas are drawn from Robinson, "Some Developments Over the Past Year in the Field of Chemical–Warfare Armament," p. 3.

[88] Don Oberdorfer, "US Curbs Chemicals to Iran, Iraq," *Washington Post,* 31 March 1984.

[89] Bernard Gwertzman, "US Includes Syria in Chemicals Ban," *New York Times*, 6 June 1986.

[90] "US tightens exports of warfare chemicals," *Chemical and Engineering News*, 10 August 1987.

[91] *SIPRI Yearbook 1985*, pp. 174–75.

[92] John Tagliabue, "Bonn Limits Export of Chemical–Arms Materials," *New York Times*, 8 August 1984.

[93] *SIPRI Yearbook 1985*, pp. 174–75. The countries involved included Belgium, Denmark, France, the Federal Republic of Germany, Greece, Italy, Luxembourg, the Netherlands, the Republic of Ireland, and the United Kingdom.

[94] *International Affairs* (Moscow), April 1986, pp. 151–52.

[95] *ADN*, 24 June 1987 as reported in FBIS–EE, 25 June 1987, cited in *Arms Control Reporter, 1987*, p. 704.B.231.

[96] Ember, "Worldwide Spread of Chemical Arms Receiving Increased Attention," p. 11.

[97] Oberdorfer, "Chemical Arms Curbs Are Sought".

[98] Oberdorfer, "Chemical Arms Curbs Are Sought," and Ember, "Worldwide Spread of Chemical Arms Receiving Increased Attention," p. 11.

[99] Robinson, "Some Developments Over the Past Year in the Field of Chemical–Warfare Armament," p. 7.

[100] Don Oberdorfer, "US Soviets May Meet Soon On Curbing Chemical Arms," *Washington Post*, 4 January 1986.

[101] "Text of the Joint US–Soviet Statment: 'Greater Understanding Achieved,'" *New York Times*, 22 November 1985.

[102] William Beecher, "US seeking action to halt chemical arms spread," *Boston Globe*, 4 February 1986.

[103] Leslie H. Gelb, "US–Soviet Pact on Chemical Arms Said to Be Near," *New York Times*, 12 November 1985.

[104] Soviet Embassy, "Statement by Mikhail Gorbachev," p. 7.

[105] "US rejects chemical weapons plan," *Boston Globe*, 11 February 1986. See also, Thomas W. Netter, "US Rebuffs Soviet on Chemical Arms," *New York Times*, 12 February 1986.

[106] A US Government Fact Sheet released at the end of 1987 simply stated that the two sides "have reviewed export controls and political steps to limit the spread and use of chemical weapons." US Information Service, "US Arms Control Initiatives (Text: US Government Fact Sheet)," London, 8 December 1987, p. 6.

[107] Iraq reportedly has used both these tactics. See, "A plaque of 'hellish poison,'" *US News and World Report*, 26 October 1987; and Steven Dickman, "Nerve gas cloud hangs over West German firms," *Nature*, 14 April 1988, p. 573.

[108] US Senate, *Joint Hearing on Chemical Warfare*, p. 17.

[109] Thomas W. Netter, "US Rebuffs Soviet of Chemical Arms," *New York Times*, 12 February 1986.

[110] Ember, "Worldwide Spread of Chemical Arms Receiving Increased Attention," p. 12.

[111] United Kingdom Mission to the United Nations, "Speech by the Rt Hon Sir Geoffrey Howe QC MP at the Third Special Session of the United Nations General Assembly Devoted to Disarmament," 7 June 1988, p. 2.

[112] Some past reports on subnational proliferation problems may be found in *SIPRI Yearbook 1983*, p. 407; and US House of Representatives, *Report on Binary Weapons*, pp. 69–71.

[113] Gary Yerkey, "Experts study threat of chemical weapons in terrorists' hands," *Christian Science Monitor*, 29 August 1986.

[114] Seymour Hersh, "US Aides Say Iraqis Made Use of a Nerve Gas," *New York Times*, 30 March 1984; and Oberdorfer, "US Curbs Chemicals to Iran, Iraq."

[115] Marie Colvin and John Witherow, "Syrian nerve gas warheads alarm Israel," *Sunday Times*, 10 January 1988.

[116] "Iraq, Saying it Hit Tankers, Threatens Chemical Strikes," *International Herald Tribune*, 30 March 1988.

[117] Aaron Karp, "The frantic Third World quest for ballistic missiles," *Bulletin of Atomic Scientists*, June 1988, pp. 14–20.

[118] These concerns are clearly articulted in Francois Heisbourg, "Missiles: Steps That Might Check Proliferation," *International Herald Tribune*, 30 March 1988.

Turmoil or Stability in the Middle East

SIR DAVID MIERS, KBE, CMG

The author is Assistant Under Secretary of State at the Foreign and Commonwealth Office. The following contribution stems from a lecture given at the RUSI. The opinions expressed are the author's own and do not necessarily represent those of the Foreign and Commonwealth Office.

THIS ARTICLE concentrates on the two chief problems in the area, namely the Gulf War and the Arab-Israel problem. It is a subjective account, giving my personal assessment of the problems as I see them. It will not necessarily reflect the views of HMG, so let nobody think that any of the indiscretions which I may volunteer can be taken down and used in evidence against the Government.

THE GULF WAR

To start then with the Gulf War, and the prospects for 1988. Of course, a lot of people say that there will be no solution to this war, no settlement, while the Ayatollah Khomeini and President Saddam Hussein remain in place. The logic for that reasoning is that the Ayatollah has insisted that the war must not stop until the idolator Saddam Hussein has been over-thrown. So therefore until one of them disappears you cannot have the war stopping. Personally, I do not think that this is a valid position from which to start because the war is too serious to be left for time to provide a solution. We have to keep trying to find a settlement. The casualties and the damage resulting from this war are not things that we can just sit by and accept as inevitable. Also, the war is showing dangerous signs of spreading. The incidence of attacks upon shipping for instance have gone sharply up. Finally, the Gulf is much too important a place, in terms of economic and strategic factors, for us to be indifferent about the continuation of the war.

In the last year, 1987, there were some encouraging features. First, we had UN Resolution 598, which was a big step forward compared with what the Security Council had been able to achieve in the early years of the war. This was a mandatory resolution, the terms very much in conformity with what the UN Charter, written more than 40 years ago now, intended. It didn't attract the veto which has prevented the Charter from being worked as intended in other theatres on so many previous occasions. Another encouraging factor was that the Iranians began, following the adoption of the resolution, to show some signs of readiness to take account of international pressures. Even if one dismisses this as being tactical, the fact that they were able to engage in detailed discussions about the implications of the resolution does show a change from their previous attitude which was one of refusing to deal with the Security Council.

Now, perhaps one should have a brief parenthetic look at why the international community has been unable to deal with this problem before. Why has there been no effective Security Council action? Apart from the problem of Superpower rivalry, which I referred to above, when the war began there was no effective Security Council response to Iran's complaint about invasion by Iraq because Iran itself was in breach of a Security Council Resolution and the International Court of Justice injunction on the US hostages. So there was not much sympathy for Iran's appeal in those circumstances to the Security Council. Subsequently, Iran lost more sympathy by being seen in many quarters as responsible for continuing the war even when the Iraqi forces which had been intruded onto her territory were withdrawn. For these reasons UN Security Council resolutions, up until last year were lacking in bite. Insofar as they had managed to achieve anything, like the mission for instance of the Secretary General's special representative, Olaf Palme, ex-Prime Minister of Sweden, UN efforts had run into the sand because Olaf Palme had been murdered and no successor had been appointed.

Security Council Resolution 598 in my view had a good effect. An important political signal was sent to the Iranians. They didn't reject the resolution, as many had expected. They accepted to get into discussion with the Secretary General. They even asked for some of their communications to be circulated as documents of the Security Council. They also agreed to parts of the resolution: in particular to a ceasefire in place if the panel which was going to take a view about responsibility for the war was set up. (This of course would have allowed the Iranians to keep the option of bargaining about withdrawal from Iraqi territory which they were then occupying at Fao and elsewhere.) Now this Iranian attitude could be a tactic or it could be indicative of genuine readiness to cooperate with the resolution. Many people think, from public Iranian statements and from an analysis of the political situation in Iran, that the Iranians are set on continuing the war and not on genuinely cooperating with the Security Council. This has not yet been incontrovertibly put to the test because the Security Council has not been able to follow up its resolution with the speed which many of the members hoped for. One Permanent Member in particular, the Soviet Union, has been slow to agree to follow up work as envisaged in paragraph 10 of the Resolution which provides that the Council shall meet again to decide appropriate measures in the event of non-compliance. It was not possible to do anything effective on this for about five months after the passing of the Resolution, and in my view this resulted in Iran being able to avoid some of the pressure which the Resolution should have brought to bear on her. However, in December, the Soviet Union agreed that work on a second, follow-up, Resolution could begin. But at the same time the Soviet Union pressed ahead with unhelpful and ambiguous proposals which it had made about a UN Naval Force and a reactivation of the Military Staff Committee of the Security Council which was provided for in the Charter. Why do I think such proposals unhelpful or ambiguous? Of course, it is not for me to explain proposals made by someone else; but there are those who think that these proposals were intended, indeed are intended, either to provide, through the Security Council or the Military Staff Committee, a capacity for vetoing the activities of warships in the Gulf: or else to provide a UN flag under which Soviet warships could enjoy port facilities in the Gulf at present denied to them. Those are two theories which have been advanced. But we have, I think, to look and see what a naval force could do; there are a number of possibilities. A naval force could be designed to separate the belligerents. The flaw in that is that the belligerents are actually carrying out their war chiefly on land, rather than at sea. So what you need is a land force rather than a naval force. Also if the UN naval force did manage to enforce a ceasefire at sea, this would be a partial ceasefire, which is not what the Resolution calls for. Finally, there would be quite a serious danger that both the belligerents, or either of them, would attack the UN naval force, or elements of it. I don't think we could have confidence that they would respect the UN flag on the ships of such a force. This would pose complicated questions about force levels, command and control of Rules of Engagement.

An alternative possibility is that the force would be put in place to preserve a ceasefire, like other, land-based, UN forces which are in existence around the world. But this begs the question of whether we get a ceasefire or not, because in these other cases a ceasefire has been in place before the force deployed. A third possibility is that you would have a force to protect non-belligerent shipping. But this again is a very difficult suggestion and in my view bordering on the impractical. The problems of providing such a force with political direction in the very dangerous circumstances, prescribing the Rules of Engagement, and deciding controversial political questions like whether it should escort non-belligerent, third party shipping into ports of both the belligerents, particularly as the Iraqi ports are up the Shatt al-Arab river and not accessible in the face of Iranian opposition, would pose some very difficult issues. Would the force be empowered to fire on aircraft of the belligerents, or would it have to refer back? If the latter, it might be too late. If the former, it would be very difficult to work out in advance exactly what the Rules of Engagement would be. Also you have a resource problem: the number of ships that would qualify for protection in the Gulf at any given moment is very high, perhaps as high as 500. It would be very difficult to muster the necessary warships, all to be placed under the command of the UN, to carry out these defensive duties.

A final possibility for a UN force would be to enforce an arms embargo once it had been adopted. But there again the precedents for a UN blockade are not very good. There is also the problem of the land frontiers of the country against which the embargo was being enforced. In practice, we have not yet got an embargo in place. The primary obligations to make one work would be on the individual UN member states who would be supporting it. The UN would then need to monitor it and identify the loopholes, and finally to stop them. The time to look at these questions of how to enforce an embargo is when you have got one in place rather than before you have even agreed on it.

So, what are the prospects? In my view we should not be too

readily optimistic. Ayatollah Khomeini still seems committed to the overthrow of the regime in Iraq. Fighting on the land front has not been heavy in 1988 compared with the massive assaults made at the beginning of 1987; but equally, there has been no lull on that front and a particularly long and deadly revival of the war of the cities. It has been suggested that even if the Ayatollah were to die, his successors would be obliged to cling to the policies he prescribed in order to validate their claims to succession. So, there are some who think that the war could continue, possibly with even more commitment, if he were to disappear.

The Iraqis appear determined to maintain their attacks on Gulf shipping. It may be that they calculate that this will provoke Iranian retaliation on Arab shipping which will in turn tend to internationalise the war. We deplore this, but realistically we cannot rule out that it will continue. Also, there is a question mark over the commitment of the Soviet Union. In the second half of 1987 the Soviet Union held up the discussion of a follow-up Resolution, and we do not know why. We do not know whether their reasons for doing that are still in force. Maybe there is some mistrust of the Americans; maybe there is insufficient pressure on them from those who feel that the second Resolution should go through. Or maybe there is an Afghan or other regional angle of Soviet policy which needs to be taken into account. What is clear is that the political factors which led to a certain rivalry between the Superpowers in the area will continue and this is going to circumscribe the scope for cooperation. So in my view, we should not be too optimistic, but equally we must continue trying. We have a clear line of action for the Security Council, to which we are committed and on which our efforts can be deployed.

THE ARAB-ISRAEL PROBLEM

I now turn to the Arab-Israel problem and the peace process. I suppose you could say that there are some links between this question and the Gulf War. But what follows is a self-standing analysis as I see it of where we stand on the peace process.

You can't assess the prospects for 1988 without taking account of the fact that it is election year in both Israel and the United States. At the end of 1987 many people were expecting little of 1988 for that reason: insufficient flexibility and freedom of manoeuvre to carry the peace process forward on the part of two crucial Governments. But, I would like to suggest, perhaps provocatively, that this analysis is not necessarily right, especially in view of the disturbances which have been taking place in the occupied territories. These disturbances have had a very high profile, they have made a great impact on television screens, particularly in America where public opinion on this question and on the amount of support that Israel enjoys, or should enjoy, from the United States can have a crucial impact on the prospects for peace in the Middle East.

To form a view on whether there really is scope for progress we will backtrack a little on the attitudes of the parties concerned. When the coalition in Israel between Labour and Likud was formed in 1984, its approach to the peace process was based on the Camp David formula, i.e. direct negotiations between Israel and each of its neighbours. At Camp David the United States had acted as a midwife and the coalition was prepared to accept this role for the future because Israel trusts the United States. However, it was not prepared to have the international community as a whole, or the Security Council, or the European Community, or the Soviet Union, or any other outsider, acting as a midwife. In other words, the coalition was opposed to an International Conference. It feared that international pressures for concessions at such a conference might prove impossible to resist.

The second important element in the Israeli attitude is the difference between Likud and Labour, two elements of the coalition, about the substance of any peace settlement on Camp David lines. Likud, as I understand it, is not ready for any significant withdrawal from the occupied territories, which many of its members regard as essential parts of a broader Israel. The most that these elements are prepared to consider is some very limited form of autonomy for the Palestinians in the occupied territories, and that in an administrative and not a territorial sense. Labour, on the other hand, accepts the principle of withdrawal (but not from all the occupied territories) provided that the Arabs give Israel peace in return, as Egypt did at Camp David. This approach is characterised as the "territory for peace" or "territorial compromise" formula. It would be subject to agreement on the precise territories from which Israel would withdraw. This approach is also, broadly speaking, the approach of the international community as embodied in Security Council Resolution 242.

On the Arab side, Egypt was ostracised after Camp David for signing a separate peace, and there are many objections to the agreement and to approaching the peace process on the basis of it. The basic objection is that Egypt conceded recognition of security for Israel without having recovered all the occupied territories, only the Sinai, and without having obtained a commitment to Israeli withdrawal. The best that Camp David provided for was the Israeli obligation to respect the legitimate rights of the Palestinian people and their just requirements, but without limiting Israeli settlement and without forcing the abandonment of Likud's claim that the occupied territories were really part of greater Israel. Then, of course, there was the invasion of Lebanon in 1982 by Israel, which further discredited in some people's eyes the agreement which had been reached before.

Also, for the Arabs the central element which Camp David ignored was the requirement for self-determination for the Palestinians, which the Arabs interpret as providing for a Palestinian state. For them, the PLO is the sole legitimate representative of the Palestinian people: at any negotiations with Israel, for example at an International Conference, the PLO must be present, and must represent the Palestinians. Israel, on the other hand, regards the PLO as a terrorist organisation and does its utmost to discredit it.

So the question of Palestinian representation, which was side-stepped at Camp David, hangs over the peace process. Israel would like to negotiate about the future of the occupied territories only with Jordan. Jordan has been unable to contemplate such negotiations without certain assurances: for example, that the Arab position on self-determination for the Palestinians would be protected, that a formula for Palestinian representation in the negotiations would be found, and that there would be adequate backing from the Palestinians or the Arabs as a whole for a Jordanian decision to enter negotiations on this basis. Of course no such backing would be available for direct negotiations on the Camp David model between Jordan and Israel. This is why the idea of an International Conference, rather than direct negotiations, is so important to the Arab side.

You might think that those two positions were irreconcilable. But, actually, since the beginning of 1985 I think that there has been considerable evolution in the attitudes of the two parties, at least amongst the attitudes of key elements on each side. On the Israeli side, while Mr Peres was Labour Prime Minister of the Coalition, he made it clear that he was prepared to accept the idea of an International Conference as a framework for direct negotiations, provided that there were safeguards, to prevent the conference turning into a means of piling pressure on Israel; and in particular to prevent a plenary session of the conference imposing solutions or vetoing what might have been agreed in direct negotiations. The key point in the evolution of the Labour position here was that the legitimacy of the Arab side requirement for an International Conference was accepted. And underlying this was the Labour Party's readiness to put to the Israeli people, within the framework of the Coalition or if necessary at elections, the vital question of territorial compromise in return for security for Israel. Mr Peres made it clear that he was confident of winning a favourable answer to that question, if only he could show that the prospect of security in return for withdrawal, i.e. peace in return for territory, was real. That's why it is so important that the PLO should denounce violence, accept Israel's right to exist, and accept UN Security Council Resolutions 242 and 338 which embody the principle of territory for peace.

On the Arab side, there was some evolution as well. In February of 1985 the PLO and Jordan went as far as a joint platform, which envisaged negotiations with Israel at an International Conference, which would lead to Israel's withdrawal and then to a confederation between Jordan and the parts of Palestine from which Israel would have withdrawn.

But, in order to make this platform realistic, it was necessary to convince Israel that the PLO was indeed ready to acknowledge Israel's right to exist, as had been suggested, or implied in successive Arab summit declarations. Despite lengthy negotiations about the conditions of making this move in 1985 and early 1986, it was not possible for the PLO in fact to make it, and the joint platform was dissolved in February 1986. However, all was not lost because there still remained on the Arab side a firm readiness to take advantage of the opportunities offered by Labour's readiness to accept an International Conference on certain conditions. Diplomatic activity continued, although with a low profile compared to what had been going on in the previous year, and continued even after the rotation of the premiership from Mr Peres to Mr Shamir. By the summer of 1987 it had become clear that arrangements for an International Conference which were acceptable to Mr Peres, to Jordan, to Egypt and perhaps to the United States, could be reached. But it was at this point that the process got stuck, because Likud adamantly refused to agree to the idea of an International Conference. It insisted that the only way forward was through a regional conference between Israel and its neighbours, which would provide direct negotiations between Israel and each of them on the Camp David model. Peres was unable to muster a majority in the Knesset to force the Cabinet's hand or precipitate an election. He had the option of forcing elections by withdrawing support for the Coalition, but that would have brought elections only in three months time—a hazardous procedure since by then the whole question might have changed and in the meantime, during the long run-up to the elections, Likud would have been in sole control of the Government without any con-straints from the Labour Party, their erstwhile Coalition partners.

At the end of 1987 no clear way of breaking the deadlock seemed apparent. There was a feeling prevailing that time had run out because of the elections due in 1988 in Israel and the United States. However, in December there came the long predicted uprising in the occupied territories. With this there definitely came a new attitude in many quarters. In Israel, there is perhaps a new realisation that the status quo cannot continue, that the occupied territories cannot remain in limbo, and that demographic and other problems threaten Israel if no political solution for the territories can be found. In the United States, too, where the disturbances have attracted very wide coverage on American television, certain existing assump-tions have been questioned—not least the conventional wisdom that nothing can be expected of US Middle Eastern policy in an election year—which has been belied by the very personal involvement of the Secretary of State himself in US Middle Eastern diplomacy in 1988.

United States activity on the peace process this year has been more inventive than in any earlier year of President Reagan's administration. By the middle of the year it had exposed very clearly the nature of the obstacles—and in particular where responsibility lay for blocking an international peace confer-ence as a framework for direct negotiations between Israel and the Arab parties concerned. All this tended to suggest that 1988, instead of being a year of stagnation because of the elections in Israel and the USA, was capable of being a year of decision. In the case of Israel there was an opportunity for national decisions of profound import, if only it could be so contrived that the question of providing for Israel's future security through territorial compromise could be put in credible terms to the Israeli electorate and the right answer obtained. In the case of the American elections, there was an opportunity for testing how far the activities of pro-Israeli lobbies would have their traditional impact on the US electoral process—and in particular on the platforms espoused by the Congressional candidates as well as the Presidential contenders —in circumstances where publicity for the uprising in the occupied territories was going to make the expression of uncritical sympathy for Israel, let alone unqualified support, considerably more controversial than usual.

By the time this appears in print both sets of elections will be over and the world will be able to judge how far the right questions were put and the right answers given. Whatever the outcome, it is likely to be clearer than ever that new and more pressing forces are at work in the region, and that reliance on immobilism or the indefinite preservation of the existing situation provides no sound basis for policy. The problems inherent in Israel's continued occupation of the West Bank and Gaza are not going to disappear and cannot be pushed into limbo. Nor can they be solved out of the barrel of a gun. So it will be necessary in 1989 to tackle the problems that were not squarely faced in 1988, and each year of delay makes them more difficult to solve in the future.

Afghanistan: Coming Home to Roost?

S. DE BANZIE

*S. de Banzie graduated from the School of Slavonic and East European Studies, University of London,
and is currently a member of the research staff on the Soviet and East European Programme of the
Institute. This contribution stems from that programme, for which the generous support of the Esmée
Fairbairn Charitable Trust is gratefully acknowledged.*

ON 14 APRIL 1988 the Soviet Union signed the accord on the withdrawal of its troops from the territory of Afghanistan after more than eight years of war. Though some have proclaimed this a defeat for the USSR, the Soviets are leaving neither as victors, nor as vanquished. There is, however, evidence to support the view that Moscow has much to gain from a withdrawal and that the political capital to be made from pulling out now is considerable, both for Mr Gorbachev, establishing his new style of leadership, and for Soviet international prestige.

THE INVASION

When the Soviets invaded Afghanistan in December 1979, they did so with an initial assault force of four motor-rifle and one and a half air assault divisions. The small scale of this force indicates that the Soviets did not expect much difficulty in the achievement of their objectives. They did, however, take the precaution of stationing another 30,000 men north of the border in case their calculations proved incorrect.[1] By early January 1980, the number of Soviet soldiers in Afghanistan had reached 85,000. The invasion forces, largely drawn from the two military districts bordering on Afghanistan (Turkestan and Central Asian), were filled out with reservists living locally. It has been estimated that 30 or 40 per cent of the original force consisted of Central Asians, most of whom would have been from Moslem families. There are four possible perceived advantages in the use, by the Soviets, of non-Russian troops to invade an area to whose population they were strongly tied by race, language and religion.[2] Firstly, using the reservists living closest to Afghanistan helped to preserve the security of the operation. Those called up were therefore mainly Tadzhiks, Uzbeks, Turkmens and Kirghiz. Using locally based soldiers also had the advantages of economy and convenience. The third point, subsequently explained by KGB defector Major Vladimir Kuzichkin, was that Moscow hoped to gain a political advantage by using troops of a similar ethnic and religious background to the Afghans. However, this proved to be a miscalculation. As Kuzichkin told *Time* magazine, "They (the Central Asian soldiers) were supposed to make our intervention go more smoothly ... It was an error ... They showed little interest in fighting "their neighbours".[3] A few (estimates put the number at less than 100) even deserted to join those they had been sent to subdue. Fourthly, stories of the fighting, by returning soldiers, would remain in the Central Asian area.

The invasion force was impeded not only by the absence of enthusiasm of the Central Asian soldiers to fight their Afghan "brothers", but also by their lack of fighting skills and poor training. It had been practice for some years to assign Central Asian conscripts to non-combat construction units and to give them only very basic military training.[4] It has therefore been suggested that Central Asian soldiers in the first invasion force were probably drawn from non-combat construction battalions.[5] Moscow quickly realised its mistake in sending such soldiers and by the middle of 1980 most Central Asians had been withdrawn from combat duty in Afghanistan and replaced by what were deemed to be more reliable troops, of mainly Russian, Ukrainian and Baltic origin.[6] Having invaded, the Soviets found themselves embroiled in a war which involved spiralling costs, repeated changes of tactics in unsuccessful attempts to subdue the resistance, and international outrage.

It was recently revealed by a Soviet official that the initial decision to invade Afghanistan, was taken by a small group of "four or five" members of the Politburo, including Brezhnev, Defence Minister Ustinov, KGB chief Andropov and Foreign Minister Gromyko.[7] It is highly unlikely, however, that the group did not include Suslov and Romanov, and in a Politburo vote the plan would very probably have received the support of hardliners Kunaev and Aliev. Clearly, the discussion prior to the decision must have taken in a wider group than just a few Politburo members. Senior officials from the KGB First

Directorate, the Middle Eastern Countries Department of the Ministry of Foreign Affairs, the Military and possibly members of the Academy of Sciences would have been involved in supplying advice and information. Frequent visits to Afghanistan by high-ranking Soviet military figures during 1979 testify to a significant degree of preplanning of the operation. In April 1979, Army General Aleksei Yepishev, Chief of the Main Political Administration of the Soviet Army and Navy, headed a delegation visiting Kabul.[8] It is worth remembering that Yepishev made a similar visit to Czechoslovakia in 1968 a short time before the Soviet invasion. In August 1979, Army General Ivan Pavlovskiy toured Afghanistan with a group of 60 Soviet officers.[9] However, it appears that of the group which purportedly took the final decision, only one (Gromyko) is still alive. The decision was not necessarily a unanimous one. Kuzichkin claims that "senior KGB officers" had warned strongly against such involvement in Afghanistan.[10]

If the invasion brought problems for the politicians, it certainly enhanced the careers of the military men involved. The man put in charge of the invasion was then First Deputy Defence Minister Marshal Sergei L. Sokolov. His Southern Theatre of Military Operations controlled two army group HQs; one based in the Turkestan Military District, and commanded by Colonel-General Yuri Pavlovich Maksimov, the other, based on the Central Asian Military District, under Colonel-General Pyotr Georgievich Lushev. The senior Soviet commander in Afghanistan from 1982 was Army General Mikhail Ivanovich Sorokin. Sokolov subsequently became Defence Minister, but was retired a few days after Mathias Rust landed his Cessna in Red Square in May 1987. Maksimov was made a Hero of the Soviet Union in 1982 and was promoted to the rank of Army General in the same year. He is currently a Deputy Defence Minister and Commander-in-Chief of the Strategic Rocket Forces. Lushev, promoted to Army General in 1981, was made a Hero of the Soviet Union in 1983. After a spell as Commander-in-Chief, Group of Soviet Forces Germany, he was made First Deputy Minister of Defence in July 1986. Sorokin was appointed Deputy Minister of Defence in July 1987.

REASONS FOR THE INTERVENTION

The coup by Hafizullah Amin, ousting the Moscow-favoured

leader Taraki in September 1979, and the fighting between the two factions of the People's Democratic Party of Afghanistan, threatened the stability and continued Soviet-friendly nature of the Kabul government. Moscow stood to lose much of its influence, and foothold in Afghanistan, which could choose to seek foreign support elsewhere. Any regime following the downfall of the current Marxist one, was likely to be hostile to the Soviet Union, since much of the Kabul government's opposition came from anti-Soviet Muslim fundamentalist circles.[11] Moscow would also have feared the danger of Muslim fundamentalist influence spreading across the Soviet border.[12] Additionally, in the decision to invade, ideology was a significant factor.[13] The principle of ensuring that the revolution was not reversed was deemed appropriate to the case of Afghanistan.[14] Soviet mistrust of the Afghan leader Amin, was also instrumental in their decision to invade. Well before the intervention there were rumours of Soviet plots to depose him.[15] Amin was never trusted by the Kremlin. He had studied for some years in US universities, and spoke English, not Russian. He was suspected of pro-Western sympathies.[16] Moreover, Amin's policies were accelerating the collapse of the army and alienating the population.

ECONOMIC INTERESTS

One factor not mentioned by the Soviets was probably Moscow's long time economic interests in, and military involvement with Afghanistan. An appreciation of these interests is necessary to understand the Soviet invasion. Soviet economic interest in Afghanistan stems back to the early post revolutionary years of the USSR, and takes over from earlier Tsarist interest. Despite this, in the period up to 1955 Moscow's attempts to bring about Afghan economic dependency were largely unsuccessful. However, from 1955 onwards the Soviets began to gain a foothold, by offering Afghanistan much needed credit, which totalled $1.265 billion in 1978.[17] Indeed in the two decades from 1955, Soviet financial aid to Afghanistan was only exceeded by their loans to Egypt and India. The USSR chose to receive repayment of these loans to Afghanistan in commodities such as natural gas, for which they were able to exact prices below those on the world market. Moscow's economic aid was, moreover, directed almost exclusively towards projects which were linked to, or of use to the

Soviet economy. Soviet policy, since long before the invasion in 1979, was to achieve Afghan dependence through selective aid (to economic programmes of potential use to the USSR) and increased bilateral trade linking the Afghan and Soviet economies more and more closely, and isolating Afghanistan from world markets as much as possible. This has been coupled with a strategy of encouraging indebtedness, and setting prices in barter trade below those obtainable on the international market.

MILITARY INTERESTS

Soviet military interests in Afghanistan date back to the early 1920s. Experts from the USSR assisted in the formation of the Royal Afghan Air Force in 1924, and around 1927 the Soviet Union began to supply Kabul with aircraft, piloted and serviced by Soviet personnel. Training for Afghan crews was provided in the USSR.[18] Following the United States' refusal to proffer military aid in 1955, the USSR and Afghanistan concluded an agreement on arms worth $25 million, for Afghanistan in 1956. The deal included MiG–17s and Il–28 bombers, helicopters, T–34s and small arms, supplied by the Soviet Union and her East bloc allies.[19] Over the next decade Soviet military trainers in Afghanistan replaced the traditional Turkish instructors and more and more Afghani personnel were sent for training to the USSR.

Following the ousting of King Zahir Shah, Soviet military supplies to Afghanistan showed a significant increase. By September 1975, the Soviets had provided Afghanistan with 400 armoured personnel carriers, additional howitzers, mortars and multiple rocket launchers, and more Il–28s. Between 1975 and 1978 a further 350 tanks (amongst them some new T–62s) and 400 guns, howitzers and mortars were supplied, as well as AT–3 anti-tank guided weapons, SA–3 and SA–7 SAMs, and AA–2 air-to-air missiles. The Afghans also received extra MiG–21, Il–14, Il–18 and Su–7 aircraft, and Mi–8 helicopters.[20] In addition, the Soviets assisted in the construction of roads, airfields and the Salang Pass tunnel. Thus, long before their invasion, the Soviets had secured influence over the Afghan military.

THE AIM OF THE INVASION

The small size of the original invasion force was indicative of the limited nature of Soviet aims. Their objectives included securing key political and military targets in the capital, and installing as the new leader, Babrak Karmal, in diplomatic banishment in Prague, a man they thought capable of forming a Soviet-friendly government able to control the country. Having put their man in power, Moscow hoped that after some time in which to consolidate its position, the People's Democratic Party of Afghanistan would be able to stand on its own and stabilise the country, and that armed intervention to maintain control would no longer be necessary.[21] As the USSR has found to its cost, such hopes were seriously misplaced.

CRITICISM OF THE INVASION

True to tradition, the Soviet media has set about looking for scapegoats, and the first recriminations and criticisms of the invasion have already begun. "Experts" have been blamed for mistakes and an incorrect evaluation of the situation. The net of criticism spreads wide, taking in specialists on Islam, diplomats, politicians and the military, although no one has yet (by May 1988) been singled out by name.[22]

Although Soviet officials attempt to portray the withdrawal in a positive light, denying that it is a defeat for Soviet power, some of those not involved in the decision to invade are now beginning to distance themselves from it. Oleg Bogomolov, Director of the Institute of Economics of the World Socialist System of the USSR Academy of Sciences in Moscow, pointed out in March 1988 that his institute had been opposed to the invasion from the outset, and that a memo criticising the intervention had been addressed to the appropriate quarters on 20 January 1980. Indeed the institute had been sending reports throughout the second half of the 1970s advising restraint and caution in dealing with Afghanistan.[23] Such criticism is not without precedent. Fairly early on in Moscow's Afghan adventure (in Spring 1981) one or two Soviet officials did dare to voice personal disapproval of the intervention.[24]

THE COST OF INVASION

There is no disputing the fact that the Soviet Union has paid dearly for its attempts to "assist" the Afghan people, in human, financial and political/propaganda terms. No Soviet figures have been released on how much the war is costing them, and Western estimates vary enormously. In the summer of 1987 Marin Strmecki (a research associate to Dr Zbigniew Brzezinski at the Center for Strategic and International Studies) suggested that the war in Afghanistan had cost the Soviet Union between $30 billion and $48 billion, and that the burden was around $12 billion annually.[25] Furthermore, with the delivery of US Stinger and British Blowpipe missiles to the rebels costs undoubtedly rose. More recently *The Economist* put the cost to the Soviets at 1 million Roubles per day.[26] Whatever the true figures, (which will probably never be published), Afghanistan is undoubtedly a serious drain on Moscow's finances.

The cost to the USSR in human terms is also difficult to ascertain. Casualty figures were not released by the Soviet Union during the war and despite recent promises to publish a full list, including names, of all Soviet fatalities following withdrawal it is highly unlikely that names will materialise. A US State Department estimate in 1987 set the number of Soviet troops killed in Afghanistan at 15,000 in stark contrast to the Afghan resistance estimates of 50,000. Privately, Soviet officials have mentioned a casualty figure of 25,000 though what precisely is included in this figure is not known.[27] A Soviet expert interviewed by the *Far Eastern Economic Review* agreed that a Western estimated 35,000 Soviet casualties was "very close".[28] Another more circumspect Western estimate put the number of Soviet fatalities over the last 8 years at 12,000 and roughly 5 times that number wounded.[29] A few days after the start of the withdrawal, Eduard Rozental of Novosti said that he believed that Western estimates in the region of 12,000 to 15,000 dead were "more or less correct", and the official Soviet figures were announced by General Aleksey Lizichev at a press conference on 25 May, just ten days after the official start of the withdrawal. He claimed that a total of 13,310 Soviet soldiers had been killed, 35,478 wounded, and 311 reported missing during the eight and a half years of war in Afghanistan. By summer 1987, between 500,000 and 700,000 Soviet soldiers and done their "internationalist duty" in Afghanistan.[30] Whether the death toll is 13,000 or 50,000, whether the number of veterans is 500,000 or 700,000, to the vast majority of the Soviet Union's population, the war in

Afghanistan remains remote and unreal. Moscow has little to fear from domestic disapproval or unpopularity of the war. Soviet public opinion, *glasnost* or no *glasnost,* still counts for very little indeed.

The international political cost to the Kremlin is far higher. The invasion has resulted in heightened anti-Soviet feeling around the world, damaged the USSR's influence within the non-aligned movement and antagonised the Muslim world, as well as making normalisation of Sino-Soviet relations even less probable than before. (Withdrawal remains one of the Chinese preconditions for improved relations.) Furthermore, the venture damaged the chances of disarmament and a reduction in international tension.[31] It also weakened the argument of unilateral disarmament movements in the West. Moreover, contrary to Soviet expectations, the West has not forgotten Afghanistan. Soviet influence in the world has, as a result of the invasion, suffered a major setback.

THE BENEFITS

Afghanistan has, however, not all been bad news for the USSR. The Afghans have undoubtedly borne much of the cost of their own subjugation, and the Soviet Union may even gain from the venture in the long run. "Defence" expenditure is bleeding the Afghan economy dry. According to a recent statement, 60 per cent of the Afghan budget is spent on defence.[32] Much of the military hardware required by the Soviets and their Kabul clients for the campaign, is purchased by the Afghans. Soviet statistics show that whereas in 1979 Afghanistan purchased "aviation technology" worth only 247,000 Roubles in 1986, this soared to 104,439,000 Roubles.[33] Machinery, equipment and transport vehicles to the tune of 1.76 billion Roubles, much of which must have been designated for military use, was purchased between 1979 and 1986. It is likely that the Soviet and Afghan regime forces found use for some of the Soviet petroleum products, exports of which to Afghanistan rocketed from 39,442,000 Roubles in 1979 to 75,650,000 Roubles in 1980 and 116,075,000 in 1983.[34] Estimates put total Afghan debt to the Soviet Union at over $2.1bn by the end of 1986.[35] Moscow's economic hold on Afghanistan is therefore a strong one, and the Soviets obviously have a clear idea of the way in which they require the payments to be made: one of the areas of the Soviet Union possessing the best oil and mineral resources is Turkestan, and across the border, the geologically similar area of Northern Afghanistan in particular, is also rich in such resources. Soviet exports to Afghanistan of drilling and geological equipment (worth 181,986,000 Roubles between 1980 and 1986)[36] indicate their longer term interests in the exploitation of Afghan mineral resources. The Soviets have been involved in developing Afghan natural gas resources in the northern area and have been the chief clients for many years, setting a price below the international market rate. It has been estimated from Soviet data, that between 1977 and 1981 the USSR bought Afghan gas at around $48 per 1000 cu m whilst the international price was, at the time, $114.78 per 1000 cu m.[37] It should also be noted that the metres on the gas pipeline from Afghanistan to the USSR are situated on the Soviet side of the border, and that the Afghans are not given access. Bilateral trade has been arranged so that raw materials constitute the majority of Afghan exports even where local processing capability is available. For example, despite Afghan refining capabilities, crude oil is exported and Soviet refined oil is imported. In the last few years, a significant number of agreements have been signed on cooperation and trade between the USSR and northern Afghanistan. The country will be paying its debt (much incurred by the war) to the Soviet Union, in Soviet-desired commodities, at Soviet-set prices, for years to come. The USSR has acquired another source of cheap raw materials.

A MILITARY TRAINING GROUND

Afghanistan has provided the Soviet military with a proving ground. Since the end of the Second World War, Soviet soldiers have had no involvement in real armed combat. A local war, using only conventional weapons, with no real third country intervention, right on their southern border, in difficult terrain, has provided an opportunity for young soldiers and officers to get first hand experience in particular of mountain and guerrilla warfare.[38] It has also given the military leadership an opportunity to evaluate the behaviour and performance of their troops, and to make appropriate changes, in training and tactics. In the 1970s, prior to the war in Afghanistan, the USSR Ministry of Defence journal *Voennyy Vestnik* published a mere 15 articles devoted to mountain

TABLE 1. SOVIET EXPORTS TO AFGHANISTAN IN THOUSANDS OF ROUBLES

	1978	1979	1980	1981	1982	1983	1984	1985	1986
Machines, Equipment and Means of Transport	72,116	92,836	115,019	164,190	220,014	180,999	335,106	294,424	361,466
Aviation Technology	247	763	698	57,806	91,993	42,416	163,581	75,666	104,439
Tractors	523	13		207	543		951	1,261	1,244
Drilling and Geological Equipment			37,295	37,064	20,267	17,186	16,676	24,148	29,350
Oil and Petroleum Products	23,436	39,442	75,650	71,749	72,350	116,075	119,621	118,925	68,006
Medicaments	408	457	747	495	653	740	991	821	516

Source: Vneshnyaya Torgovlya SSSR

warfare. However, since the war began, *Voennyy Vestnik,*
Sovetskiy Voin and *Aviatsia i Kosmonavtika* have featured over
100 articles on training for mountain warfare.[39] In 1988, in-
depth coverage of the subject of mountain warfare training for
all branches of the army is to be provided by *Voennyy Vestnik.*[40]

With the Soviet media publishing articles in praise of the
decision to pull out, it is interesting to note that General Yazov
has yet to comment on the move.[41] Absence of praise could
tend to indicate disapproval. Indeed for the Soviet military,
the Afghanistan venture had many pluses. Besides giving the
Command an opportunity to review tactics and providing the
soldiers with experience of real combat, it has been a road to
advancement of their careers for officers who served there. At
last, too, officers can point their troops to recent examples of
heroism, rather than rely, as has been the case up to now, on
reciting stories of Great Patriotic War heroes, old men now, in
battered suits with rows of medals, with whom it is extremely
hard for 20-year-old soldiers to identify.

In the Soviet military press in January of this year, a
"polemical letter" by the editorial board, rejected the comparison of their Afghan venture with the US Vietnam campaign.[42]
The letter also indicated the military's desire to stay and fight it
out in Afghanistan.[43] In a radio interview in March 1988, a
member of the editorial board of the military daily *Krasnaya
Zvezda,* Colonel Viktor Filatov, denied the suggestion that the
intervention had been a mistake.[44] There is evidence therefore,
that the Soviet Command perceives an advantage in their
involvement in Afghanistan.

THE ISLAMIC THREAT

The Kremlin has long been concerned about the influence of
Islam and particularly the spread of fundamentalism, and its
influence on Soviet Central Asian Muslims, and there are
indications that the Soviets perceive a certain degree of success
in stemming the tide of fundamentalist Islam in Afghanistan. A
recent article in the Soviet press claimed that the invasion had
prevented the possibility of the emergence of such a regime.[45]
In his radio interview, Viktor Filatov claimed that the invasion
had prevented the appearance of a Khomeini-style regime in
Afghanistan, and had thus removed a threat to the USSR's
southern border.[46] The fact that the withdrawal agreement was
signed shortly after a flare up of nationalist tensions in the

southern USSR, (albeit not in Central Asia, but in another hot spot—Azerbaijan) indicates that the success in preventing the spread of fundamentalist Islam is perhaps perceived throughout the top Soviet decision making circles. Indeed it must be clear to Moscow, that fighting is likely to continue in Afghanistan for many years after a withdrawal, thus preventing the emergence of any fundamentalist Islamic regime able to influence Soviet Muslims to a damaging degree. It should also be noted that those areas of the USSR bordering on Iran do not seem to have experienced any particular upsurge in Islamic discontent.

THE WITHDRAWAL

As has often been pointed out, a Soviet withdrawal from Afghanistan would be the first voluntary Soviet pull out from territory invaded in war in 33 years. The comparison with the withdrawal from Austria is inappropriate since military resistance played no part in the Soviet decision to pull out on that occasion.

Mention of the Soviet desire and conditions for a pull out of Afghanistan began shortly after the invasion. In mid January 1980 the Soviets declared that since the assistance of their troops had been requested to help counter "external agression", they could be brought out only when "the reasons for the Afghan leadership's request for them disappear".[47] The Soviet troops remained. Shortly before the Olympics in 1980, the Soviets announced that they were pulling out some of their troops. They turned out to be units unsuited to the type of warfare, and were quickly replaced by more suitable forces.[48] When Mikhail Gorbachev came to power in 1985, one of the priorities he set was solving the Afghanistan problem. In February 1986 at the Soviet Party Congress, Gorbachev reiterated Moscow's desired aim of bringing out the forces "in the very near future". He talked of the problem as one of border security, and pointed out that its regulation was a "vitally important aim of our foreign policy". Five months later in Vladivostok, Gorbachev promised the withdrawal of six regiments from Afghanistan. Those pulled out were either sent in after the July speech ready to be withdrawn, or they were of no use in Afghanistan and were brought out only to be replaced by more useful forces shortly after.

What makes the agreement signed on 14 April 1988 (on the

withdrawal of all 115,000 Soviet troops by 15 February 1989) different? Prior to the invasion, the regime in Kabul was in danger of collapse and the various factions were fighting bitterly. Hatred of the foreign invader helped to unite the rival groups with a common purpose, but after withdrawal that cohesive force will almost certainly cease to exist. The first enemy of all the resistance groups will be Najibullah's regime. In the event of a resistance victory over Kabul, the struggle for power will then be between the main rebel factions. The Jamiat-i-Islami group, to which guerrilla commanders Massoud and Ismail Khan are affiliated, receives only the third largest allocation of weapons from the West. However, the group has probably the best fighters and has captured significant quantities of weapons from Soviet and regime forces. The Hesb-i-Islami faction, with which the guerrilla leader Abdul Haq is connected, receives the second largest supply of weapons. Haq specialises in attacks on Kabul. A rival Hesb-i-Islami group led by Heckmatyar, which receives the most Western arms, despite being anti-Western, is the third strongest fighting force. The balance of strengths and weapons supplies of these and other groups will ensure that the fighting continues for some time to come, and that no one group is likely to emerge in control, or form a stable government in the near future. Whilst engaged in a bitter power struggle, the various Afghan factions will have little time to devote to attempts to influence their brothers north of the border. Moscow can rest assured that the Afghans will remain locked in feud for the foreseeable future. Incursions over the border such as the one which took place in March 1987[49] are containable, and cease to be as significant for the Afghans once they are left to determine their own future. For the present, they are unlikely to attempt to transfer their holy war to Soviet territory. Whether or not the sight of Russian troops withdrawing from Afghanistan after more than eight years of war will fan flames of nationalism and religious fervour in Soviet hotspots of discontent,[50] remains to be seen. What is clear, however, is that the Kremlin is still prepared to send in the troops when disturbances within its borders threaten to get out of hand. The troubles in Nagornyy Karabakh were an eloquent demonstration of Moscow's priorities. When faced with a choice, *glasnost* goes, and Soviet power remains.

Post Withdrawal Afghanistan

The growing number of agreements on cooperation between Afghan northern provinces and Soviet republics (some of which were signed as recently as March 1988), the recent announcement of the formation of a new province in the north of Afghanistan and Soviet pull outs in April 1988, following the signing of the accord, from garrisons in the south, all point to a preparedness on the part of both the Soviet Union and the Afghan regime to abandon southern Afghanistan, over which they never had any real mastery, to the rebels.[51] The regime failed to maintain control in the region with Soviet assistance. Without, they have no chance. Their priority then remains maintaining power in the north, and consolidating their forces in the area to defend themselves against the opposition. The Soviet Union has invested heavily in the development of resources in this area, and will expect the continuing repayment of Afghan debts in commodities available in the north, such as natural gas. The USSR has also said it intends to honour agreements already made with the Afghan regime. These will include the supply of such military hardware (and almost certainly advisors) as should be necessary. Under such circumstances it seems probable that following a withdrawal, some form of partition of Afghanistan will take place, with the south (largely desert) abandoned to the "rebels", and the north, tenuously held by the regime, propped up with Soviet military hardware. The Soviets will, at least for some time, have a Soviet-friendly, Soviet-dependent regime on their southern border.

A Change of Policy?

That Moscow means business this time over the withdrawal is perhaps demonstrated by the appearance in Soviet publications of articles deliberating on the benefits and drawbacks of involvement in Third World liberation struggles, and on the precise nature of a marxist revolution. Recent questioning of the effectiveness of military intervention in countries fighting for "liberation", suggest that a change in Soviet policy may be on the way.[52] It remains to be seen whether the debate takes in other areas of Soviet Third World activity. The theoretical debates in the media have tried to present an ideological explanation and justification for the withdrawal. As the true revolution is irreversible, according to Leninist doctrine, the question to be argued by ideologists is whether or not the Afghan revolution was truly socialist.[53] A withdrawal from Afghanistan must not be seen as a renunciation of the Brezhnev doctrine, which would almost certainly apply were the situation in an established socialist Soviet satellite state to get out of hand. However, in countries of the Third World, where socialism has not yet taken firm root, it could be that the Soviets will apply a different principle, and decide to avoid direct military engagement, both for fear of getting their fingers burnt, and to avoid the serious financial burden involved, for the promise of very limited returns at best.

The Benefits of Withdrawal

Although Afghanistan undoubtedly represents a defeat for the Soviets, pulling out the troops, or at least most of them in a highly visible manner, has a number of advantages:

Afghanistan is now in even greater debt to the USSR than before the invasion, and any Soviet friendly regime in Afghanistan (or more likely the northern part), will be dependent on continued Soviet support. Moscow is thus assured a continuing source of cheap raw materials in payment of these debts, at least for the duration of the present regime, although probably only in the north of the country. When the United States signed the accord with the Soviet Union, it assisted Moscow in its attempts to present the withdrawal as an honourable move agreed upon by all parties concerned. By making the withdrawal into a multilateral agreement, the true nature of Soviet presence in Afghanistan has been glossed over and the public presentation of the pull out enhanced.

Withdrawing now has an additional advantage. Moscow can present the move as a break with the years of "stagnation" under Brezhnev, and the present leadership can disassociate itself with the original decision to invade.

The political capital to be made by pulling out of Afghanistan in a blaze of publicity is considerable: much of the Western media and public opinion, ever eager to see in Mr Gorbachev the great reformer, will quickly forgive and forget Afghanistan—it was after all not of his making. Western attitudes towards the Soviet Union as a result are likely to be warmer than they would have been had the USSR not invaded in the first place. A withdrawal may also help smooth the way for, and boost Western cooperation over, further arms cuts. A

large and influential slice of Western public opinion will see it
as proof that the Soviet Union is no longer an aggressor, and
pressure will be put on NATO governments to concede much
in arms negotiation.

The Soviets leave behind a country in debt to them for the
privilege of being invaded, a ravaged and poverty stricken land,
thousands of dead, and millions of displaced Afghans who
must now attempt to rebuild their lives. The West should be
careful not to reward Moscow with unnecessary concessions
and effusive congratulations.

NOTES

[1] In comparison, about 20 divisions, totalling 250,000 men, were sent in to
Czechoslovakia in 1968.
[2] S. Enders Wimbush and A. Alexiev, "Soviet Central Asian Soldiers in Afghanistan",
Conflict, Vol. 4, Nos. 2/3/4, 1983, p. 335; and Alexandre Bennigsen and Marie
Broxup, *The Islamic Threat to the Soviet State*, (New York, St. Martin's Press, 1983), p.
113.
[3] "Coups and Killings in Kabul: A KGB defector tells how Afghanistan became
Brezhnev's Viet Nam", *Time*, 22 November 1982, p. 34.
[4] J. Bruce Amstutz, *Afghanistan: The First Five Years of Soviet Occupation*, (Washington
DC, National Defense University Press, 1986), p. 169.
[5] Wimbush and Alexiev, op. cit., p. 332.
[6] *Daily Telegraph*, 26 August 1980.
[7] "Oh what a horrible war", *The Economist*, 16 April 1988, p. 51.
[8] *Kabul Times*, 7 April 1979, in M. Urban, *War in Afghanistan*, (London, Macmillan
Press, 1988), p. 31.
[9] *Washington Post*, 2 January 1980, in M. Urban, op. cit., p. 39.
[10] "The KGB tried to explain tactfully that a communist takeover in Afghanistan
presented hair-raising problems . . . An open communist regime would arouse hostility
that would then be directed against the Soviet Union" Kuzichkin reported that his
boss, a KGB general said after the invasion "Now we are bogged down in a war we
cannot win and cannot abandon. It's ridiculous. A mess" *Time*, 22 November 1982.
[11] According to one Soviet official, "the Afghan state was on the verge of
disintegration . . . To leave the Afghan revolution without internationalist help and
support would mean to condemn it to inevitable destruction and to permit access to
hostile imperialist forces to the Soviet border": Ulyanovskiy, "The Afghan Revolution", USSR Report, 20 July 1982, quoted in H. S. Bradsher, *Afghanistan and the Soviet
Union*, (Durham NC, Duke University Press, 1983), p. 154.
[12] "We had either to bring in troops or let the Afghan revolution be defeated and the
country turned into a kind of Shah's Iran . . . We knew that the victory of counter-
revolution would pave the way for massive American military presence in a country
which borders on the Soviet Union and that was a challenge to our country's security",
Izvestia, April 1980, quoted in *Dawn*, Karachi 10 August 1983, in Joint Publications
Research Service, *Near East/North Africa Report*, 7 September 1983, p. 80–81.
"When the red flag over Kabul began to be threatened by trouble, when it was shot
at . . . with American recoilless rifles, Chinese machine guns, and English infrared
missiles and when expansion from Pakistan became a reality, and began to threaten to
overthrow the system, we sent in the troops . . .": A. Prokhanov, "Afganskye
Voprosy", *Literaturnaya Gazeta*, No. 7, 17 February 1988, p. 1.
[13] R. Medvedev, "Interview", *New Left Review*, No. 121, May–June 1980, p. 93.

[14] "When external and internal forces hostile to socialism try to turn the development of a given socialist country in the direction of restoration of the capitalist system, when a threat arises to the cause of socialism in any country—a threat to the security of the socialist commonwealth as a whole—this is no longer merely a problem for that country's people, but a common problem, the concern of all socialist parties": H. Bradsher, op. cit., p. 137.

[15] In cable No. 199533, from Kabul on 11 August 1979, the US mission in Kabul warned the State Department of such rumours: M. Urban, op. cit., p.37.

[16] According to Kuzichkin they "had doubts about Amin from the beginning. Our investigations showed him to be a smooth talking fascist who was secretly pro-Western . . . we also suspected that he had links with the CIA but we had not proof": *Time*, 22 November 1982.

[17] O. Cooper and C. Fogarty, "Soviet Economic and Military Aid to the Less Developed Countries, 1954–78" in US Congress, Joint Economic Committee, *Soviet Economy in a Time of Change: A Compendium of Papers* (Washington, DC, US Government Printing Office, 1979), pp. 648–662.

[18] American University, Foreign Area Studies Division, *Area Handbook for Afghanistan 1969*, (Washington DC, US Government Printing Office, 1969), p. 52; and G. Nollau and H-J. Wiehe, *Russia's Southern Flank*, (New York, Praeger, 1963), p. 103.

[19] L. Dupree, "Afghanistan's Big Gamble, Part II", *American Universities Field Research Reports*, South Asia Series, IV (4), pp. 10–11, in P. J. Garrity, "The Soviet Military Stake in Afghanistan: 1956–79", *RUSI Journal*, September 1980, p. 32.

[20] *The Military Balance*, (London, International Institute for Strategic Studies, 1973–1979), yearly.

[21] A. Prokhanov, "Afgansye Voprosy", *Literaturnaya Gazeta*, No. 7, 17 February 1988, p. 9.

[22] Ibid.

[23] O. Bogomolov, "Kto zhe oshibals'a?", *Literaturnaya Gazeta*, No. 11, 16 March 1988, p. 10.

[24] In March 1981, Vitaliy Kobysh, deputy head of the Central Committee International Information Department, told an audience in Cincinnati that the invasion was "a mistake".
In April 1981, Yuriy Velikanov, a Soviet diplomat in the Seychelles Islands stated: "for us, Afghanistan is an embarassment. There were mistakes when we went in and we are looking for ways to get out": J. J. Collins, "The Soviet-Afghan War: The First Four Years", *Parameters,* Summer 1984, p. 561.

[25] M. Strmecki, "Gorbachev's New Strategy in Afghanistan", *Strategic Review,* Summer 1987, p. 39.

[26] "Oh what a horrible war", *The Economist,* 16 April 1988, p. 52.

[27] M. Strmecki, in *Strategic Review,* Summer 1987, p. 39.

[28] S. Quinn-Judge, "A costly adventure", *Far Eastern Economic Review,* 21 January 1988, p. 15.

[29] *The Economist,* 16 April 1988, p. 52.

[30] Strmecki, op. cit., p. 40.

[31] O. Bogomolov, "Kto zhe oshibals'a?", *Literaturnaya Gazeta,* No. 11, 16 March 1988, p. 10.

[32] M. Shukeir, (a member of the Palestinian Communist Party), "Peace to You, Afghanistan", *Problemy Mira i Sotsializma,* 4 April 1988, p. 77.

[33] *Vneshnyaya Torgovlya SSSR,* 1980–1986, Moscow, yearly.

[34] Ibid.

[35] M. Siddieq Noorzay, "Soviet Economic Interests in Afghanistan", *Problems of Communism,* May–June 1987.

[36] *vneshnyaya Torgovlya,* SSSR, relevant issues.

[37] M. Siddieq Noorzay, op. cit.

[38] In 1985, journalist and author Alexandr Prokhanov commented that Soviet soldiers were finally getting combat experience in Afghanistan: "Zapiski na Brone", *Literaturnaya Gazeta,* 28 August 1985.

[39] V. Konovalov, "Afghanistan and Mountain Warfare Training", *Radio Liberty Research,* Munich, RL 118/88, 17 March 1988, p. 3.

[40] *Voennyy Vestnik* No. 1, 1988.

[41] By April 1988.

[42] "Perhaps somebody might believe that there is no difference between the Americans in Vietnam and our (troops) in Afghanistan . . . As if the difference between imperialist robbery and international assistance were not obvious . . .": *Krasnaya Zvezda,* 16 January 1988.

[43] As the author of the article put it: "A commander of the Red Army does not surrender!": *Krasnaya Zvezda,* 16 January 1988.

[44] *The Economist,* 16 April 1988, p. 52.

[45] "Despite everything that was not achieved, fundamentalism of the Iranian type is no longer possible in the country—the country would not accept it—and the threat of the emergence on the USSR's borders of an extremist Muslim regime, prepared to carry over its propaganda and practice, on to the territory of our Central Asian Republics—there will be no such threat". A. Prokhanov, "Afganskye Voprosy", *Literaturnaya Gazeta,* No. 7, 17 February 1988, p. 9.

[46] *The Economist,* 16 April 1988, p. 52.

[47] *Pravda,* 13 January 1980.

[48] "Afghanistan, Seven Years of Soviet Occupation", Special Report, No. 155, Washington DC, United States Department of State, Bureau of Public Affairs, December 1986.

[49] *Krasnaya Zvezda,* 27 March 1988.

[50] The Baltic Republics of Estonia, Latvia, and Lithuania; Armenia and Azerbaijan; and the Central Asian republics of Turkmenia, Uzbekistan, Tadzhikistan, Kazakhstan and Kirghizia.

[51] *Pravda,* 18 March 1988.

[52] An article in *Krasnaya Zvezda* on 22 March 1988, warned that in countries "fighting for their freedom" the arm can act either to accelerate or to slow down progress.

[53] As Prokhanov asks: "Is a stable political structure of socialism possible in a country where a countless number of tribes, nomadic peoples, agglomerations, chiefs . . . form a constantly seething, pliable, social soup, always exploding in bursting bubbles? In this medieval melting pot, forms appropriate to today were only just taking shape, and the task was to build, on this marshy swamp, a socialist edifice": Afganskye Voprosy, *Literaturnaya Gazeta,* No. 7, 17 February 1988, p. 9.

The Army and Egypt

FRANCIS TUSA

Francis Tusa is Middle East analyst at the Royal United Services Institute for Defence Studies

IT IS not taken for granted in Western countries that the armed forces should have a leading role in the affairs of the nation. In Western Europe, armies have been generally pushed to a more subordinate position in affairs, and now obey without question their political-civilian masters. There is thus a tendency to regard countries where the military either runs, or takes a great part in the government as abnormal and despotic. But especially in the Arab World, historical and religious factors have led to a situation where the military play a leading role in civilian matters, if any such divide can actually be made. Two of the most obvious examples of Middle East countries today where the military are prominent are Syria and Iraq.[1] In both these cases, the regimes came to power as a result of coups led by the army, coups in the end led by nationally ethnic and religious factions who had found advancement only in the socially inferior armed forces. But Egypt is perhaps the most important Arab country in which the power, both military and political, of the armed forces has been manifested, in this case by the Free Officers coup, and the subsequent rules of Gamal Abdul Nasser, Anwar Sadat and Hosni Mubarak.

Certain factors seem to be common in all the military coups, and subsequent military regimes in the Middle East, which have taken place in the post war period. For Egypt, as with Iraq and Syria, the assessed corruption of the civilian government, the massing of wealth and prosperity in a few hands[2] and in some cases its failure to act decisively over the Arab-Israeli wars of 1948–49 shocked the officer classes.[3] The civilian regimes which had taken over after Independence, whether from the French or British, are little different in social class and outlook from those that had aided the colonial or mandate power, and to a certain extent showed little social change from the nature of administration that was seen under the Ottomans. On the other hand, the social standing of those officers pre-eminent in such military movements as the Egyptian Free Officer Movement was generally not high, that is to say they had often joined the army to attempt to better their position, as other avenues to social advancement were difficult without a good education, money and influence.[4] The motives of the civil governments were very conservative, and little change to the economic and political system was deemed to be necessary.

Another reason for Middle Eastern military political precociousness was that historically the military had been more exposed to modern ideas than the civilian governments. With the advent of industrial warfare in the nineteenth century, armies had to absorb new weaponry, and new, radical tactics to use the equipment. Egypt was a leader in the nineteenth century in the field of such change. Under Mehmet Ali, the army was heavily and rapidly modernised, and foreign officers were brought in to train local troops. The modern outlook of the military, while not necessarily always in the political field, was retained by the army into the twentieth century, and onwards to the times when foreign officers were the exception rather than the rule.[5] It was the lower echelons of officers (Captains and Majors) which became politically involved in Egypt in the fight for Independence from the British during the 1930s, and it was these groups (now often promoted to the rank of Colonel) who were so scarred by the defeats in the wars against Israel in 1948–49. This last episode was crucial in the minds of officer groupings in the confirmation of the belief that their governments and systems were selling them down the river.

In Egypt, the Free Officer Movement, as established by Nasser, saw itself as the defender of the people, and the bringer of new prosperity, new freedoms, and true freedom from colonial influences. Manifestos and exhortations were privately published and distributed claiming such things as, "The homeland is in danger . . . rally to the Free Officers, and thus you will triumph, you and the people of which you are an inseparable part" and "We demand that the people be granted all the freedoms, since the people cannot struggle against imperialism when it is chained by laws which limit its freedom".[6] The Free Officers saw themselves as the only force that could end the internal crises, both economic and political, which Egypt was going through with monotonous regularity, and solve the foreign policy problems of the continued British presence in Egypt, which in turn affected the political situation. The problem of relations with Britain was being dealt with very ineffectively by the civilian governments, and little progress was made towards an equalisation of the situation between the two countries; even the British withdrawal to the Canal Zone was seen as being not enough, and nothing short of a full departure of British troops was deemed to be sufficient. In late 1951 and early 1952, there was a great deal of terrorism in and around British Canal bases, and in retaliation, British troops closed off the Canal Zone, thus isolating the city of Suez and Egyptian troops in Sinai from the rest of Egypt. On the question of Israel, it was widely believed that the government had been deliberately lax in coordinating the assault on Israel with other Arab states.

The coup in 1953 was rapidly carried out as the conditions seemed ripe, with widespread civil disorder as a result of clashes with British troops in the Canal Zone, and a power vacuum in the Palace and government. Tradition was maintained even when King Farouk was deposed, as his son was declared King, and a Regency Council was appointed. But very soon monarchist aspirations were dropped and the head of the Council, General Naguib, was made President, the old constitution was abrogated and a commission to establish a new one was set up. With power firmly in Free Officer hands, officers played an important part in the rule of Egypt. Up until 1967 (a period of some fifteen years), never less than 30 per cent, and in 1967 some 65 per cent of the government were officers.[7] While such percentages fell at later stages in the 1970s, it is worth bearing in mind that Anwar Sadat was of the Free Officer Movement. Likewise, President Hosni Mubarak, having succeeded to the Premiership from the Vice-Presidency, is not strictly a civilian, as he was head of the Air Force in 1973, and an architect of the campaign which did so much to restore Egyptian pride. Even now, there are significant posts in the present government which are held by military men, chief of which must be that of the Deputy Prime Minister and Minister of both Defence and Military Production which is held by Field Marshal Abu Ghazala, head of the Armed Forces, and potentially a ruler of Egypt should something untoward happen to President Mubarak. The Interior Minister (in charge of security) is also an army officer, Major General Badr. As will be seen, it is the Armed Forces in Egypt which are continuing to provide political stability, improve links with the Arab World broken by the signing of the Camp David Treaty, increase the importance of Egypt in the region and, surprisingly, are important in the economic and industrial growth of the country.

AEGYPTA RESURGA

On 11 November 1987, the final statement of the Arab League meeting at Amman permitted individual members to restore diplomatic relations with Egypt.[8] After President Sadat signed the Camp David Treaty with Menachem Begin in early 1979, Egypt lost its role as the leading Arab nation, and diplomatic relations with all but a handful of Middle East countries were broken. But within ten days of the end of the Amman summit, nine countries (including most of the wealthy Gulf states) reopened their embassies in Cairo. The reason for such a volte-face on the part of the Arab states was not in any measure due to new perceptions of the Egyptian role in the Palestinian problem, but was a result of increased fears about a possible spill over of the Iran–Iraq war into the weakly defended states of the Gulf. The Iranian Scud-B missile attacks on Baghdad during 1987 showed clearly that Iran had the power to strike over long distances, and the "Silkworm" missile hits on Kuwaiti oil export terminals brought the meaning of support for Iraq home to the Gulf Cooperation Council (GCC) states. The GCC armed forces, while often armed with some of the most modern equipment, were small and dependent in many cases on foreign staff as the level of local training was often deficient. Thus a security guarantee was deemed necessary. But the country had to be a regional one as non-regional nations, such as the Superpowers, were either unreliable like the United States, or were not sympathetic to the specific needs of the Gulf states, as was the case with the Soviet Union. With Syria, the most powerful Arab state in military terms, still perversely supporting Iran, and all the other moderates too small or too far away to be of value, the GCC states had to look for some regional power to help act as a threat to Iranian ambitions. The only country in any proximity and with any reasonable strength was Egypt. Such a view is backed up by President Mubarak, who said in a speech to the Second and Third Field Armies on 24 February 1988 that Egypt's leading foreign policy role was due to its military strength, and that

Egypt's security was intertwined with the Arab World's security and thus dependent on the power of the armed forces.[9]

THE EGYPTIAN ARMED FORCES IN 1987

It is thus ironic that Egypt should be seen as crucial to the Arab cause at a time when her armed forces were at such a low ebb. In the late 1960s and early 1970s, Egypt had possessed without doubt the largest and most powerful Arab armed forces, and was for that reason the greatest danger to Israel. The Soviet Union had been its major supplier, and had been quick to supply equipment and then to replace destroyed items, as was seen after the 1967 war. This single supply source produced commonality of equipment and tactics throughout the armed forces, thus easing supply and logistical problems. But in March 1976, President Sadat, as part of his rapprochement with the United States, abrogated the Treaty of Friendship with the Soviet Union and cancelled dept repayments on Soviet military loans. Although weapon supplies did not stop instantly, any long-term plans for modernisation of the Soviet-based forces would have to be reassessed. The hope, after the signing of the Camp David Treaty, was that US weapons would flood in to modernise the Egyptian forces, weapons "bought" or donated on the same basis as they were for Israel. This expansion has yet to be fully realised. With the US commitment to keeping Israel supplied with enough weaponry to protect itself against all Arab armies together, supplies to Egypt, regardless of the principle that Israel and Egypt are equals, have been less than those to its neighbour. Thus Egypt has been left with armed forces which are a mixed bag of Soviet and US equipment, which can produce logistic nightmares in wartime, and which do not help coherent tactics.

Although the actual size of the Egyptian Army was smaller in 1973, in terms of the proportion of the population taken into some form of military service it was larger, and the actual number of vehicles and planes were proportionally greater, and were also larger in comparison to the Israeli forces at the time. In terms of modernity, although the mainstay of the Egyptian armed forces in 1973 was the even then venerable T–54/55 (approx 1,500), it still represents nearly 40 per cent of the 1987 tank force. In terms of the air force as well, the present 42 F–16A fighters (with potentially up to 80 more on

TABLE 1: EGYPTIAN FORCES IN 1973 AND 1987

	1973	1987
Manpower	260,000	320,000
Reserves	500,000	500,000
Tanks	2,000 (100% Soviet)	2,250 (65% Soviet, 35% US mainstay old)
APCs/MICVs	2,000 (100% Soviet)	3,000 (40% US, 50% Soviet, some domestic built)
Arty	Numerous (100% Soviet)	Numerous (90% Soviet, some US, some domestic)
Combat Aircraft	400+ (100% Soviet)	375 (33% Soviet, 24% China 20% US, 23% France)

Source: IISS, *Military Balance*, London, 1973 and 1987.

the way) from the United States, and the 14 Mirage 2000s from France represent a fraction of the air force. The majority are still old Soviet MiG–21s and slightly more advanced Chinese J–6s and J–7s. When this is compared to the advances made in the Israeli air force on its own, it is a minimal improvement, and when compared to the modernisation of the Syrian air force, it still looks moderate.

Egypt is not rich enough to buy the necessary weaponry to modernise her armed forces rapidly to a sufficient level. The cost of modernisations outside the American assisted Foreign Military Sales (FMS) programme is causing her severe financial problems already. During 1987, the French government stopped deliveries of Mirage 2000 fighters until the Egyptians actually produced the money they promised.[10] In the end, it was several months before deliveries were resumed, and the option to buy further models might well not be taken up. The same problems have often been encountered in fields outside that of combat aircraft. Thus while the Egyptian command is planning a reduction in the number of men under arms, for equipment they have several choices; either the size and capabilities of the armed forces has to fall (and thus the number of men cannot fall without a reduction in capabilities), or some parts can be modernised slowly in line with US deliveries, or Egypt can attempt to produce its own equipment. Since the Egyptian government and High Command are unhappy with the first two options, as they lead to Egypt losing its position in the Arab World even further, it is the third which is receiving more attention.

HOME-MADE?

The Egyptian arms industry was originally set up in 1948 to make the country as independent as possible of foreign weapons supplies. With relatively free and cheap access to Soviet equipment in the 1960s and early 1970s, there was no economic incentive to take such production to its limit. But with the breaking of the link with the Soviet Union, ready access to weaponry ceased, and the cost of modern tanks and aircraft on the open market was exorbitant.

The Arab Organisation for Industrialisation was set up in 1975 to try and fill this gap, not only for Egypt, but also for the Gulf States. It was the latter countries (Saudi Arabia, the United Arab Emirates and Qatar) who provided the majority of the funds to start the industrialisation of Egypt, at that stage somewhere in the region of $1 billion initial capitalisation with perhaps $9 billion in reserve. Unfortunately for Egypt, with the Camp David Treaty, further funds were not forthcoming, and the original rapid growth ceased. But Egypt persevered with her building. At the end of the 1970s, updates for Soviet RPG–7s were starting to be made, and the WALID APC based on the Soviet BTR–40 was filling the motor pools of the Egyptian Army, as well as being exported. Small arms and large ranges of ammunition were equally seeing Egyptian production.[11]

But such projects merely scraped the surface of equipment costs. To modernise the armed forces, the large systems were needed. Unfortunately, these are also the most expensive; tanks, airplanes, artillery and sophisticated sensors and electronics associated with such weaponry. There was not, and still is not, the technical or industrial know-how in Egypt to produce modern, effective, advanced weaponry, therefore European and American companies had to be asked to participate in Egyptian development. Examples of such cooperation include British Aerospace with licensed production of the Swingfire Long Range Anti-tank Missile and Dassault-Breguet/Dornier for building the Alphajet. But even so, such contracts tended to be for items which were not necessarily at the technological edge of weapons development; the systems involved were simple and ran little risk of technology transfer to the Soviet Union.[12]

Even so, such limited production paid ample dividends. It is estimated, for want of publicised figures, that such sales produce approximately $300 million per year, and in certain boom years—like 1985—over $1 billion of equipment and ammunition was sold, much of it to Iraq. A good deal of the money seems to have been used to buy more weaponry for the armed forces, as well as for more factories for further production. At a time when revenues from staple products such as oil, Suez Canal dues, tourism and expatriate remittances are far from steady, and US economic and military aid are not growing rapidly, such money is very useful if the armed forces are to be modernised, and in future could be vital to the economy as a whole.

The most recent attempt to improve Egyptian industrial capabilities involves a possible contract with the United States for licensed production of the M–1 Abrams tank, possibly in cooperation with Turkey. As yet, the status of this project is uncertain; at a speech to troops on 29 February 1988, Field Marshal Abu Ghazala, Head of the Armed Forces and Minister of Defence and War Production claimed that such a contract had been agreed, and while the US Administration has not quashed such rumours, Congressional approval has yet to be given.[13] Not only would such production significantly update the ageing Egyptian tank force (as long as Congress did not demand that too many items were deleted for export purposes), but potentially, the tank could also be produced for Saudi Arabia, also in the throws of selecting a modern armoured force. Such a contract could be worth several billion dollars net, a significant boost for the Egyptian Armed Forces modernisation plan, and perhaps not least for the appalling debt situation. But the normal Congressional hurdle of the threat to Israel has yet to be crossed, and in election year, that may prove thorny.

Failing that, there is still a potentially profitable area for Egypt; that of upgrading the tanks they possess already, and using the same processes to perform similar operations on other countries' old tanks. Royal Ordnance has already won a contract to provide update kits for the T–55, all parts for such refits to be eventually produced in Egypt, while the General Products division of Teledyne Continental Motors has produced prototypes of an upgraded T–54 with a new engine, a 105mm gun instead of the 100mm, laser range finder and battlefield targetry computer. This would significantly improve the fighting capabilities of the T–54, extending its life for a few further years until a modern tank can be purchased.[14] Also, the upgrade uses as many M60 parts as possible to maximise commonality between the two vehicles. Egypt hopes to eventually procure the licence to produce such an upgraded vehicle, and then sell the product to other countries.

The aim of all this work is not simply the modernisation of the armed forces. It has always been hoped that the spin-off effects of such projects will benefit Egyptian industry as a whole in terms of plant sophistication, acquired skills and training and capital gained from such work. In a speech by President Mubarak in February, Egyptian industrialisation was praised, and it was hinted that defence production was seen as being part of the developing infrastructure required for Egypt.[15] Certainly, the expansion of plants such as those dealing with electronics would mean that a significant number of engineers and workers would have to gain skills which they would otherwise not possess, and could theoretically transfer such skills to directly civilian projects. The expanded military industrialisation programme can also be seen to be of use in another area, namely that of skilled unemployment. Egypt used to suffer greatly from the loss of her skilled workforce to expatriate jobs in the Gulf states and similar areas. But with the fall in the price of oil, and the subsequent fall in Gulf State revenues, many Egyptian workers have had to return home. Not only has this meant a spectacular fall in the revenues obtained by the government from expatriate earnings, but skilled unemployment can lead to vocal dissatisfaction in such groups. But with a growing defence industry, some form of "worthwhile" work could be found, or even created with the trained people only now available.

THE EGYPTIAN ARMY AS A DEVELOPMENT TOOL

The lack of skilled workers in Egypt has been mentioned in the previous section, but the armed forces have always played a role in trying to fill this gap, and are still doing so. There is a general policy in Egypt that graduates are guaranteed jobs. This has led irrevocably to a rapidly expanding civil service. But since there are also degree courses in Military Science, there is a ready supply of officers and warrant officers for the army. The Military Science course is one of the more prestigious courses to take, and since historically armed forces officers have become politically powerful, the quality of students taking the course is high. With the history of incompetent and

corrupt civilian officials, the management skills prevalent in
armed forces training, and the constant swapping of uniform
for civilian clothes means that many may spend part of their
military career in government ministries not directly concerned with armed forces business. This means that there is a
cadre of military personnel in administration to whom the
executive (largely military itself) can look to in times of trouble.

The army especially is looked to for help with infrastructure
projects such as building roads, irrigation and other engineering projects; recent examples of this can be seen in Army
engineering support for the Cairo Metro and Upper Egypt
irrigation projects. By and large, it is the army that can provide
cheap, reliable and relatively incorruptible manpower to
implement government schemes. While these types of projects
are often vexing to the armed services, perks are given in the
form of free, quite luxurious housing, cheap shops and food
and so forth, which elevates the social status of the officers and
senior ranks in the army, thus perpetuating the tradition of the
army as a source of social mobility. Likewise, with education
provided for poorer troops who could not afford it in civilian
life, the history of the armed forces being at the head of social
improvement is also continued. It is in ways such as these that
the armed forces attempt to become slightly less of a burden to
the economy, and slightly more of a benefit.

The Army as Guarantor of the System

It was the lack of support for King Farouk in the army that
lead to his eventual downfall. He had tried to control both the
civilian government and the upper echelons of the armed
forces, but failed. The political parties were either unwilling to
be controlled or were deeply unsatisfactory to the country and
thus had to be replaced. In the armed forces, the Free Officer
Movement's growing discontent and challenge to the existing
command structure in the end lost the King his throne. It is the
convergence and concord of the military and civilian arenas
which now guarantees the regime.[16]

The army stood firmly behind the government during the
violent food riots of 1977, allowing Sadat time to rescind the
subsidy cuts. The cohesiveness of the armed forces after the
assassination of President Sadat in 1981 by Muslim extremists,
even though the ring-leader of the assassination party was a
lieutenant in the army, allowed the peaceful transfer of power

to the Vice-President Hosni Mubarak. The chances of a Muslim-inspired revolution at the time were quite high, especially when one remembers that Sadat's death was followed by extensive fighting with Muslim extremists in the southern town of Assiut. In Cairo, one foreigner can remember seeing troops at all the major buildings and intersections in the city ready to combat further insurrections. Perhaps the most notable example of military support for the regime can be seen in the way the army stepped-in to stop the riots of the militia in Cairo in 1985, when many of the conscripts rebelled over poor conditions and started to burn symbols of Western affluence on the outskirts of Cairo. The army reaction in this case was swift and effective, rounding up all the conscripts, and arresting or shooting other associated rioters. Bearing in mind the personal history of President Mubarak, and the position of Field Marshal Abu Ghazala, support from the armed forces of the regime will maintain the present system of government and lend political stability to a country short on economic stability.

But the lessons of Sadat's assassination exemplify the problem of dependence on military support. Islamic Fundamentalism has been present in Egypt in the active form of the Muslim Brotherhood since the 1920s. There was great tension between the Brotherhood and Nasser during the early days of military rule, and more under Sadat. The Iranian revolution of 1979–80 gave extra impetus to Fundamentalist activities in the Arab World, and has created a sub-culture inside Egypt. The mosques have become an area where the government dare not go in force, and in which the Fundamentalists can debate their opposition to the regime and organise themselves. There are now four large Islamic investment houses which hold billions of dollars of investments for common people; there are also Fundamentalist hospitals and food stores which provide cheap and rapid service, better than that provided by the government. It is this area which presents more of a challenge to the government than the opposition political parties. Not only are the Fundamentalists proving popular with many, but with the present conscription system there will be a certain proportion of followers of Fundamentalism who will find their way into the armed forces. Before the repression by Nasser, there were known Brotherhood members of the armed forces and some help was gained during the Free Officer coup. But since then, anyone with such affiliations has tended to keep them secret. The presence of Fundamentalists in the armed forces could lead to a situation where elements of the army become politically unreliable and unwilling to protect the existing order, endangering the regime. To counter this, regular purges of the officer class have been held, though they have been more regular since 1979–80. The plans for the future of the armed forces will reduce manpower to possibly 300,000 in total which, it is hoped, will make screening of candidates easier, and thus the military more dependable. In the meantime, President Mubarak alternately arrests the most violent Fundamentalists while allowing the Brotherhood (theoretically outlawed) to enter Parliament in an alliance with other parties in the 1987 election. To back-up the limited liberalisation in political matters, the State of Emergency enacted after the assassination of Sadat has been extended regularly, the most recent occasion being in March 1988 for three years. It is the security services (run by the military) which then act as the immediate guarantor of the government.

KEEP THE POWDER DRY

While the reliability of any countries' armed forces dictates the stability of the regime, it is perhaps especially true in Egypt. The regime, while trying to broaden its support in many areas, faces a multiplicity of complex problems. The country faces a never-ceasing external debt problem which eats up much of its foreign currency earnings, revenue sources from all areas (Suez Canal, tourism, oil, remittances) are stagnant or falling, and growth is not rising appreciably. The population grows apace, outstripping the present infrastructure, and causing food to be imported. The Nile has suffered from several years of minimal rainfall and the Aswan Dam is silting-up as a result, entailing a possible loss of hydro-electric power. To maintain the inflow of foreign aid capital, Egypt may well be forced to cut the large system of subsidies which provide cheap food and power to the mass of the population which, as the riots of 1977 show, could affect the stability of the regime, especially with the growing support for Fundamentalist groups among all walks of life. It is in this case that the role of the armed forces as guarantors of the government should really be seen. Other roles for the military have included enabling Egypt to regain some of its prestige in the Arab World, and the growth of domestic defence manufacture is managing to produce foreign currency, if as yet only some $300 million a year on average.

Whether the army is able to keep the lid on violent Egyptian dissent, while acting as a leader in industrial development, or whether it will slowly become more and more riddled with Fundamentalists, remains to be seen. But in the short term, the military will remain to be important in government, and may well grow in influence and importance.

NOTES

[1] Kessler, *Syria: Fragile Mosaic of Power*, Washington 1987, National Defence University Press, pp. 28–31.

[2] Be'eri, *Army Officers in Arab Politics and Society*, London, Praeger 1970, p. 81; See also *Syria: Fragile Mosaic of Power*, op. cit., p. 60.

[3] P J Vatikiotis, *The History of Egypt from Muhammad Ali to Mubarak*, London, Weidenfeld 1985, p. 366.

[4] Heikal, *Autumn of Fury*, London, Corgi 1983; *Army Officers . . .*, op. cit., p. 321.

[5] *Army Officers . . .*, op. cit., p. 318.

[6] Ibid, pp. 93–94.

[7] Ibid, see diagram at pp. 29–30.

[8] INA, Baghdad 12 November 87, in *BBC Summary of World Broadcasts*, (thereafter cited as SWB), 16 November 1987, ME/0001 A/3.

[9] *Mena*, Ismailia, 24 February 88, *BBC SWB*, ME/0085 A/3, 26 February 88.

[10] *Military Technology*, November 1987, p. 65.

[11] *Military Technology*, November 1987, p. 65. *Defence and Armaments Heracles International* No. 69, January 1988, p. 18.

[12] *Military Technology*, February 1985, pp. 50–56, *Military Technology*, February 1988, pp. 80–83.

[13] *Mena*, Alexandria, 29 February 88, *BBC SWB* ME/0089 A/7, 2 March 88.

[14] *Military Technology*, February 1988, p. 81. *International Defence Review*, February 1988, pp. 185–187.

[15] *Mena*, Ismailia, 24 February 88, *BBC SWB*, ME/0085 A/3, 26 February 88.

[16] *History of Egypt . . .*, op. cit., p. 441.

The Superpowers in the Middle East: Lessons from the Persian Gulf

PROFESSOR JAMES BROWN

James Brown is currently Professor of Political Science at the Southern Methodist University, Dallas, Texas. He was formerly Special Assistant in the Office of the Deputy Under Secretary of Defense (Planning and Resources).

A LITTLE shy of eight years and one million dead, the dangerous and very brutal conflict between Iran–Iraq may now basically come to a close. The beneficiaries of this tragic ending are many: the war-weary peoples of both Iran and Iraq, the Persian Gulf states whose sovereignty and commerce the war threatened, and the Western nations who are dependent on the region's oil supply. However, for the foreseeable future the presence of the United States and the Soviet Union in the Persian Gulf region will continue.

It was the summer of 1987 which witnessed a major build up of Western naval forces in the region of the Persian Gulf.[1] The United States and several Western European nations deployed an unprecedented number of naval units into this area to ensure the safe transit of ships navigating the Gulf waters. On the other hand, the Soviet Union's Indian Ocean squadron which serves as Moscow's political instrument in the region did not substantially increase its presence in conjunction with the marked increase in Western naval power. All of these naval deployments raised interest about the nature and purpose of US, Soviet, and European intentions in the Persian Gulf region. Although all of this naval activity appears as an aberration, in many ways these are manifestations of trends that have shaped naval deployments to this region in the post Second World War period.[2]

THE UNITED STATES PRESENCE

The naval activity of the United States in the Persian Gulf in the last 20 years has progressively increased in its level, intensity, and scope, from a position of detachment in the 1960s to direct involvement of its naval forces as a central policy tool in the 1980s. This incremental change in intensity and purpose basically reflects differing perceptions of the nature of the threat held by successive administrations since President Truman. The perception of the Soviet "threat" appears to be the deciding factor as to why Washington expanded its naval commitments to the region.

The policy of the United States for much of the post war era relied upon the British and then regional actors such as Shah Reza Pahlavi of Iran to contain Soviet expansionism, maintain regional stability, and ensure continued access to the region's oil reserves. Under these circumstances, a small Middle East Force (MIDEASTFOR) was the only permanent American naval force in the region from 1949–69. It consisted of two small warships and a command vessel. Its primary function was to show the flag, generate good will, and underscore the American commitment to protect its interests. The US presence was augmented by the infrequent dispatch of units of the Seventh Fleet (Pacific) to the Indian Ocean on goodwill cruises.

The first major change in US naval activity came about in the early 1970s when Washington began to deploy an aircraft carrier or cruiser task force to the Persian Gulf. This was initially precipitated by Moscow's deployment in the late 1960s of naval units to the region. Although a paucity of information exists regarding Soviet deployment decisions during this period, most analysts believe that limited out-of-area sustainability, indecision over the role of the Soviet fleet, and the view that the strategic nuclear balance remained unfavourable, were the primary reasons the Soviets did not deploy naval forces to the Persian Gulf region prior to 1968.[3] Soviet ships that appeared were scientific and fishing vessels, and warships transiting between the Soviet Union's European and Pacific bases.

Also, about this time the British withdrew their forces from East of the Suez.[4] Until that withdrawal Washington had believed that defence of the region was primarily a British responsibility resulting from Britain's long history of involvement in the region's affairs and the network of British bases and facilities along the Indian Ocean littoral. Washington maintained that if London curtailed its overseas commitments, it should do so in Western Europe rather than in the Persian

Gulf region, because the United States was unwilling to extend its defence obligations, particularly in the wake of its growing involvement in South-east Asia.

While this region has always been regarded as a possible target of Soviet expansionism, the dependence of the West and much of the Third World upon Persian Gulf oil is what makes this area's defence vital. In the 1970s price rises, an oil embargo, and producers' instability dramatically demonstrated Western vulnerability on this score. The collapse of the Shah's Iran and the resulting chaos including the seizure of the US Embassy in Tehran in April 1979, and the Soviet invasion of Afghanistan later that year are what focused this issue squarely for Washington and precipitated the decision to deploy almost continuously an aircraft carrier in the region.

The Carter Doctrine enunciated in the State of the Union Address in January 1980 stated that

(A)ny attempt by an outside force to gain control of the Persian Gulf will be regarded as an assault on vital strategic interests of the United States of America, and such an assault will be repelled by any means necessary, including military force.[5]

The practical application of this doctrine depended from the outset on US capability to project substantial military power in the Gulf region with maximum speed. But formidable obstacles confronted Washington planners, including distance, an absence of treaty relationships to facilitate joint contingency planning with regional states, political constraints against obtaining permanent military bases, and competing demands on US military resources elsewhere.

The rationale behind the Carter Doctrine was the desire of the then Administration and subsequently the Reagan Administration to contain Soviet influence, and to demonstrate support to Washington's moderate Arab allies against regional threats. Subsequently, assignment of units to United States Central Command (USCENTCOM) and actual naval activities reflected the concerns that underline this doctrine.[6] The US forces became more directly involved in the implementation of policy through such actions as the evacuation of persons from Iran in early 1979 and lending support to the Iran hostage rescue attempt in 1980.

It was the outbreak of the Iran–Iraq war in 1980 that further provided a test case for an extension of the Carter doctrine to intra-regional conflicts. Administration officials at that time voiced concern that Iran and Iraq would launch strikes against each other's oil facilities and that the Straits of Hormuz might be closed. They also expressed apprehension over possible Soviet intervention should Iran disintegrate.[7]

Iraq began the so-called "tanker-war" in May 1981 by attacking merchant ships steaming to and from Iranian ports at the extreme northern end of the Gulf. In March 1984, Iraq increased the frequency of its attacks, and in May 1984, the Iranians responded by initiating their own attacks. As the tanker war intensified, Washington began to emphasise its determination to keep open the shipping lanes of the Persian Gulf, although it remained neutral in the land war.[8] (See Table 1).

TABLE 1: ATTACKS ON SHIPS IN PERSIAN GULF

Ships Attacked	1981	1982	1983	1984	1985	1986	1987
By Iran	0	0	0	18	14	45	91
By Iraq	5	22	16	53	33	66	88

Source: "Persian Gulf: US Military Operations" by Ronald O'Rourke, *Congressional Research Service*, 1988, p. 3.

By the beginning of 1987, both Iran and Iraq had greatly intensified their attacks on each other's shipping. Iran, in particular, began to stop numerous merchant ships of various flags and search them for war material bound for Iraq. Also, Iran attacked ships serving ports on the Arab side, particularly singling out Kuwaiti ships in order to intimidate Kuwait for its support of Iraq's war effort.[9] In May 1987, Iran apparently began to lay mines surreptitiously in a concerted way.[10] At first, Iran appeared to employ them solely in the waters off Kuwait's ports, but by mid-summer they were being found in various places in the Persian Gulf and in the Gulf of Oman. It bears noting that the tanker war to date has neither significantly reduced oil shipments nor substantially curtailed shipping traffic.

Throughout the period 1980–86, the Reagan Administration policy was one of caution. Secretary of State Alexander Haig warned that the Iran-Iraq conflict might lead to "unforeseen and far-reaching changes in the regional balance of power, offering the Soviet Union opportunity to enlarge its influence". He repeated Washington's commitment to neutrality and its refusal to supply military equipment to either side, and he pledged to be more active "with other concerned members of the international community" in efforts to end the conflict.[11]

The final step in the direct utilisation of US forces was taken by the Reagan Administration, fearing that the Soviet Union would take advantage of Iranian threats against Kuwait, when it offered to reflag 11 Kuwaiti tankers.[12] This came about at the end of 1986 when the Kuwaiti government sought both Washington and Moscow's protection for its tanker fleet. Kuwait's concern was the increasing frequency of Iranian attacks on her flag tankers since September 1986, which jeopardised Kuwait's secure flow of oil from the Persian Gulf. It has also been suggested that the Kuwaitis saw the reflagging issue as a method to make the expanding Gulf War an international issue by drawing the Superpowers into a confrontation with Iran.

At first, the Reagan Administration was reluctant to take up the Kuwaiti request. It appears that when Washington learned of the progress in Soviet-Kuwaiti talks, it became determined not to allow the Soviets an important political and military opening in the Persian Gulf. Thus when Kuwait requested to place six of its tankers under the US flag on 2 March 1987 (the other five under the Soviet), the Reagan Administration responded on 7 March with an offer to reflag all 11 tankers. In return, Washington received a commitment from Kuwait that it would limit the Soviet escort operation to three leased tankers, and that Kuwait, along with the other Arab gulf states, would not offer the Soviet Union use of its port facilities as long as the US operation continued.[13]

Another rationale for the reflagging policy was the perceived need for Washington to shore up waning influence among the Arab gulf states. These needed reassurance about US reliability only because their confidence in Washington's judgment had been so badly shaken by the Reagan Administration's sale of arms to Iran.[14]

Washington believed that its reflagging policy would counter Iranian intimidation of Kuwait and other moderate Gulf states, and deny Moscow the opportunity to become the guarantor of the free flow of oil from the region. Although the Reagan Administration recognised the danger inherent in the policy, it felt that Iranian "prudence" in the past, and not Iranian rhetoric, should be the prime indicator of an Iranian response.

Recognition of the role that Kuwait plays in supporting Iraq did not alter Washington's belief that US neutrality was in no way altered by the reflagging policy.[15]

The Reagan Administration's requirements for reflagging were completed in June and the first two ships—the *Bridgeton* and the *Gas Prince*—were convoyed by US naval units on 21 July. Three days later, the *Bridgeton* struck and was damaged by a large contact mine about 18 miles west of Iran's Farsi Island. It was now time for Washington to reassess its Persian Gulf policy.

The 24 July mine attack on the *Bridgeton,* coming on the first convoy, embarrassed the Navy and reinforced many of the doubts of on-scene commanders that they had little capability to counter mines, which had damaged ships of Kuwait's coast since May 1987.[16]

The *Bridgeton* incident and the growing tensions in the region led the Reagan Administration to enlarge the US naval presence.[17] To coordinate these forces more effectively, Washington announced the formation of the Joint Task Force Middle East (JTDME) to take direct control of US forces in the Persian Gulf-Arabian Sea area.[18]

President Reagan declared that the reflagging operation would not end until the Iranian threat to the Kuwaiti ships is no longer present and freedom of the seas is ensured. By early 1988, the convoying operations had become so routine that Secretary of Defense Frank Carlucci scaled back the size of the US Persian Gulf force.

But any sense that there was a detente in the Persian Gulf was shattered on 14 April when the frigate *Samuel B. Roberts* struck a mine injuring 11 crewmen. Washington was quick to retaliate.[19] It has been suggested that Iran intended its mines to sink a US warship, hoping that American public opinion would then force Washington to withdraw from the Persian Gulf and thereby relieve pressure from Iran, which has suffered a recent series of military setbacks at the hands of Iraq.[20] However, the *Roberts* incident places Washington in a position of "fleeing forward." As a result, President Reagan ordered the US Navy to expand its duties to include the protection of neutral, non-communist ships that seek assistance if attacked, along with vessels flying the US flag.

"We are not the policemen of the Gulf nor do we wish to be," declared Secretary of Defense Frank Carlucci. "We cannot stand by and watch innocent people be killed or maimed by malicious, lawless actions when we have the means to assist and perhaps prevent them."[21]

What began as a limited policy initiative, designed to curtail Soviet influence and to demonstrate support for Kuwait and other moderate Arab states, became, instead, a high profile test of wills between Tehran and Washington.[22] The refocusing of the initiative on deterring an Iranian attack on US forces has replaced earlier concerns of Soviet expansionism and the former has become the driving force behind the size, composition, and activities of US military forces in the Persian Gulf.

WESTERN ALLIES IN THE PERSIAN GULF

For the last decade, many American observers have argued that the Western allies should do more to assist the United States defend Western interests in the Persian Gulf.[23] Formally, no arrangements exist for military operations outside of NATO territory, including concerted action either to protect shipping or to undertake military actions against another nation's territory. The Japanese case is a bit different. Her constitution requires that her armed forces be used only for self-defence. This has been interpreted as prohibiting the use of Japan's naval forces outside of home waters.

Unlike the United States, the Western allies have been driven primarily by either economic or historical ties to the region, and not strategic competition with the Soviet Union. Great Britain and France became involved in the region's affairs in the 18th and 19th centuries. As a result of their historical attachment, both have developed an intricate web of economic, political and social ties with peoples in the area—ties that continue to affect the pattern of naval deployments of the two countries.

In the case of London, important interests and close ties developed with the Persian Gulf sheikdoms. Although Great Britain, a North Sea oil producer, no longer relies on imported oil, she continues to maintain close economic ties with the nations of the region. Reflecting these ongoing interests, London's "Armilla Patrol", has since 1980, re-established a permanent naval presence.[24] In fact, Great Britain's naval vessels have operated a protection service for her shipping in the Persian Gulf since 1981.

France also has substantial economic interests in the area.

However, Paris is hypersensitive regarding its Persian Gulf policy. It does not want to appear to have been influenced or swayed by Washington. A French diplomat characterised his country's policy as "Une Evolution, non pas unchangement". France continues to import some 30 per cent of its oil from this region, even with diversification of her energy supplies. In order to "recycle" her energy expenditures back into the French economy, Paris has become a major supplier of arms to the sheikdoms of the region and Iraq, whose indebtedness to France reputedly amounts to billions of dollars. The dispatch of the aircraft carrier *Clemenceau*[25] plus other naval vessels reflects the intense diplomatic disputes that exist between Tehran and Paris, that culminated with the break in diplomatic relations in July 1987.[26]

In addition to Great Britain and France, several other European countries (Belgium, Italy, the Netherlands) are performing high profile roles in the waters of the Persian Gulf.[27] An impressive 32 ship European naval force is deployed in and around the region. Minesweeping is the key allied contribution.[28] Although the European navies lack effective air cover and pack less firepower than the US naval forces, their minesweepers are far more modern and effective than their US counterparts. In addition, the Europeans protect their own commercial ships. European patrols and minesweeping operations have been a major deterrence to Iranian attacks. However, none of the allies has agreed to formal joint naval operations with the United States in the Persian Gulf. There is a sharing of intelligence information on the region and all naval contingents communicate with one another routinely.[29]

The French tend to operate independently, not coordinating their operations with Allied naval units. They also refuse to integrate their 13 ship fleet into any type of intra-European force. French diplomats point out that they have different interests than the other Europeans and the United States. France is a privileged ally of Iraq, its second largest arms seller after the Soviet Union. While British commercial interests are stronger in Iran, Washington, on the other hand, has the largest geopolitical stakes in the area and sees long-term interests with Iran, Iraq, and other Persian Gulf states.

These deployments also signal a change in European cooperation outside of the NATO area. The Gulf experience has been a big plus to Atlantic solidarity. After the frigate

Samuel B. Roberts was damaged, the Europeans re-emphasised their solidarity in a communiqué issued by the Western European Union (WEU). The communiqué affirmed the right of free navigation in the Persian Gulf and called for "an immediate end to all mining and other hostile activities".[30] It stressed that such activities require ships to act in self-defence.

The European presence also provides diplomatic cover for Washington. It has helped convince the world that the United States is not acting alone and bullying small states in yet another far-off regional conflict.

THE SOVIET FACTOR

Historically, the Soviet Union, because of its geographic proximity, has always considered the Persian Gulf region of vital interest. Moscow's designs upon this area long predate the predominance of oil in international politics. They were tellingly expressed by Soviet Foreign Minister V. M. Molotov in 1940, at the time of the Ribbentrop-Molotov Accord, by indicating "the area south of Batum and Baku in the general direction of the Persian Gulf is recognised as the centre of the aspirations of the Soviet Union".[31] Unsuccessful Soviet efforts after the Second World War to extend their hegemony in that direction by annexing territories in eastern Turkey and western Iran provided one of the opening volleys of the Cold War. The rise of oil has only added an important dimension to this longstanding Russian aspiration for direct access to the warm waters of the Persian Gulf and Indian Ocean.

Washington's policies in this region, as discussed earlier, have largely responded over time to the perceived Soviet threat. The desire to contain and minimise Moscow's influence has been intensified by the Soviet naval build-up in the region. Two important points need to be made regarding the nature of Soviet naval deployments. The first is that the Soviet-Indian Ocean squadron, established in 1968, has had primarily a political rather than a military mission. Deployment of these vessels is a way that Moscow can expand influence by gaining recognition from the littoral states and outside powers of the Soviet Union's "legitimate" interests in the region.This recognition enables Moscow to acquire access to naval facilities in the area from which they can better monitor US activity, counter US naval diplomacy, and more effectively attempt to erode Washington's influence in general.

The Soviet's offer to reflag five Kuwaiti tankers, and the acceptance in March 1987 of a leasing arrangement for three of these, along with the provision of Soviet escort, are the most recent examples of a pattern of naval diplomacy that is designed to win such recognition.

The second point is that the Soviets did not substantially change the size of its naval deployment in spite of the expansion of US naval presence. This directly contradicts previous patterns of Soviet naval activity established during the 1970s.[32] This lack of similar response in 1987 suggests that, in this region, General Secretary Gorbachev is perhaps pursuing a more modest approach to military competition with Washington.[33]

Suffering from a series of military defeats, and growing economic hardships in 1988, Iran's momentous decision to accept the United Nation's Security Council Resolution 598 and end the Persian Gulf war appears to set the country on a more moderate and realistic course, with potentially profound consequences both internally and for the region as a whole. The road to a comprehensive agreement between Iran and Iraq, however, is likely to be a long and very arduous process involving complicated border disputes and competing claims to the Shatt al Arab waterway that were the focal point of friction even before this conflict began in 1980.

Throughout this conflict, the Soviet Union has attempted to maintain its links with both belligerents, providing military assistance to both, while attempting to position itself to play a mediating role. There is no doubt that strategically Iran is much more important to the Soviet Union than Iraq. However, Moscow has been unable to gain any telling influence in Tehran.[34]

The late 1986 decision by Kuwait to invite the Superpowers to protect Kuwaiti shipping proved decisive for Soviet interests in the Persian Gulf. Not only did Moscow's acceptance reflect the activism of Soviet foreign policy, but the role of the Soviet Union became a central factor in the Persian Gulf and in United States policy toward it.

The invitation by Kuwait is an important milestone in Moscow's efforts to improve relations with many of the Persian Gulf states. By agreeing to protect Kuwaiti oil exports, for the first time, Moscow was given an active role in defending these states—a role that in the past exclusively belonged to Great Britain and the United States. Moscow may also have hoped that this involvement would lead to the expansion of Soviet arms sales to Kuwait and the initiation of such sales to other countries in the region.

The Soviet Union, however, has kept its arrangements with Kuwait in perspective. Moscow has not attempted to compete with Washington to be the Superpower with the most naval vessels in the Gulf. The Kremlin realised that a rapid Soviet naval build-up would only lead to an equal or greater American naval build-up. Even more importantly, the Soviet Union was not interested in improving its relations with the several Arab Gulf states at the expense of its longstanding goal of improving ties with Iran.[35]

Tehran, angered by the Soviet Union's willingness to protect Kuwaiti vessels, proceeded to attack Soviet freighters.[36] The Soviet Union, however, did not retaliate; instead it played down the incidents. Soviet minimisation of the risks of conflict with Iran and its restraint after the attacks is in stark contrast to Washington's behaviour toward Tehran.

In the spring of 1987, Moscow proposed a joint East-West security arrangement for the region. This was similar to an offer made by Leonid Brezhnev in 1980 that had been rebuffed by both Presidents Carter and Reagan. Indeed, if the United States could not provide for Persian Gulf security without the involvement of the Soviet Union, its credibility would be further eroded. The Soviet diplomats then proposed that an Iran–Iraq peace conference be convened in Moscow. Iraq might have been amenable, but Iran was not.

From Washington's perspective, however, the greatest risk lay in the benefits that could accrue in major military actions against Iran. But the risk of major Soviet influence remained. Throughout 1987 Soviet diplomats were assiduous in politically courting Tehran. This underscored the basic nature of the United States' gamble in the Persian Gulf. Not in control of the pace of events in 1987, the United States risked actions that could provide the Soviet Union with unprecedented opportunity in the region—the possibility that it could realise a centuries-old dream of decisive influence.

As the war of nerves between Washington and Tehran escalated in mid-1987, the Soviet Navy maintained a low profile in the Gulf. It was in August 1987 that relations between Moscow and Tehran warmed up with the announcement of a major economic cooperation accord.[37] This active flirtation with Moscow by Tehran was a realisation by Iran that

Soviet military aid to Iraq was one of the principal obstacles preventing an Iranian victory in the Iran–Iraq war. By holding out the prospect of stronger Soviet influence in Iran, Tehran sought to provide Moscow with an incentive to avoid increasing aid to Iraq or cooperating with the United States in its attempt to cut Iran off from its external arms supplies. On the otherhand, the improvement of Iranian-Soviet relations only provided a further incentive for the moderate Arab gulf states to rely on Washington, even if their hopes had not completely vanished that Moscow could somehow restrain Tehran.

In a broader context, Moscow sought to persuade Tehran that while the United States was its enemy, Moscow was its friend. They also attempted to persuade all states in the region that American actions against Iran only heightened the prospects for increased conflict, but that the Soviet Union (not the United States) could help bring peace to the region. Moscow's position is that peace between Iran and Iraq is necessary so that the Muslim world can focus its attention on the common enemy—Israel.[39] The continuation of the Iran-Iraq war serves Washington (and Israeli) interests and detracts Muslims from the Arab-Israeli conflict.

While strengthening the Soviet position in the Persian Gulf, Gorbachev has not pursued policies that differ from those of his immediate predecessors.[40] But if Soviet policy towards the region has not changed much, political conditions in the area certainly have. These changes have led many states, which opposed a greater Soviet role in the region just a couple of years ago, to welcome a greater Soviet presence, or at least to reduce their objections to it.

Can the Soviets transform their greater presence and acceptability in the Persian Gulf into long-term influence that might permit them to expand their role while diminishing Washington's influence? Serious obstacles abound. While the fear of an Iranian victory has led the Arab Gulf states to welcome an increased Soviet role in the region, this fear has also led them to seek an even greater American and Western role in the area. These states have not heeded Soviet claims that Washington, the main source of tension in the region, should withdraw its forces. On the other hand, they have no desire to see either Tehran or Moscow become the strongest military force in the area. The rapprochement between the Arab Gulf states and Moscow has taken place because these nations perceive that they share common anti-Iranian interests with the Soviet Union.

Gorbachev's policies have led to greater influence, but the Soviet Union is not in any position today to transform this greater influence into predominance in the region. It also signals to Washington that Moscow will now play a more active role in the Persian Gulf, where Soviet interests historically have been primary. It further suggests that Gorbachev believes that the Soviets are less exposed than Washington and, therefore, the Soviet Union can use its lower military profile to perhaps enhance its image as a "peacemaker" and to draw a sharp contrast with the United States.

Today, it is not feasible to ask "Where else can the regional states of the Persian Gulf go?". Gorbachev's Soviet Union— neither loved, nor trusted—is making gains as an active player in the region, and with some success. This does not mean that Moscow will soon, if ever, supplant Washington as the key outside arbiter of events. But it does suggest that the United States is not alone; its diplomacy and actions are being compared to that of the Soviet Union, and its position is now being challenged.

NOTES

[1] The area includes not only the Persian Gulf but the peripheral waters of the Gulf of Oman, Gulf of Aden and the Arabian Sea.
[2] Robert J. Ciarrocchi, "US, Soviet, and Western European Naval Forces in the Persian Gulf Region," *Congressional Research Service Report to Congress,* December 8, 1987, p. 1.
[3] *Ibid.,* p. 22.
[4] The majority of the British forces were withdrawn, including garrisons in Singapore and Southwest Asia, and an aircraft carrier group from the Indian Ocean.
[5] *US Department of State Current Policy,* No. 132, January 23, 1980, p. 2.
[6] USCENTCOM is responsible for responding to contingencies in Southwest Asia. In addition to US Army and Air Force units, Navy and Marine Corps units include three aircraft carrier battle groups, one surface action group, three amphibious ready groups, and five anti-submarine warfare patrol squadrons.
[7] The Carter Administration initiated measures to improve Saudi Arabian defence capabilities (e.g. F–15 sale) against the event war might spread. At the same time Saudi Arabia and Oman were requested not to permit Iraqi use of their airfields for attacks which might encourage Iran to widen the conflict. *The New York Times,* October 12, 1980.
[8] In late 1984, US Navy ships began to escort ships of the Military Sealift Command.
[9] In mid-March 1987, it was reported that Iran obtained shore-based Chinese made HY–2, "Silkworm" anti-ship cruise missiles and had constructed several launch sites on the Iranian side of the Straits of Hormuz. Iran is believed to have presently as many as 200 missiles. In the most recent conflict between Iran and the US, on 18 April, 1988, it was reported that "Silkworms" were fired at US naval vessels without hitting their targets.

[10] On 17 May, 1987 an Iraqi Mirage F–1 fired two Exocet missiles at the US frigate *USS Stark*, killing 37 crewmen and wounding 21. Iraq apologised for the incident and said it would pay damages.

[11] "Peace and Security in the Middle East." An address by Secretary of State Alexander Haig before the Chicago Council on Foreign Relations, 26 May, 1982. *US Department of State Current Policy*, No. 395.

[12] In April 1987, US navy ships were ordered to increase the amount of time spent in the Persian Gulf, and deployment of the aircraft carrier *Kitty Hawk* to the Arabian Sea was extended. Also in the same month, Secretary of Defense Casper Weinberger ordered that aircraft carrier battle groups begin spending their full deployment to the Persian Gulf region in the Arabian Sea. *The New York Times*, 7 August, 1987, p. 10 and 3 May, 1987, p. D2.

[13] *The New York Times*, 23 August, 1987, p. A12.

[14] The Iran-Contra affair came to light in early November 1986, when a pro-Syrian newspaper in Lebanon revealed that the United States had been selling arms to Iran. This was in contradiction to the declared US policy of cutting off the arms supply to the regime of Ayatollah Khomeini. Even more, it soon transpired that the US arms sales, at least in part, were designed to help secure the release of several Americans who were being hostage in Lebanon.

For additional details see Robert E. Hunter, "United States Policy in the Middle East," in *Current History*, Vol. 87, No. 526, February 1988, pp. 49–50.

[15] "US Policy in the Persian Gulf," *US Department of State Special Report*, No. 166, July, 1987, p. 11.

[16] Vietnam era coastal minesweeper and numerous smaller craft, plus RH–53D Sea Stallion helicopters for use in minesweeping were delivered to the Persian Gulf after the *Bridgeton* incident. Subsequently, on 16 October, 1987, the *Sea Isle City* was struck by an Iranian missile after it left the company of its US naval escort.

[17] The War Powers Resolution, 1973, was an assertion of the US Congress power to declare war (and by implication, to refuse to do so) as a counterweight to the long-asserted presidential authority over when and where to deploy US forces. In 1987, the Reagan Administration expended considerable political capital arguing that, although the crews of the warships escorting the reflagged tankers would receive "danger pay," the ships would not be at risk of "imminent hostilities" as specified by the War Powers Resolution. The US Congress thus far has been stymied in their efforts to invoke this Resolution. Some members who are proponents of a strong congressional check on presidential freedom to deploy forces have abandoned the legislature arena in search of a judicial remedy.

[18] As of April 1988, the major US units in the Persian Gulf consist of the command ship *LaSalle*, six to eight major surface combatants, one amphibious dock ship, and six ageing ocean-going minesweepers. In addition, the Navy has 10 to 16 smaller patrol boats manned by special operations personnel, and an Explosive Ordnance Team with a team of four specially trained dolphins who are used to detect mines. Also, there are four US and five Saudi Arabian Airborne Warning and Control System (AWACS) aircraft. It is estimated that a total of 4,000 or so personnel are manning these units. The overall cost of this operation is put at $10–15m per month.

Most of the Gulf states friendly to the United States allow port visits by US ships and all of them deny them to the Soviets. Bahrain provides limited port facilities to US and British forces operating in the Gulf and some prepositing of equipment. Oman allows US P–3 patrol planes operating out of Diego Garcia to refuel at one of its air bases, and in 1980 signed an agreement with the United States permitting US access to Omani bases, with Omani permission, in certain contingencies. In early January it was also reported that Oman made available to the United States its facilities on Massira Island. Other Gulf Arab states are sometimes rumoured to have secret base-access agreements with the United States.

To aid the escort operation, the Saudis have agreed to use their five new US-made Airborne Warning and Control System (AWACS) planes to maintain a patrol over the southern half of the Gulf (four US AWACS planes based at Riyadh maintain a patrol over the northern half). The planes are routinely protected by Saudi F–1 fighters. The Saudis have also allowed two of their four minesweepers to help sweep Kuwait's waters.

The United Arab Emirates has agreed to allow the Saudi AWACS planes to overfly UAE territory. Kuwait in July 1987 reportedly agreed to (1) grant landing rights to US aircraft involved in protecting the reflagged tankers and (2) supply free fuel to the US ships participating in the escort operation. It has assisted in minesweeping operations in its own waters. Kuwait has not allowed US warships escorting the reflagged tankers into its territorial waters in the past, but may have relaxed this position. In early December, it was also reported that Kuwait had approved a US request to station a third barge fortress near Kuwait's offshore oil terminal, in Kuwaiti territorial waters.

The Gulf states are also reportedly assisting the US effort in many small, often unreported ways. For example, they are believed to be allowing US minesweeping and other helicopters to use their air bases for occasional "emergency" stops.

[19] US forces attacked two oil platforms at Sirri and Sassan in the southern gulf. These platforms were used as command and control radar stations. The Iranians engaged the American forces and US ships sank a missile patrol boat, severely damaging two frigates and sank or badly damaged three barge armed speedboats. This was the worst confrontation between US and Iranian forces since the *Bridgeton* incident. For more details see *The New York Times,* 19 April, 1988, pp. 1 and 4–5. Most recently, in July 1988, the *USS Vincennes* mistakenly downed Iran Air Flight 655, killing all 290 people aboard.

[20] *The New York Times,* 20 April, 1988, p. 8.

[21] *The Washington Post,* 30 April,1988, p. 1.

On April 29, 1988, President Reagan ordered the US Navy to expand its duties to include protecting neutral, non-communist merchant ships that seek assistance if attacked, along with vessels flying the US flag. *The Washington Post,* 30 April, 1988, p. 1.

[22] For a detailed discussion of US Congressional views on the Persian Gulf operations, see *Congressional Quarterly Weekly Report,* Vol. 46, No. 17, 23 April, 1988, pp. 1051–1058.

[23] As of the end of April 1987, an average of about 25 tankers a day pass through the Straits of Hormuz, carrying more than 7 million barrels of oil. Western Europe receives some 3 million barrels per day from the Persian Gulf or 25 per cent of its total needs, while Japan receives 2.5 million barrels per day or 58 per cent of its total needs. The United States is less affected, importing about 5 per cent of its consumption from the region.

[24] Presently the "Armilla Patrol" consists of one frigate, two destroyers, three minesweepers, and three auxiliary vessels. The patrol usually operates in the southern gulf region. To date, the Royal Navy has accompanied over 200 vessels safely through the Straits of Hormuz. *The Christian Science Monitor,* 2 May, 1987, p. 9.

[25] The French naval forces consist of one aircraft carrier, four frigates, two destroyers, three minesweepers and three auxiliary ships. *The Christian Science Monitor,* 2 May, 1988, p. 9.

[26] There are several issues that divide Paris and Tehran. One is for France to limit the sale of arms to Iraq and begin selling to Iran, the restoration of diplomatic relations broken off in the summer of 1987 over the Vahid Gorji affair, and Tehran would like the repayment with interest of the final slice of a $1 billion dollar Iranian loan to a state-owned French nuclear reactor company. See *The Economist,* Vol. 307, No. 7549, 7–13 May, 1988, pp. 48–49.

[27] For a detailed discussion of deployment efforts by these nations, see *The Christian Science Monitor,* 2 May, 1988, p. 9.

[28] The constitution of the Federal Republic of Germany forbids the use of its armed forces outside NATO territory, but Bonn announced that ships of her navy would fill gaps in NATO areas created by allied deployments to the Persian Gulf. Spain, likewise, will use its ships to fill NATO gaps. Norway has deployed a minesweeper to NATO's English Channel command.

[29] The United States and its European allies have decided to coordinate more closely their naval operations in the Persian Gulf to eliminate wasteful duplication of efforts searching for Iranian mines. *The Washington Post,* 28 April, 1988, p. 33.

[30] *The Christian Science Monitor,* 2 May, 1988, p. 9.

[31] R. J. Sontag and S. S. Beedie (eds.) *Nazi-Soviet Relations, 1939–1941* (Washington, D.C.: United States Government Printing Office, 1948), p. 259.

[32] As the US would increase its forces in response to regional crises, the Soviet Union would do likewise (e.g.: 1971, Indo-Pakistani war; 1973; Arab-Israeli war; 1977, Somali-Ethiopian conflict).

[33] As of April 1988, five to seven Soviet military vessels are in the immediate vicinity of the Persian Gulf, including two or three destroyers or frigates (Krivak I Class), three minesweepers (Natya I Class), plus a support ship and a Vytegrales class command and control vessel.

The Soviet Union has access to the port at Aden in the Peoples Democratic Republic of Yemen. It is used for liberty and ship replenishment, and anchorages at Socotra and Perim Islands. The Soviets' only naval communications and intelligence gathering facilities in the region are located in Yemen. When not under steam, most of the Soviet naval squadron lies at anchorage off the Socotra. Also, they began flying IL–38 reconnaissance aircraft out of Aden International Airport in late 1978, and also have access to Al-Anad airfield. It is reported that four Il–38 May aircraft, and sometimes TU–95 Bear bomber/reconnaissance aircraft, operate out of Aden, providing aerial reconnaissance for the squadron. In addition, the Soviet Union also has access to facilities in the Seychelles Islands and Ethiopia. In the latter country, they have important naval maintenance and repair facilities, which are capable of handling submarine as well as surface facilities. In order to protect the Ethiopian facilities, a small contingent of Soviet Naval Infantry stand guard. Ciarrocchi, *op. cit.,* pp. 41–42.

[34] Khomeini brutally suppressed the Tudeh Party and has provided some aid to the *Mujahadin* fighting Soviet forces in Afghanistan.

[35] "Soviet Policy in the Middle East," by Mark N. Katz, in *Current History,* Vol. 87, No. 5, 26 February, 1988, p. 59.

[36] On 6 May, 1987, a Soviet merchant ship the *Ivan Koroteyev* was attacked by an Iranian speed boat. This attack was followed on 17 May by the incident of the *Marshal Zhukov,* one of the three Soviet tankers leased to Kuwait, which struck a mine in the main channel leading to Kuwait's oil ports.

[37] The Soviet Union has agreed to build an oil pipeline to carry Iranian crude oil to the Black Sea, and to add links to the Soviet-Iranian railroad system.

[38] Katz, *op. cit.,* p. 59.

[39] *Ibid.,* p. 60.

The Soviet Scene—Review and Prospect

AIR COMMODORE F. S WILLIAMS, CBE, MPHIL, RAF (RETD)

The author is a Soviet Studies Associate at the RUSI and has been closely connected with Soviet studies since the 1950s. In 1964 he was appointed Assistant Air Attaché in Moscow, where he served for three years. He spent another four years as UK Defence Attaché in the Soviet Union until late 1981. His latest work, The Soviet Military: Political Education, Training and Morale *was published by the Institute in association with Macmillan in January 1987. This contribution stems from the RUSI Soviet Union and Eastern European Programme, supported by the Esmée Fairbairn Trust.*

VERY FEW years since 1917 can claim to have been more exciting for Soviet watchers than 1988. In the course of 1986 it could be seen through the glass darkly that Mikhail Gorbachev's radically progressive policies were not universally welcome in certain strata of Soviet society, both high and low. It was not difficult to predict that the already brisk dialogue between the zealous Gorbachev supporters and the more conservative, "softly-softly" reformers epitomised by Mr Ligachev would continue to sharpen and become more positively polarised through 1987, into 1988 and beyond; that it would become the pivotal issue for the future of the Soviet Union, for the whole socialist commonwealth and indeed for the rest of the world.

It was, however, difficult for some to perceive Ligachev as quite the formidable counterweight he clearly is—wishful thinkers in the West found it hard to accept that the relatively young, exciting and supposedly popular Mr Gorbachev, who for them had come to represent communism with a human face, did indeed have an opposition and might not necessarily retain his position for ever. It was impossible for them to believe that the General Secretary's position might even be precarious. But the dialogue has expanded into an almost open brawl conducted on a number of fronts and Mikhail Sergeyevich has not always won the day. As one dramatic example of this, he was forced to sack Boris Yeltsin, his favourite and protegé, ostensibly because he was urging the introduction of reform measures too quickly. Yeltsin may have been guilty of wanting actually to implement *perestroika*, not just to talk about it.

THE PROGRESS OF *GLASNOST* AND *PERESTROIKA*

This of course is still the nub of the problem. First, how much of *perestroika* is mere sloganeering and how much represents actual on-the-ground reconstruction? Second, how much of the vaunted *glasnost* is genuine open debate and, if it has any, where do its real limits lie? Above all, how much of *glasnost* is merely a new and more subtle element of the admitted ideological struggle with capitalism, and how much is a confession that the system really has to undergo a complete root and branch change in both its internal structure and in its relations with the rest of the world?

Obviously it is difficult for an outside observer to answer these questions convincingly. Sovietology used to be a relatively easy pursuit in the days when nothing much changed. This was particularly so during Stalin's long reign and even though some dramatic things happened in the 30 years plus that followed, this continued to be true in essence. Sovietology remained a soft option for those who found social studies too demanding. It was simply a matter of reading *Pravda*, of claiming that you could read between the lines, and of bandying about the names of Politburo members. One could make a decent living out of it without too much sweat. And, although then as now sovietologists hated each other, they nevertheless all came to the same conclusions, so one could not go far wrong.

Now all is very different. Gorbachev has shot the fox. In what might still prove a desperate gamble he called the 19th All-Union Communist Party Conference in late June 1988— the first for almost half a century. It seems more than probable that he had a twofold aim in taking this momentous step. First it would be a once and for all test of strength with Ligachev, a clearing of the ground in order that a solid new social and economic state structure could be built on the already established foundations of *glasnost*. This was a test that could not wait until the next Party Congress, not due until 1991. Second, to put his own political future at stake would be the loudest declaration he could make to the West that his policies are indeed genuine.

Suddenly in consequence there is more information than

ever before—certainly since the end of the Second World War. An information explosion has occurred. More data is now available on all sectors of Soviet activity, political, military, economic, social and historical, than can be easily assessed. No longer is there a single consensus view of the Soviet Union or, at the most, two. Now a range of widely differing interpretations may have validity. Another disconcerting thing is that for the first time one has to read what is actually on the lines in *Pravda*; there are not many enigmatic messages written between them anymore. Not only Westerners but also the Russians themselves are finding this more than a little confusing.

In one important aspect, that of gauging what Soviet policies actually are, this information explosion means we know less about Soviet intentions than we might have done before. The patent collapse of the internal monolith suggests that foreign policy intentions are equally mixed. That could be dangerous when trying to interpret Soviet signals. At present NATO governments appear well aware of the danger and the Reagan-Gorbachev Moscow Summit of May 1988 reassured the rest of us that we are not going to be saddled with thoroughly bad strategic and conventional arms agreements concluded with unbalanced concessions made merely to gain short-term political advantage—as many in Europe still regard the conclusion of the now ratified INF Treaty. The ghost of Reykjavik will haunt the scene for a long time to come.

Future governments might, however, choose to forget this danger, blinded by the fascinating cornucopia of revelations that *glasnost* strews in front of us daily. Suddenly we know much for certain about those facets of Soviet life we have hitherto only suspected, for example the extent of ethnic unrest, industrial inefficiency, crime, corruption, youthful disaffection, apathy, prostitution and much else. We also now know for sure about many things we did not suspect, the extent of the drug problem, for instance, and the appalling state of the health organisation. But of course these revelations come as a greater surprise to Westerners than they do to the Soviet citizen because to him they are not really revelations. He has known the score, or most of it, all the time. The comrade-in-the-street has always been aware of the crimes of Stalin, and the Gulag. It is only the novelty of reading about them in the newspapers that shocks him. He still cannot believe what appears to be happening and finds it difficult to adjust. Those who meet foreigners still tend, at least initially, to talk from behind those strong individual carapaces which have always protected the Soviet system, right or wrong, for better or for worse. In any case they cannot accept yet that foreigners are not what they always were. *Glasnost* and *perestroika* promise to radically change, and hopefully to improve, Soviet life but the enemy, capitalist-imperialism, remains unchanged and continues to threaten their existence from without.

Now is the time for all good Soviet watchers to examine the state of the Party. They are flocking to Moscow to assess the present condition of the USSR and test the validity of their earlier pre-*glasnost* perceptions. Their experiences are strikingly similar. They expect only to have their prejudices updated and to an extent that is in general what happens. They find the same old semi-efficient Soviet Union and the same lovable Russians they have known for years. Superficially nothing appears to have changed, nor does it seem to be changing. *Glasnost* is there for all to see, it fills the media, but if *perestroika* has produced anything at all, it is not yet very obvious to the foreign eye. The city appears unchanged from five years ago. Life for a Soviet consumer has not yet noticeably improved. The shops are still dreary beyond belief. Not even the window dressing has improved. Gorbachev's experiment with private enterprise is scarcely noticeable. The much discussed new private enterprise "cooperative" restaurants are as yet no advertisement for *perestroika*. The entrepreneurs who are trying to run these establishments are still heavily reliant on the state for their wholesale supplies and services with the result that they are all too frequently closed. They fear they will go out of business long before *perestroika* arrives. Only the opportunities for relieving foreign visitors of hard currency keep them afloat. The leader of one such "cooperative" said recently, "it seems as though 'they' don't want us to open".

If that is typical it is not surprising that private enterprise is slow to take off. The sole and puny result of its introduction so far has been to legitimise some of the small trading operations conducted traditionally by peasant families or urban moonlighters. There has scarcely been a public stampede to take advantage of the doubtful franchises on offer. In fact, they seem not to appeal at all to the vast majority of orthodox citizens long schooled to abhor any form of capitalism, however petty. Ironically it seems that 70 years of Soviet conditioning have been all too successful in creating this particular Soviet ethic in the nation's soul. If the British disease is a tendency to scorn trade and commerce, then the Russians scorn them entirely.

But that is the visual impression only. Beyond this there comes the consciousness that *glasnost* has already created a new dynamic in all walks of life—an excitement that has not been felt since the early days of Lenin. It is especially noticeable among the higher echelons of the regime. If he has achieved nothing else, Gorbachev has restored a faith in the future among the vast majority of the population. It is a future still mistily discerned but rosy nonetheless, based on science, space and the perfectibility of man. The prophet Gorbachev points the way to that future, his popularity and support, as predicted in these pages last year, dependent on his ability to deliver it. "The reform policy cannot be halted now", he told the Central Committee in January 1988. "During this year, 1988", he said, "we shall begin to see results from the groundwork already accomplished. It would be disastrous to stop it at this stage".

Who would want to stop it? The opposition of course, the flies in the ointment for, in spite of the popular euphoria, the wider debate on the economic system, as reflected in Soviet media throughout 1987 and the early months of 1988 has been ominously negative. The gist of the argument critical of Gorbachev's proposals seems to have been that central planning does not permit the existence of a free market sector of any significant size and that any measure of devolution to management will not work out. It has to be all or next to nothing. The choice seems therefore, again as ever before, to be between staying with the existing, inefficient but politically controllable system, or its complete abandonment, with all that this would entail, ie, the introduction of competition, full currency convertibility and entry to the world market. The implications of adopting this latter free market course are awful to contemplate. Both options seriously threaten the stability of the regime, the latter amounting to nothing less than the renunciation of Marxist-Leninist ideology.

A compromise was expected to emerge at the 19th Party Conference. It did not. Instead, the battle between Gorbachev and Ligachev was fought on political grounds. Spectacular innovations to streamline the unwieldly dual government of Party and Ministries were proposed and other novelties introduced to give a more democratic look to the system.

Gorbachev appeared very much the glossy winner with a mandate to continue the restructuring process but after the tumult and the shouting died down, it was noticed that economic problems had scarcely been mentioned. But they will refuse to go away of their own accord and there is little doubt that Ligachev, having received his own impressive affirmation of support from a large sector of the delegates, will soon return to this dismal subject.

It ought, in fairness, to be said that Ligachev is not simply a powerful reactionary champion of the new "bourgeoisie". The so called "opposition" to Gorbachev has understandable and legitimate fears, not merely for its own privileges, but for the continued stability of the regime. It is not the hidebound bureaucracy that the cartoonists of *krokodil* see sharpening its pencils to fight innovation with regulation. Nor is it a strong dam of inertia holding back the flow of modernisation. If anything it sees itself as quite the opposite—as a dam holding back a flood that threatens to sweep the whole state away and leave only chaos. It may well be right and many an average Soviet citizen thinks it is, those in the constituency represented by Mr Ligachev who continues to say that he is by no means an opponent of *perestroika*. It is just that he is fearful lest the baby be thrown away with the bath water. And on a careful count it is not at all clear whether Gorbachev really did consolidate his authority at the June conference. Ligachev needed to say little at the conference to warn of the perils of precipitate reform. The daily communiqués from Nagorny-Karabakh said it dramatically for him.

Gorbachev's recent foreign policy successes have nevertheless enhanced his domestic standing as an international statesman. In the eyes of the Soviet population, the summits and the INF Agreement are purely Soviet triumphs gained by their champion in the teeth of Western opposition. These are being exploited at home for their own glory; every day across the range of their media we see how they endeavour to exacerbate the intra-NATO fears and dissensions that the agreement has caused. *Tass* announced in December after the General Secretary's brief stop at RAF Brize Norton that the climate between the UK and the USSR had never been warmer. One would not judge so from the unabated virulence of their sustained media attacks against Western policies. Although governments themselves are receiving less direct criticism, their agents, the Wall Street puppeteers, the City, the revanchist fascists and the neo-colonialists, serve equally well as targets, represented as they are as the enemies of disarmament and peace.

A NEW MILITARY DOCTRINE?

Such attacks are unlikely to lessen in number as long as their much advertised "new political thinking" is expanded and expounded further. Whether this truly represents a new approach to relations with the rest of the world or merely a refurbished version of the peace offensive is hard to determine. It is possibly best examined in the context of the new Soviet military doctrine published in May of 1987, a doctrine that has become associated with Dmitri Yazov, the new Defence Minister. Whether the "new political thinking" is indeed something new or whether there really is a new Soviet military doctrine in existence or in the making, there is no doubt that a radical re-appraisal of Soviet military requirements is underway in the general staff. It appears, as far as personal Western contacts indicate, to address the implications of the INF Treaty, the "50 per cent" proposal for strategic nuclear forces and perhaps in the course of time a significant reduction in conventional weapons. The difficulty lies again as always in isolating the genuine from the propaganda of the ideological struggle but it does seem more than probable that an element of motivation in the framing of this new military doctrine was a genuine desire for an improvement in relations for whatever purpose. A lot of pudding will have to be eaten, however, before this can be proved, although it is quite clear from recent assessments that the Soviet economy would find it extremely difficult both to satisfy domestic consumer demand *and* sustain the East–West confrontation at the current level. The military speakers at the conference ungrudgingly supported Gorbachev's programme, implying that the only way for the Soviet Union to enter the 21st century is through restructuring—the only way to ensure that the Soviet armed forces can hope to meet the technological challenge of Western forces in the years to come.

The opening theme is another reminder that the aim of the "new military doctrine" is universal peace with guarantees of security for all nations. Its terms are a blend of the naïve, the obscure and the cunning, all reminiscent of those massive propaganda offensives of a few years ago aimed at the more

sanguine opinion-formers in the West in the hope that they
would squash the "neutron bomb", which they did, and the
two-track INF policy, which they failed to do. Reminiscent, but
more so, since the new emphasis is no longer confined to
nuclear weapon issues but equally to conventional war.
Conventional war, they claim, is not an option any more. It is
now just as much a threat to the continued existence of
mankind as nuclear war. Not only would modern conventional
weapons cause unimaginable casualties far beyond the scale of
Passchendaele or Stalingrad, but the inevitable destruction of
nuclear power stations would in turn destroy the world ecology
no less finally than would nuclear missiles. The effects of the
Chernobyl experience are obviously very much in their minds
when they claim this, with an innuendo that the West is
unaware of this threat.

Thus far, this scarcely adds up to a new military doctrine and
clearly it is not. It appears more like a bid to attain, or retain,
the hegemony of the world peace cause—the Soviet Union still
fearlessly confronting the war-mongering politicians of
Western reactionary circles. This time, the aim is higher and
wider but still in similar style to the discovery of peace, the
Soviet Union has now discovered the ecology, the world
environment and the cosmos surrounding it. They are striving,
it would seem, to capture the propaganda high ground this
presents. They have had considerable success already insofar as
many voices in NATO countries are already demanding large
force reductions in the light of their perception of a lower level
of threat.

All that appears to have been added to make a military
doctrine of this are the terms "sufficiency", ie, sufficiency in
armaments to preserve "equal security", the other term. They
accept at last that verification will be necessary in the wake of
any arms reduction agreements concluded. This of course they
have conceded in principle since Stockholm, 1986, and in the
formulation of the INF Agreement, but it is already plain that
the very levels and systems of verification themselves will not be
easy to implement although during this year several practical
steps have been taken in this direction including exchange
visits to INF installations and chemical warfare establishments.

One of General Yazov's first acts on taking up his new
appointment was to write an article elucidating this new
doctrine. He emphasised the underlying point that the Soviet

Union is not seeking to raise its security at the expense of other nations but will not settle for less than it considers essential. As long as general peace is in danger from attempts to break it, the Soviet armed forces must be maintained in reasonable shape to give the USSR adequate defensive capacity. And, he continues, lest other nations feel insecure within their own borders the Soviet Union wishes to make it clearly known that it requires its armed forces only in order to uphold its own security and that of its allies.

In astonishing egg sucking fashion he then tells us that the true purpose and character of a state's military doctrine can be measured not so much from the strength of its political and military leaders' official statements and everyday rhetoric, as from the actual steps they take in such things as military programmes and budgets, the declaration of strategic intentions, their attitudes to arms reduction and slowing down the arms race. To what extent his tongue was in his cheek when he wrote this cannot be known, but on all counts he concludes that Soviet military doctrine is undoubtedly defensive while NATO's is agressive.

As his readers begin to wonder what exactly the article is about he reiterates that their's must be a defensive doctrine because of their proposal to eliminate all nuclear weapons and other means of mass destruction by the year 2000—this is supplemented by the Warsaw Pact's proposal that conventional forces be reduced drastically in the region between the Atlantic and the Urals, with commensurate cuts in defence spending for the benefit of humanity.

It might therefore be said that nothing new is coming out of Moscow in spite of *glasnost* and in spite of a declared military doctrine proclaiming itself to be new. This might just be wrong, because there have been some hints that in the wake of a conventional agreement they would have to re-shape their military art from the current "offensive defence" concept to one of "defensive defence". Senior Soviet officers in personal contacts with Westerners have suggested the need for some form of positional defence with "smart" weapons capable of annihilating an aggressor the moment he set a foot across the socialist frontier. Note, the *socialist* frontier, indicating that come what may, those Eastern European countries are not going to be given a freer rein militarily. They are still to be the buffer zone and if necessary the battlefield. In the immediate future Eastern Europe will be exciting to watch. Each of those countries has in varying degree shown disquiet about the INF Treaty and the possibility of further agreements which might expose them to greater danger from NATO attack. In this they fear for their security in almost a mirror image of the Western European concern over their increased vulnerability to the Warsaw Pact. In view of economic strains and *perestroika* rising expectation knock-on effects, the Soviet leadership must have grave doubts about the longevity of the alliance.

If the Moscow summit and the ratification of the INF Treaty were the predictable highlights of 1988, the most amazing event was surely the Soviet decision to start pulling out of Afghanistan on 15 May. Amazing because such a retreat goes against the dialectic grain and, although Gorbachev himself first announced the intention to withdraw over two years ago and Soviet military authorities have informally spoken of it for even longer, it still came as a surprise when the columns formed up and actually began to leave. It confounded most analysts, especially those of the opinion that time was on the Soviet side and victory inevitable. Moscow, of course, claims victory but this fools nobody, not even Soviet citizens. General Gromov, the ground forces commander, in a speech on 1 July at the 19th Conference denied it was a defeat, but it manifestly was.

> We went to Afghanistan on a mission of the greatest goodwill in order to defend the people, women and children, peaceful villages and towns and, finally, the country's national independence and sovereignty. And we carried out this task, at a high cost, but with honour.

This thought will now be reiterated sufficiently often enough to ensure it enters the history books.

How the decision to withdraw came about will long be debated but the natural starting point must be the Soviet policy to restrict the size of their "international contingent". Throughout the campaign they never exceeded a presence of approximately 105,000 men, but not because they could not field more. At that force level the war was unwinnable but they could not raise it without repeating the United States' escalation experience in Vietnam and encountering the increased opprobrium from the court of world opinion which would have ensued had they attempted an all-out win. There was also the opprobrium of their own ethnic minorities to consider. The events of spring 1988 in Armenia and Azerbaijan detonated by the climate of *glasnost* and its effect on the long-standing Nagorny-Karabakh dispute proves the oft-repeated claim that "the Soviet Union now has no internal enemies" to be an empty boast. To have stayed longer in Afghanistan, campaigning with greater ferocity, would have had unpredictable effects on the Central Asian and Transcaucasian populations.

By whatever process of reasoning the decision to withdraw was reached, it had to be cold-blooded. The most optimistic forecast for the future in Afghanistan without the Soviet Army is a state of low intensity civil war between rival factions, with an embattled government managing to hold on in Kabul and perhaps some other towns. Why then should Gorbachev decide to go at this very moment when, with a little more patience, a better deal might have been wrung out of Geneva? A deal offering the prospect of a more stable neighbour on the southern flank of the USSR. Several possible reasons suggest themselves, all relating to *perestroika*, but whatever the true story might be, it is plainly evident that the Soviet Union has washed her hands of the Afghanistan affair. Whatever happens there once the withdrawal is completed will not be her responsibility. Although Gromov maintains that the Soviet task has been successfully concluded, he recognises that the country is far from pacified and foreshadows the essence of future Soviet policy in his statement that:

> The enemies of positive change in Afghanistan, both within and beyond its borders, still nurture plans to return the people to their semi-feudal past by force of arms. Any means will be used to this end—terror, shelling, the senseless destruction of population centres, the mining of lines of communication etc. This process . . . is given full support by the USA and Pakistan governments.

In other words the Soviet Union is covered whether Najibullah stands or falls in the coming months. Some prophesy his fall before the end of 1988. Others cite increasing animosities among already difficult Mujahedin bedfellows which might give him better survival chances than are commonly reckoned. We can only wait and see. What is certain is that the Soviet Army was found sorely wanting in Afghanistan. Gromov has thrown in his lot with Gorbachev and calls for a total *perestroika* of the armed forces but, like the United States after Vietnam, it will be some time before confidence is restored.

It is unlikely in the coming months that Yazov will supply a more concrete expression of his new military doctrine nor will any reflection of it be seen in their procurement programme but it may be expected that other authors will expand and interpret it. This year will be notable for difficult bargaining on verification despite the enthusiasm Gorbachev displayed in his calls for ratification of the INF Treaty. We shall also see many statements and counter-statements concerning conventional arms reduction proposals. General Tatarnikov revealed the Soviet position in his important *Red Star* article of 5 January 1988. He emphasised that tactical nuclear weapons cannot be separated from conventional weapons in any reduction agreement. We shall see and hear much more of the same. The road ahead is still going to be rocky.

Finally, we should expect Mr Gorbachev to continue his overtures to China for a summit with Deng Xiaoping. Relations with the East for him loom as large as do relations with the West. He has made painful concessions already but unless he is prepared to concede the "three obstacles" entirely to Chinese satisfaction he is unlikely to get that Summit.

Glasnost and the Soviet Military

NATALIE GROSS

Natalie Gross is currently Professor of Political Military Studies at the US Army Russian Institute, Garmisch–Partenkirchen, West Germany.

DURING THE past few years Mikhail Gorbachev's *glasnost* (public openness) policy has begun to challenge the public image of the Soviet Union as an expansionist power seeking military superiority over the Western world. Yet, after the signing of the INF Treaty and a series of Soviet-initiated confidence-building proposals, Western observers of the Soviet scene continue to debate whether and to what degree the Soviet leadership has extended *glasnost* from domestic politics into the military and international security realms. Clearly, if *glasnost* is to have an enduring effect on Soviet security policy, it should enjoy the support of the military establishment, which plays an important role in shaping military doctrine. This article examines the role of the Soviet military in implementing Mikhail Gorbachev's *glasnost* reform in international and domestic contexts as a case study of civil-military relations under the new leadership.

GLASNOST AND ARMS CONTROL

The reshuffle of the Soviet High Command in May 1987, which followed the embarrassing landing of a young West German pilot in Red Square, has heralded the beginning of the Soviet military's involvement in Gorbachev's *glasnost* initiatives abroad designed to smooth the arms control negotiation process. Prior to the shake-up of the Soviet military establishment, senior officers of the Soviet General Staff took part in Gorbachev's arms control negotiations primarily as military experts, the role they had traditionally played during the SALT negotiations in the Brezhnev period.[1] Before Mikhail Gorbachev reaffirmed party control over the military by appointing Army General Dmitri Yazov as the new Minister of Defence, Soviet generals had publicly voiced disagreement with Gorbachev's unilateral moratorium on nuclear tests and opposed on-site inspections of the signatories' compliance with the agreement.[2]

By spring 1987 a new approach to arms control diplomacy, which linked arms reductions with military *glasnost*, began to take shape. In April 1987 a group of American and Soviet retired generals met to discuss methods of preventing nuclear war.[3] This was the harbinger of what were later to become high-level inter-alliance consultations between Soviet and US military leaders. Following this initial bilateral meeting, Marshal V. Kulikov, Commander of the Warsaw Treaty Organisation Forces, proposed an open discussion of the European military balance with his Western counterpart, NATO's Supreme Allied Commander Europe, General Bernard Rogers. These initial proposals set the stage for an official endorsement of military *glasnost* announced in the 29 May 1987 Berlin Communiqué of the Political Consultative Committee of the Warsaw Treaty Organisation.[4] The communiqué, which contained a comprehensive programme of arms control and confidence-building measures in nuclear and conventional areas, emphasised open exchanges of military information and free discussions of military doctrines and force development between the alliances.[5]

The Berlin Communiqué, however, did not indicate that Soviet openness on international security issues would mean a fundamental change of Soviet political values in the conduct of international relations. Like *glasnost* in the civilian area, which the leadership sees as a political instrument designed to stimulate political participation and accelerate technological development in their best interests, military *glasnost* is an expedient political tool for implementing arms control policies and enhancing Soviet security interests.[6] Vitalii Zhurkin, Gorbachev's civilian adviser on national security issues and director of the new Research Institute for the Study of Europe, has endorsed this approach to military *glasnost* in the leading party journal *Kommunist:*

> In the past closedness and secrecy in foreign and military policies were considered an important avenue for strengthening national security . . . . Today it is becoming more apparent that this secrecy has become counterproductive. Only openness allows us to bring home to people of other countries our policy goals, convince them of our peaceful plans and intentions, and helps to identify reactionary and militarist groups. For this reason measures for

expanding openness in our foreign policy and military activities play an enormous role in strengthening our national security.[7]

According to this authoritative writer, military *glasnost* is designed to influence Western public opinion in order to restore Soviet international prestige eroded in the post-detente years as a result of the continuing military build-up and undue reliance on military power in international relations. By winning favours with Western public opinion and influential political groups, Zhurkin hopes, greater openness on security issues will help slow down the pace of Soviet military technological competition with the West, check the Western lead in new emerging technologies, maintain the current balance in conventional capabilities and secure for the Soviet economy a desperately needed *peredyshka* (breathing spell). In other words, Gorbachev's top expert stresses the connection between openness in the national security area and its long-term pay-offs for the domestic restructuring programme:

> . . . Imperialism intends to use America's leading edge in science and technology in order to degrade Soviet military capabilities in a permanent and systematic fashion. It hopes to force the Soviet Union to engage in new military expenditures and then again devaluate them . . . Attempts have been made to throw our country to the periphery of the world economic system . . . The problem of carefully coordinating foreign policy and domestic goals and the methods for achieving them become all the more important.[8]

During critical phases of the INF Agreement negotiations the Soviet military have played a prominent role in managing military *glasnost* to promote Soviet arms control initiatives. In August 1987, Colonel-General Nikolai Chervov, the top military expert on arms control at the Soviet General Staff, gave an interview to the Japanese news service in which he promised that the military establishment will extend *glasnost* to the military sphere and proposed an exchange of data on nuclear weapon systems with the NATO Alliance.[9] General Chervov's statement was a self-fulfilling prophecy. In early September 1987 Soviet air defence commanders escorted a group of US congressmen to the controversial Krasnoyarsk radar site and allowed them to take photographs of the once highly secret military facility.[10] A week later an international delegation took a guided tour of the Soviet chemical weapons production facility at Shikhany and inspected a special installation designed for future destruction of chemical weapons.[11]

Interestingly, these military *glasnost* initiatives appear to be directly linked to the Berlin Communiqué proposal to ban nuclear and chemical weapons. The Soviets have skilfully exploited this new military openness for propaganda purposes: later during the month, the Soviet news agency TASS used these visits as a propaganda theme to praise to foreign audiences the openness of the Soviet military system to inspection by international organisations.[12] Predictably, in the same statement Soviet propagandists charged the United States with bad faith in arms control negotiations and with barring foreigners from inspecting US military installations. The obvious message of this standard propaganda exercise was to demonstrate to the Western community Soviet readiness to comply with verification procedures and on-site inspections stipulated in the INF Agreement, while diverting the public's attention from continued Soviet violation of the 1972 ABM Treaty at Krasnoyarsk and the use of chemical weapons against the civilian population in Afghanistan. In this sense, military *glasnost* paved the way for the INF Agreement by restoring the credibility of the Soviet Union in the West as a reliable partner in arms control agreements.

One cannot fail to notice the high visibility of Soviet generals in Gorbachev's military *glasnost* and arms control initiatives, although under the new leadership the actual political power of the military in decision-making has apparently declined.[13] During the final stages of the INF Agreement, Marshal Akhromeev visited his US colleagues at the Pentagon to resolve the remaining verificiation issues.[14] In March 1988, in Switzerland, Soviet Defence Minister Dmitri Yazov discussed with the US Defense Secretary Frank Carlucci Soviet doctrine and force development, as well as possibilities of future sports competitions and exchange lecture programmes between the American and the Soviet armies.[15] Therefore, even if the military establishment had in principle disagreed with military *glasnost*, it has, nevertheless, lent its support to Gorbachev's policy as part of his broader arms control strategy.[16] For his part, Mikhail Gorbachev has clearly relied on the military's expertise in arms control diplomacy and their public relations talents to further his foreign policy goals. At least in this respect, the Soviet military has contributed to Gorbachev's political strategy and the success of his "new political thinking", which emphasises political over military instruments for achieving security for the Soviet state. It appears, therefore, that the Soviet military propaganda writer Alexander Prokhanov was not far off the mark, when he said that the military establishment's contribution to the INF Treaty is as great as that of the political leadership.[17]

LIMITS ON MILITARY GLASNOST

Up until the present time, military *glasnost* was successful primarily as a confidence-building measure which expanded and improved communication channels between alliances and senior military leaders. Another benefit of military *glasnost* to the West has been an exchange of views and information on problems of doctrine and force development between the military establishments, and access gained by Western policy-makers to Soviet military facilities. However, it should be stressed that this access has so far been sporadic, carefully timed to coincide with Soviet arms control initiatives, and limited to influential political players.

At the same time, the Soviets have refused to disclose to the public at large specific information pertaining to their strategic and operational concepts, force structure and weapon systems. The quality of statistical reporting in this area has not improved: the figures related to the defence budget, allocations for defence programmes, and arms sales to the Third World remain shrouded in secrecy. The defensive military doctrine of reasonable sufficiency, that the Soviets have presented to the Western public as evidence of their new openness in the age of reform, was formulated in characteristically vague and general propaganda terms.[18] Since no specific information on military deployments and strategic and operational concepts has been released, the declaratory statements about defence and war prevention have enhanced the apprehensions of Western observers that military *glasnost* is primarily a propaganda ploy aimed at weakening the NATO alliance.

Foreseeing the unfavourable effects of limited military openness on Western public opinion, civilian representatives of the science community inside the Soviet Union have opposed the leadership's reluctance to expand *glasnost* in the military sphere. They have argued that release of defence-related information and an open debate on defence matters will benefit the USSR's security interests, yet, at the time of writing, their pleas have fallen on deaf ears.[19] A possible explanation for these limits on *glasnost* is that the military establishment, as well as the conservative groups within the party apparatus, who support the traditional penchant for secrecy in the military sphere, continue to exert leverage on the Gorbachev leadership.

The coverage of international security issues to domestic military audiences provides a good indication of the limits of military *glasnost*. Prior to Gorbachev's shake-up of the military establishment, the military press expressed in its cryptic fashion its disagreement with the party's arms control initiatives and policies of accommodation with the West. At the time of the summit meetings between Mikhail Gorbachev and President Reagan in Geneva and Reykjavik, the military press chose not to discuss in any detail the arms control proposals, and emphasised the Western military build-up and the US Strategic Defense Initiative (SDI). During this period, the anti-American propaganda campaign in the military press reached its peak with the publication of an article vividly describing plunder and sadistic atrocities allegedly inflicted by the American expeditionary force on the Soviet civilian population during the Allied Intervention of 1918.[20] Consistent with enforcing a negative view of the West, Soviet political officers instructed junior commanders to instill their men with hatred for the enemy by, for example, recounting the atrocities committed by the Germans during the Second World War.[21]

At the time of the INF Treaty, the coverage of the summit meeting expanded, and blatant anti-American propaganda subsided. The focus of the propaganda campaign has, however, shifted to the NATO Alliance as the main source of opposition to Soviet arms control policies and the threat to Soviet security interests.[22] It is important to note that Mikhail Gorbachev's visit to Washington has received incomplete coverage in the military press, compared to the full coverage it was given in the civilian media.[23] Omissions made by editors of the military newspaper were designed to play down to an average Soviet soldier the US-Soviet rapprochement and reduction of the military threat. For example, the military censor has deleted passages from President Reagan's speeches which portrayed the United States as a free and open society friendly to the Soviet Union. The military paper made no mention of Reagan's reminder that the American side was the first to initiate the zero option proposal. Likewise, quantitative comparisons of US and Soviet nuclear forces showing Soviet superiority have not found their way into the military press. Remarkably, the military press has also omitted Mikhail Gorbachev's statement that his political ideal is a violence-free world of disarmament.

The INF Treaty had little effect on threat assessments made in the military press. Furthermore, in the post-INF period Soviet military publications have stressed the increased threat to the Soviet Union from the NATO alliance. The journal for political officers, for example, has repeatedly accused NATO and the United States of a soaring military build-up which at present threatens the security of Soviet borders. Material recommended to political officers for use in political study classes emphasised the growing opposition to the INF treaty in the West which increases the threat of a new war.[24] Furthermore, a military commentator argued that the INF Agreement mandates an improved force posture and conventional modernisation, in order to offset possible force imbalances resulting from the agreement.[25]

It should be noted that despite *glasnost,* personnel and generational changes within the Soviet military establishment, no significant change between the reactions of Brezhnev's generals to the SALT agreement and those of Gorbachev's generals to the INF Treaty can be observed. As in the past, once the agreement has been signed, the military have verbally supported a stronger military posture as their way of rationalising the need to expand defence programmes and to check future far-reaching arms control agreements.[26] This continuity in the attitudes of the military regarding East-West security relations suggests that "old political thinking" still prevails in the military establishment and is perpetuated in the ideological training of young Soviet soldiers and officers. This leads one to believe that under the new leadership, the military have enjoyed the freedom to deviate from Gorbachev's new policy line in troop indoctrination.

DOMESTIC ASPECTS OF MILITARY *GLASNOST*

In Russian history *glasnost* in the military, as in civilian society, was traditionally designed to direct an exchange of opinions and ideas in the best interest of the leadership. In mid-nineteenth century Russia under Nicholas I, the champions of *glasnost* among officers promoted critical debates to correct failures of the bureaucracy and thwart corruption which thrived among Russian officers of the time. The Grand Duke Konstantin Nikolaevich, who sponsored such discussions in the naval establishment, believed that an artificially induced debate (*iskusstvennaia glasnost*) would promote conflict of opinion during debates about the new naval regulations.[27] These debates—held within limits strictly defined by the central government—contributed to Russian naval professionalism and made the military system of the time more effective.

Not unlike its predecessor in Imperial Russia, *glasnost* in the military press today means discussions critical of bureaucratic mismanagement and corruption. During the *glasnost* campaign in the military and civilian press, senior military officers and the Ministry of Defence as an institution have been criticised for inefficiency and misappropriation of funds.[28] The Soviet public has learned, for instance, that their revered two-star generals build private saunas and spas at the army's expense and make profits on the side by sending cadets to work on local farms. By castigating these social maladies, Soviet military reformers believe public openness will assist in correcting some of the army's present discipline and morale problems.

Another aspect of *glasnost* has been to encourage grass-roots initiative in order to improve military hardware and training procedures—changes designed to make the military system more cost-effective. For instance, within the framework of *glasnost,* Soviet logistics experts are encouraged to improve efficiency in the areas of resource allocation, cargo transportation and more extensive incorporation of computer technology.[29] Indeed, *glasnost* has been used to promote discussions in military units on topics ranging from awards and admonitions to shortcomings in training and exercises. Commanders are now requested to solicit recommendations from junior personnel on issues related to education and training.[30] Admiral A. Sorokin, the First Deputy Chief of the Main Political Administration, also recognises the role of public debate in facilitating the decision-making process, namely, in making the military bureaucracy more responsive to inputs from below.[31] He has emphasised the need to keep the soldier informed about command decision-making—a prerequisite for developing low-level initiative in peace and wartime. According to the Chief of the Political Administration of the Air Force, General L. Batekhin, public openness should be used to discuss possible improvements in training standards, namely, to introduce tighter combat readiness standards.[32] A new emphasis on training (*obuchenie*) over indoctrination (*vospitanie*) means that Soviet military commanders can tailor *glasnost* to promote *perestroika* (restructuring), that is, to improve training methodologies and the quality of Soviet manpower, especially its junior command component.

Restructuring for the Soviet Army means some decentralisation of the system on lower levels, more initiative for lower ranks, and closer interpersonal relations between officers and men.[38] Not unlike Western military experts, under *perestroika* reform-minded Soviet commanders stress realistic and flexible training, "accessible leadership," and self-motivated commitment over blind subordination and obedience. Traditionally, the Soviets have regarded the highly centralised senior command authorities implementing elaborate operational plans as the linchpin of combat power.[34] Today Soviet military reformers, without relinquishing the core principle of centralised command and control, pay increasing attention to smaller combat units, junior leaders and individual combatants as critical elements of success on the fluctuating modern battlefield. In this respect, the rationale for restructuring is to protect survivability and effectiveness of tactical sub-components of the Soviet military system, which, given the accelerated pace of operations, surprise situation changes and massive disruptions in command and control in a future war, will be operating independently, on detached axes, and in isolation from the main force and its command centres. In planning for this war scenario, the Soviets have come to recognise the relationship between a more accessible military system in peacetime and more effective leadership, decision-making, and improved soldier motivation and initiative on the battlefield. In re-examining leadership and training concepts, the Soviets have responded to Western technological and doctrinal developments, for example, precision weapons, target acquisition and strike systems, the Air-Land Battle and FOFA, which will fundamentally change the nature of the future battlefield. As the First Deputy Minister of Defence, Army General P. Lushev, has noted:

The human element is the main component (in combat readiness) ... Achieving high training standards is a difficult task ... This is due to changes in military affairs, the conduct of operations under conditions of enemy use of precision weapons, when defenses against fire, strike and target acquisition systems will have to be created.[35]

This shift in Soviet thinking is generally consistent with the gradual transition of the Warsaw Pact armies to the regimental/brigade structure as their "building blocks".[36] The new military establishment installed under Gorbachev favours *perestroika* precisely because it recognises its potential benefits in making the Soviet military system more effective on the technologically complex modern battlefield vis-à-vis NATO forces.

Although the Soviet High Command may find *perestroika* and *glasnost* compatible with the Army's military and technological requirements, Gorbachev's policy has encountered resistance from military bureaucrats with powerful stakes in the old system. Like their civilian counterparts in the party and state bureaucracies, groups of senior officers who owe their careers to the good old Brezhnev years have resented a more open military system where their performance is subject to scrutiny and their professional fortunes jeopardised. The right granted to lower ranks to criticise senior command decisions has naturally provoked angry complaints from seasoned officers that *glasnost* is eroding the sacred unit of command.[37] The Chief of the Political Administration of the Ural Military District, for instance, has warned military personnel that unrestricted criticism of commanders and their decisions would not be tolerated, but party members among soldiers and junior officers can use authorised party channels to criticise their superiors.[38] To mitigate the emerging conflict between senior and junior military personnel, the Defence Minister Dmitri Yazov has reassured his officers that the Marxist dialectical approach can reconcile criticism of superiors with the unity of command.

Initially, the new policy produced tensions in those units where low-ranking personnel petitioned senior authorities to investigate misconduct of their commanders. Military personnel reportedly suffered reprisal for initiating grievances or for voicing criticism. For instance, a navy captain stationed at the Leningrad Naval Base was reprimanded for informing senior military authorities that his commander employed enlisted men in his illegally run private souvenir workshop on post as well as in menial jobs in the home.[39] For quite some time, for fear of reprisal the majority of enlisted men and NCOs did not engage in critical discussions. Military writers reported that during public meetings military men were reluctant to criticise the army's political departments or their representatives.[40] As the First Deputy Minister of Defence, Army General P. Lushev, admitted at the time "since criticism is not respected in all military units, criticism from below is expressed in the form of timid suggestions, with caution".[41] However, slowly and gradually Soviet troop commanders are beginning to appreciate the benefits of *glasnost:* they report that candid discussions and closer personal relations in military units are helpful in enforcing discipline, correcting failures in training and in soliciting recommendations from qualified junior personnel.[42]

Glasnost on internal military issues has substantially changed the Soviet military press, which is clearly more open today than it has ever been since the 1920s. It discusses a plethora of social problems which the Soviet Army shares with many other modern military systems: alcoholism and drug abuse, nationality conflicts and draft-dodging, violence between first and second year draftees and AWOLs, corruption among senior officers and illegal arms trading in units stationed in Central Asia. Some truthful reports about the war in Afghanistan and candid discussions of unofficial veterans' organisations and their demands for more benefits and public recognition have found their way into the military press. The media has acknowledged reluctance among conscripts to risk their lives in combat, and disclosed methods used by parents to keep their children from being drafted.[43] Yet, in stark contrast to the galvanisation of public opinion in democratic societies, for example, in the United States during the Vietnam era and in Israel during the war in Lebanon, military *glasnost* has not developed into an open policy debate about the costs and benefits of the Soviet invasion.

Another aspect of *glasnost* in the military press has been the new candour in assessing Soviet military performance during the Second World War. Though criticism of selected aspects of Soviet operations, for example, the initial period of the war, organisation of logistics and the medical service, appeared in the military press during the late 1970s and the early 1980s, the recent discussions scrutinise Soviet military failures during all phases of the past. The Soviet Military Historical Journal has provided, for instance, an in-depth analysis of Soviet failures during offensive operations in 1944. Since the Soviets view military history as a model for refining their operational concepts for a future war, their military theory is likely to benefit from this manifestation of *glasnost*.

However, compared to relatively free current discussions of sensitive political subjects in the civilian media, the reporting of controversial political issues in the military press has not markedly changed. The military press has been haphazardly reporting the economic reforms proposed by Gorbachev. His speech at the January 1987 CPSU Central Committee Plenum, which called for broad reforms and attacked opposition to his programme, appeared in an abridged, sanitised version. Criticism of Stalin as a military commander and his use of terror against the officer corps in the military press has been restrained and limited to the more academic journal for senior officers. The civilian press, on the other hand, has engaged in an unprecedented destalinisation campaign which blames current Soviet economic and political failures on Stalin's dictatorship.

Since in the era of *glasnost,* the military establishment, as well as other bureaucracies, for example, the party, the ministries, the KGB, law enforcement authorities, have drawn fire in the civilian media, it has naturally found Gorbachev's policy of public openness detrimental to its status and institutional interests. Significantly, it was the political leadership, exasperated with the military's incompetent handling of the Cessna incident, which has set the tone for critical attacks on the military in the civilian media. In June 1987 Gorbachev accused his generals of a lack of professionalism, and of having compromised the USSR's international prestige as a military power, whereas Boris Yeltsin, the former First Secretary of the Moscow Party Organisation, took the command of the Moscow Military District to task for insubordination to the political leadership.[44] To protect its tarnished reputation in civilian society, the military press now regularly attacks liberal civilian publications such as Moscow News for misconstruing Soviet Second World War failures, overstating the extent of morale and cohesion problems in today's army and, last but not least, for discrediting the military profession and military officers in the public eye.[45]

Other military spokesmen have bluntly stated that the ongoing devastating criticism of the military in the civilian press is damaging to the army's mission:

> Individual publications which question the significance of army service and discredit the constitutional duty of Soviet citizens

seriously harm the education of Soviet soldiers and their readiness to defend their country. Love for the motherland and hatred for its class enemies are inseparable.[46]

Other officers blame *glasnost* for the army's continuing discipline problems and the growth of pacifism among the 1988 pool of draftees.[47] The Defence Minister Dmitri Yazov, in less alarmist tones, has shared the concerns that *glasnost* may soften the traditionally stringent Soviet assumptions of an endemic conflict between opposing social systems and have an unfavourable effect on soldiers' morale and fighting spirit.[48]

As we can see, the Soviet military have mixed feelings about *perestroika* and *glasnost*. On the one hand, senior officers have supported military *glasnost* as an instrument of Soviet arms control policy. They have also recognised the benefits of openness in making the military system more flexible, competitive and effective vis-à-vis Western technological and doctrinal challenges. On the other hand, the Soviet military have opposed *glasnost* when, in their view, it has threatened to change traditional threat perceptions, disrupt morale and ideological reliability of the troops, and question the legitimacy of the military institution and its vested bureaucratic interests in society. In this respect, *glasnost* has exacerbated the tensions between Gorbachev's leadership and the military.

On balance, Western defence planners should be aware that the ultimate goal of *perestroika* and *glasnost* for the Soviet high command is to correct some of the army's current problems in troop training, morale, motivation and unit cohesion. If successful, these changes in the long term may make the Soviet soldier a more formidable opponent. Yet, since bureaucratic control and inertia are deeply entrenched in the Soviet military system, it will take some time before Gorbachev's reforms will take root in the Soviet Army. In the meantime, as long as Gorbachev stays in power, he can be expected to delegate authority to and receive support from his generals for his foreign and national security policies, while fighting battles with them at home over how far *glasnost* can be permitted to go in the media, scholarship and the arts.

NOTES

[1] Raymond Garthoff, "SALT and the Soviet Military," *Problems of Communism*, January-February 1975, pp. 21–37.

[2] *Voennyi vestnik APN,* No. 1 September, 1986 and *Sovetskaya Rossiya,* 23 August 1986.

[3] *The Defense Monitor,* vol. 16, No. 5, 1987.

[4] *Pravda,* 30 May 1987.

[5] Jacob W. Kipp, "Reflections on the Berlin Communique of the Political Consultative Committee of the Warsaw Treaty Organization," unpublished manuscript.

[6] Natalie Gross, "Glasnost': Roots and Practice," *Problems of Communism,* November–December 1987, pp. 69–80.

[7] V. Zhurkin, S. Karagonov, A. Kortunov, "New and Old Challenges to Security," *Kommunist,* No. 1, 1988, pp. 42–50.

[8] *Ibid.,* p. 50.

[9] KYODO, Tokyo, in English, 16 August 1987 in FBIS, 17 August 1987, AA 2.

[10] "Radar Visit Proves Doubly Astonishing," *Washington Post,* 10 September 1987 and "Breaching the Wall at Krasnoyarsk," *New York Times,* 10 September 1987.

[11] Jesse James, "Glasnost Spreads to Chemical Weapons," *Arms Control Today,* 1 November 1987, p. 22.

[12] Moscow TASS International Service, 21 September 1987, in *FBIS,* 22 September 1988.

[13] Natalie Gross, "Mikhail Gorbachev's Reforms and the Soviet Military," paper presented at the conference "Gorbachev and the Soviet Military Institution" at the French National Foundation for Political Sciences in Paris, January 1988.

[14] *New York Times,* 12 October 1987, p. 9.

[15] *Los Angeles Times,* 17 March 1987.

[16] Phillip Petersen and Notra Trullock, "Gorbachev and the Soviet Force Development Process," paper presented at the conference "Gorbachev and the Soviet Military Institution" at the French National Foundation for Political Sciences in Paris, January 1988.

[17] *Pravda,* 17 December 1987.

[18] *Pravda,* 27 July 1987, and *Armed Forces Journal International,* December 1987, pp. 48–50.

[19] *Moscow News* (English), 6 September 1987, p. 3.

[20] *Krasnaya zvezda,* 14 February 1987.

[21] *Voenno-istoricheskii zhurnal,* No. 3, March 1987, pp. 62–68.

[22] *Krasnaya zvezda,* 4 December 1987.

[23] Cf. *Krasnaya zvezda* and *Pravda,* 9 December 1987.

[24] *Kommunist vooruzhennykh sil,* No. 2, 1988, p. 82.

[25] *Krasnaya zvezda,* 14 February 1988.

[26] Raymond Garthoff, "SALT and the Soviet Military," *Problems of Communism,* January-February 1975, pp. 21–25.

[27] Natalie Gross, "Glasnost: Roots and Practice," pp. 69–70.

[28] *Krasnaya zvezda,* 22 March 1986 and *Pravda,* 21 March 1987.

[29] *Tyl i snabzhenie,* No. 11, 1986, pp. 17–21.

[30] *Kommunist vooruzhennykh sil,* No. 16, 1986, pp. 18–26.

[31] *Kommunist vooruzhennykh sil,* No. 22, 1986, pp. 9–18.

[32] *Kommunist vooruzhennykh sil,* No. 21, 1986, pp. 17–24.

[33] Cf., for instance, *Krasnaya zvezda,* 19 August 1987 and 23 January 1988.

[34] John Hines and Phillip Petersen, "NATO and the Changing Soviet Concept of Control for Theater War," *Signal* (May 1987), pp. 125–139.

[35] *Voenno-istoricheskii zhurnal,* No. 6, 1987, p. 8.

[36] *Gorbachev and the Struggle for the Future,* memorandum prepared by the Soviet Army Studies Office, Fort Leavenworth, Kansas, for Assistant Deputy Chief of Staff for Operations and Plans for Force Development, p. 10.

[37] *Krasnaya zvezda,* 9 June 1987.

[38] *Kommunist vooruzhennykh sil,* No. 18, 1986, pp. 52–59.

[39] *Krasnaya zvezda,* 17 March 1987.

[40] *Krasnaya zvezda,* 14 January 1988.

[41] *Kommunist vooruzhennykh sil,* No. 5, 1987, p. 17.

42 *Kommunist vooruzhennykh sil.*, No. 24, 1987, p. 13 and No. 1, 1988, pp. 44–49.
43 *Pravda,* 18 May 1987.
44 *Pravda,* 26 June 1987.
45 *Krasnaya zvezda,* 7 February 1988.
46 *Kommunist vooruzhennykh sil,* No. 2, 1988, p. 12.
47 *Sotsialisticheskaya industriya,* 5 December 1987.
48 *Krasnaya zvezda,* 17 December 1987.

The Defence Establishment at your Fingertips

BRASSEY'S BRITISH DEFENCE DIRECTORY

Editor: Air Commodore D H Sutton CBE

The quarterly, computerised, regularly updated directory of senior service and civilian personnel in the Ministry of Defence, Royal Navy, Army, Royal Air Force and NATO. Comprehensively indexed.

An essential reference work for defence equipment manufacturers and all who need up to date and accurate access to the Defence Establishment.

"...very good and fills admirably a long-felt want."
Major General R Britten, CB, MC, Defence Consultant.

"...extremely useful, both in my liaison with MOD and also in briefing senior members of Lucas Aerospace/Industries."
R F Mudge, Marketing Manager for Lucas Aerospace Limited.

"...previous editions of the directory have been excellent."
Robert J Scott, Managing Director, MEL.

Free specimen pages available on request

Subscription Information
Published quarterly
Annual Subscription (1988)   £210.00/US$355.00

 **BRASSEY'S DEFENCE PUBLISHERS**
(A Member of the Pergamon Group)

Orders to: **Brassey's Defence Publishers**, Headington Hill Hall, Oxford OX3 0BW

East European Economies: Facing Inwards, Outwards or Backwards?

JONATHAN EYAL

Dr Jonathan Eyal completed a Doctorate at Oxford University in East European relations and is currently the Soviet and East European Research Fellow at the Institute. This work stems from the research programme which the Institute undertook in the problems of this area and which is supported by the Esmée Fairbairn Charitable Trust.

IN THE spring of 1988, Moscow's traffic police took delivery of its newest and most visible item of equipment: West German-produced BMW motorcycles which will slowly replace the Soviet-made Zhiguli bikes. The reason? The Soviet capital's youngsters are fond of joy-riding around town in Czecho-slovak-built Jawas to which the Soviet Zhigulis are no match. This story, buried amongst the wealth of information now percolating out of the Soviet Union, neatly encapsulates the state of economic relations between the Kremlin and its East European allies: in order to match the quality of some of the East Europeans' industrial items, the Soviet Union has had to resort to Western products. The automotive industry is but a small sector on which Mikhail Gorbachev has set his sights since 1985. His main aim remains much grander: to reduce his country's financial commitment in the region while maximising the political benefits inherent in the maintenance of an empire. The Soviet domination of the region was not due to economic considerations, but rather to strategic and political ones. Possession of Eastern Europe automatically conferred on the Soviet Union a visible accolade of Superpower status, despite the fact that its economy remains that of a Third World state. Establishing communist regimes in the heart of Europe also seemed to confirm Marxist interpretations about the "inevita-bility" of socialist revolution while at the same time assuring the Soviet Union its leading role in the communist movement. Most importantly, Eastern Europe provided a buffer zone of states which would ensure that any future conventional conflict in the continent would not be fought on Soviet territory. The

175

East Europeans therefore provided legitimation to the land of the Soviets as well as considerable strategic benefits. The converse must also be true: the failure of East European governments to attain legitimacy in their own countries—to be accepted by their own populations—and the social and economic decay which resulted from the application of wrong economic priorities was bound to have an effect on the Soviet Union itself.

At the beginning, Stalin's USSR was content to extract the maximum economic benefits from the region while insisting on the strict application of the Soviet model of economic development. Factories were dismantled and transported to the Soviet Union; "joint" companies were established with the intention of exploiting Eastern Europe's not inconsiderable raw materials and heavy industrialisation, and the nationalisation of agriculture was implemented. The results were not slow in coming. For Czechoslovakia, an industrial state before the Second World War, the application of Stalin's orders resulted in a drop in economic rank, rather than an increase. In all countries, the deliberate neglect and suicidal mismanagement of agriculture resulted in massive shortages of food and privations. As long as local regimes aped Moscow and relied on terror to sustain their power, the significance of these economic developments mattered little. The situation changed with Stalin's death and history from the 1950s onwards is essentially one of a Soviet attempt to square the circle: how to maintain regimes favourable to Moscow without social explosions, while at the same time reducing the cost of maintaining the empire to the Soviet Union. All leaders since Khruschev had their own answers and all, until today, failed. Eastern Europe's economic decline could not be arrested, the crises and challenges to local communist regimes continued with alarming regularity and Soviet financial support rose inexorably. Gorbachev's present policies towards the region display some important characteristics. First, they are the most coherent and wide-ranging measures attempted by any Soviet leader since Khruschev. Secondly, Gorbachev has learnt from previous mistakes and is proceeding with his attempts with great thought. Thirdly, the aim is much grander than ever before: Gorbachev's wish is not merely to limit his country's subsidies to Eastern Europe and consolidate its regimes, but also to bind the region to the Soviet Union with the most durable ties possible: those of economic self-interest and cooperation which would not be easily severed in the future. Should these measures succeed, they will amount to a complete overhaul of Moscow's empire in Europe of a kind not seen since the 1940s. Gorbachev is taking great risks and may yet fail. Yet there is little doubt that he believes he has precious few other alternatives.

EAST EUROPEAN ECONOMIES UNTIL 1985

Stalin failed to understand that as long as the East European states were not a Soviet "family of nations", the independence of their regimes and their legitimacy both inside and outside their countries was important. Khruschev grasped this point and created two institutions with the aim of managing Eastern Europe. The first was the Warsaw Treaty Organisation (WTO) which identified a common "enemy" in the shape of the capitalist West in general and the "revisionist" Federal Republic of Germany in particular. The second was the Council for Mutual Economic Assistance (CMEA) which attempted to present an alternative to the Marshall Plan and revive local economies. The real reason behind Khruschev's efforts was the hope of institutionalising normality in the region and channelling relations with Moscow towards more constructive paths. The attempts ultimately failed and both Moscow and the East European governments ultimately opted for a policy of political legitimation by tolerance: a slow but inexorable rise in the standard of living of the local populations, coupled with the regimes' predictability and use of terror as a means of social control, reinforced by impermeable frontiers should, it was argued, drive home the point that the regimes were there to stay, that they did not ignore the needs of the population but also would not allow a viable alternative. The region's nations would be bought into submission by a mixture of carrot and stick which would take time but which avoided radical initiatives and unnecessary dislocations. As the examples of Hungary, Poland and the GDR show, how that mixture of policies was applied did not concern Moscow greatly, as long as it was sure that basic rules were not broken: the control of the party over all walks of life and adherence to the alliance with the Soviet Union were beyond question. One country which perceived the pecking order of Soviet priorities and used it to its advantage was, of course, Romania. It is also Romania which currently presents the best example of the ultimate failure of Soviet policy to manage its alliance through a mixture of *diktat* with national independence without laying solid foundations based on joint economic and security interests. It was a half-way house which produced little integration, perpetuated inherently unstable regimes and still portrayed the Soviet Union to the world as the colonial power. Despite the "socialist internationalism" of the Warsaw Pact, everyone in Eastern Europe knew that the only danger to Eastern Europe came from the East and not the West. And, despite increasing Soviet subsidies, most East Europeans refused to show any gratitude or greater loyalty to Moscow.

It would be unfair to suggest that the Soviet leadership under Brezhnev was not aware of these facts. Nevertheless, the East Europeans' ingratitude—the Kremlin hoped—was a phenomenon limited to local populations and not shared by the ruling elites who still had every interest in clinging to Moscow's coat-tails. Brezhnev still genuinely believed that long-term economic growth could be achieved and that this would still result in greater political cohesion without embarking on unacceptable political reforms. This strategy, which grew directly out of the invasion of Czechoslovakia in 1968 entailed the pursuing of *detente,* which brought the region financial and technological assistance from the West and allowed the expansion of their economies, coupled with half-hearted attempts at integration within the CMEA. Some of the region's economies did indeed register spectacular industrial growth rates which did promise full employment and a seemingly unstoppable rise in local standards of living. Hungary is the most cited example of a country which benefited from the ability to shape its own economic policies while not straying outside the bounds of permissible behaviour but the GDR equally adopted a different economic mechanism in the 1970s. In fact, the Brezhnev solution unravelled by the end of the 1970s with more ominous signs. First, the economic fruits of *detente* did not lead to radical economic approaches and corrections in previous economic priorities. It merely perpetuated structural deficiencies and bought Eastern Europe more time. Furthermore, as a direct result of *detente,* East European economies were not shaping the future prosperous socialist community but were growing further apart as each country attempted to capture foreign markets, usually at the expense of the other. This, in turn, resulted in the worst possible situation from the Soviet Union's viewpoint. On the one hand, no genuine alliance was achieved. On the other, economic prosperity could not be long lasting and ultimately not merely the loyalty of East European societies was in doubt, but the loyalty of local regimes could equally become questionable.

Poland is the prime example of this failure. The Polish government accepted massive credits from Western banks but handed them over to the bureaucracy. This bloated body of party and state activists proved incapable of absorbing new technology or directing resources to the industries of the future—an essential step, for most East European states aimed to repay their debts through increased exports. White elephants sprung up everywhere across the region, industrial production still proved unable to compete in open markets, hard currency could not be obtained and debts mounted. The party was over by the end of the 1970s and resulted in the explosion in Poland, but Hungary and Romania were also on the brink of economic collapse. The Soviet Union had two options. The first was to allow the East European economies to cope as best they could with the situation, partly by restricting local consumption and imports in order to repay their debts and partly by diverting more products to exports. This could risk further social crises. The alternative for Moscow had was to step in and redeem the East Europeans' debt. That would have proved difficult for the Soviet economy, which was itself under severe strain. The solution which Brezhnev adopted was typical of the man's character and lay somewhere in between these two options. Assistance, mainly in the form of raw material supplies, was provided on a selective basis. At the same time, the East Europeans were expected to undertake their own belt-tightening policies. Nevertheless, Moscow gave no indication of having perceived the irony of the situation: yet another of its policies was falling apart and the problem was back where it started. Governments could not assure prosperity and at the same time could not fulfil their societies' political aspirations. Eastern Europe remained an economic liability for Moscow and at the end of Brezhnev's rule CMEA countries owed almost US$54bn in hard currency.[1] In 1973, Eastern Europe ran a surplus of US$600m in its trade with the USSR. By 1980, the deficit stood at US$1,600m and was growing rapidly. Moreover, Soviet subsidies to the region reached a staggering US$18bn by 1982. Clearly, this situation could not be allowed to go on for much longer, yet, as a sick leader succeeded a dying one in the Kremlin, little was done until 1985, although the problem was clearly on the minds of every Russian ruler. Alarmed at the pace of East German-West German economic and political contacts, Moscow ordered East Berlin to place a quarantine on inner-German relations. Worried about the fluid political situation in Poland, the Soviet leadership assented to more economic help. Suggestions abounded but most of them involved even greater cash just at the time when the Kremlin could ill afford more financial commitments. This was Gorbachev's inheritance.

GORBACHEV'S SUCCESSION

No easy solutions dictated prudence and Gorbachev's rule was marked, in its first year, by a period of stalling. The new leader showed that he was aware of the lesson that there was little to be gained from handling social and economic crises in Eastern Europe and everything to be lost. Every reform in the Soviet Union can have an impact in Eastern Europe and usually not in the direction Moscow expected. Stability was therefore the prerequisite for future change and Gorbachev's first 18 months in power were devoted to that. He personally attended East Germany's 11th Communist Party Congress and congratulated Erich Honecker on his country's achievements at a time when Berlin felt particularly insecure about his attitude to inner-German relations. Similarly, in June 1986, the Soviet leader attended the Warsaw Pact summit meeting in Budapest and assured the Hungarians of his support for their own economic mechanism. Indeed, Gorbachev's increased emphasis on more frequent consultations within the Pact pleased everybody, although the GDR, Czechoslovakia and Romania in particular suspected that the price of greater consultation might be greater integration and waited for Moscow to name that price. The price was slow in coming for Gorbachev appeared to remember another lesson. If any reform is to succeed, the cooperation of local regimes should be achieved first. As Khruschev's experience showed, there was little point in alarming local leaders with a series of economic demands without being prepared to listen to their grievances and make a determined effort to address them. That was precisely what the new Soviet leaders set out to do.

EAST EUROPE: FROM TEACHERS TO PUPILS

The East Europeans started enjoying greater freedom in their foreign affairs. The tactic worked well in presenting yet another facet of the new leader's apparently benign intentions; Gorbachev was prepared to allow this as long as his long-term objectives remained unimperiled. The ban imposed on communist relations with China was lifted as part of Gorbachev's aim of normalising relations with his Far-Eastern neighbour and the East Europeans were encouraged to lead the way. Honecker travelled to Beijing, relishing this apparent Soviet acknowledgement of his country's relative primacy in Eastern Europe.[2] The East Germans were also congratulated for signing an agreement with the Federal Republic's Socialists on the creation of a "nuclear-free corridor" in Central Europe.[3]

The GDR and Hungary felt genuinely flattered at persistent hints that the Soviet Union was actively studying the possibility of implementing their own methods of economic reforms. Their joy and pride was certainly premature: hints that the Kremlin looked with interest upon the GDR and Hungary's economic mechanisms dated back to Andropov's rule and were revived in 1986 by Abel Aganbegyan, one of Gorbachev's closest economic advisers.[4] It is unlikely, however, that the possibility of copying East European economic reforms was ever seen as a viable proposition. While it remains undeniable that the Soviet Union's new leaders did devote some attention to the GDR's and Hungary's economic experiments, this was probably due more to intuition rather than a serious economic analysis.[5] The fact was that the GDR forged ahead to the top of the East Europeans' economic performance during the 1970s and Hungary's reforms succeeded in creating greater prosperity while assuring the population with a reasonably decent standard of living and increased personal consumption. Both results were therefore enticing. What probably attracted some leaders in the Kremlin even more was the fact that the GDR's economic performance was not accompanied by a dilution of the party's rigid control over society or by a relaxation in central planning. In fact, Berlin proceeded to liquidate the last relics of private enterprise just before launching its economic reforms in the early 1970s. In Hungary, some private enterprise was allowed but, again, this did not appear to result in great dislocation or any diminution of the party's role. Surely, some members of the Politburo felt, the GDR and

Hungary were living examples of the possibilities still available to central planners. Unfortunately for them, they were not, for two reasons. First, the reforms introduced in the GDR and in Hungary could not be translated or applied in the Soviet Union. Secondly, even these reforms were slowly grinding to a halt and threatened greater dislocation in the future.

The GDR model owes its existence to the *Kombinat* structure, which united producer, supplier and seller in approximately 157 vertically integrated trusts. This achieved specialisation and concentration in certain branches of the economy and was applied to a society which was historically and linguistically homogeneous and was used to discipline and the ethos of hard work.[6] In short, success in the GDR relied on all the basic factors which the Soviet Union did not possess. In any case, the Soviet Union did not enjoy the fruits of this success. In 1973, the USSR exported two billion roubles worth of goods to East Germany, of which a quarter was heavy machinery and 10 per cent oil. Ten years later, Soviet exports to the GDR did indeed grow almost four-fold, but no less than 60 per cent of the total was in the form of gas and oil.[7] Increasingly, the Soviet Union was relegated to the status of a supplier of raw materials, but that was not all; the GDR in return sold to Russia only the goods which could not be sold in the open markets of the West for hard currency. From Gorbachev's more radical viewpoint, it did not matter that the GDR's example made command economies and the socialist system more respectable and even workable; what mattered very much was the fact that as the years went by, the CMEA economies did not become compatible with each other but competitive instead. They all competed in the production of some items but especially in the penetration of hard-currency markets. Only success in these markets assured them a steady opportunity to buy necessary raw materials and technology which the Soviet Union was unable to supply. The Soviet Union was at the bottom of the export priorities list for all of them, both in quality and quantity. With the possible exception of Bulgaria which prudently and—in retrospect in the late 1980s also successfully—continued to rely on the Soviet Union for most of its finished goods markets, all other East European states delivered their lowest quality goods to the Soviet Union. There are fairly clear indications that despite all these facts, Gorbachev genuinely studied the applicability of Hungarian-style reforms and their attractions to him were quite obvious. Given his own personal career background, he found Hungary's ability to increase agricultural production and assure a sizeable export surplus of food particularly enticing but it is reported that members of his party's Central Committee were strongly against applying them,[8] although the debate lasted a long time.[9] The conclusion therefore was that East Europe could not teach the Soviet Union any useful lessons and Gorbachev's attention turned to the economic difficulties in the region the moment it became clear that even the two "model" East European economies of Hungary and the GDR were faltering rapidly. Hungary was jolted by the general drop in credit from the West in the wake of the Polish debt crisis; its inefficient industries were nearing breaking point and fairly radical reforms were still needed; and the GDR's economy was also contracting. The alarm bells came soon enough—even if Berlin still refused to recognise them—in the shape of sharply lower growth rates for key industries, such as coal, energy and heavy machinery. While in the 1981–86 period the coal and energy sectors grew by an annual average of 8 per cent, this dropped to no more than 2.1 per cent by 1987 and the heavy machinery sector's growth rate halved over the same period. This was coupled with stagnant exports and increased competition from West German producers who themselves encountered difficulties in exporting to the Third World and who increasingly regarded their own country as their only hope for growth. The GDR's combines did not remove the impediments to technological innovation. Being responsible for a great part of scientific research, they tended to retain any achievements instead of allowing these to be shared among the economy at large. Researchers were still confined to strict and centrally determined targets.[10] To put it differently, there was, after all, a price to be paid for maintaining rigid central planning while striving for efficiency and, in the absence of radical changes, the East German economy could still follow the path of Czechoslovakia's: an industrial country which became an economic dinosaur. With Romania on the point of starvation and Poland a basket-case economy crippled by foreign debts and an undisciplined labour force, the time for learning had passed; it was now Gorbachev's turn to do the teaching and the honeymoon with Eastern Europe was at an end.

The problems which Gorbachev intended to tackle were very great. His general and ultimate goal is to assure stability and prosperity; to transform the region into a genuine alliance.

This requires social and economic reforms and would encounter great opposition. His middle-term goal is to achieve all these with the political and financial commitment to his own country. His short-term solution was to begin integration in specific areas between different East European countries. The problems which he had to face were both related to the presentation of these new priorities and their actual application, for the CMEA was and continues to be riddled with difficulties which hinder genuine cooperation. In command economies, prices for all products are usually decided not by supply and demand or costs of production but by administrative fiat. There are usually two broad price bands: one for the producer, the so-called wholesale price, and one for the consumer, the retail price. The structures of these two prices are usually unconnected because of the different levels of subsidisation and taxation. Thus, the essential basis for trade cooperation was not available. Furthermore, there was the question of currencies. Within the CMEA's transactions, the so-called "transferable rouble" was introduced in 1964. At that time, that rouble was equivalent to a Soviet domestic currency unit, but the rate has remained unchanged ever since. In practice, the transferable rouble was no currency at all but a book-keeping device. It is issued only in response to imbalances in bilateral trade within the CMEA and sees no circulation whatsoever in any of these countries. As a consequence, most of the trade between CMEA states was reduced to the level of barter. It was a barter system which did not benefit the Soviet Union. Most of the USSR's exports consisted of energy products and other raw materials. In return, Eastern Europe, heavily dependent on sometimes cheap and always reliable supplies of oil, paid with products of the lowest possible quality after having satisfied all possible orders in hard-currency markets. Hard currency promised access to advanced Western technology and, by being freely convertible, also allowed local governments the flexibility to chose their own future economic priorities. Thus, Gorbachev lacked the means to establish better economic collaboration quickly. The CMEA lacked proper prices, proper currencies to conduct trade and the political will to do this. Gorbachev nevertheless had considerable advantages. First, the sharp downturn in trade with the West and the performance of East European economies made them more dependent on Soviet supplies of raw materials although economic dependency could not always be translated into political dependency, as the example of US-Israeli relations shows. In the 1970s, the Soviet Union also experienced problems with translating the East Europeans' reliance on its economic aid into greater political loyalty, but Gorbachev was better placed to correct this. The second advantage which the Soviet leader enjoyed was that, due to the five-year moving scale which the CMEA operated in fixing the price of oil, Soviet oil started becoming cheaper. The Soviet leader could thus appear to be magnanimous by responding to his allies' needs and improving the terms of trade, while at the same time, increase political demands. As even the GDR's foreign trade turnover dropped by 10 per cent in 1987 alone, the advantage of relatively cheap oil was considerable.[11]

THE FIRST SHOCK

At the 41st Special Meeting of the CMEA, the Soviet Prime Minister, Ryzhkov outlined the Soviet priorities very clearly: the "principal feature of the current period" he said, was that all East European countries had "essentially exhausted the possibility of extensive growth and where these opportunities still exist, they must be used more efficiently with the maximum possible return". The point was quickly driven home: at the beginning of November 1986, the 42nd CMEA session was held in Bucharest and from the first day, it was clear that the Soviet position has changed radically. The Poles and East Germans treated the delegates to long expositions of their "achievements"; Moscow's representative curtly replied that more coordination of these "achievements" was needed, in the form of "stable economic, scientific and technical ties between states and the effective utilisation of the possibilities of social and economic integration".[12] The plans which were unveiled went further than anything the East Europeans expected, mainly because they underestimated Gorbachev as a leader. From subsequent pronouncements, it is clear that Gorbachev learnt the lessons of Khruschev's integration failures thoroughly. Khruschev sought to establish a "division of labour" in the socialist community. Particular countries, with strong industrial traditions and good infrastructures were allotted the tasks of industrial production Others, such as Romania, were given the job of supplying the socialist community with raw materials and food. The idea was misconceived, for it attempted to impose a unified plan on the entire socialist community. Secondly, it was offensive to the East Europeans' national pride and to their own economic aspirations. Most of the region's leaders, graduates of the Stalinist school of economics, still believed that more steel smelters, more heavy industry and more furnaces was what communist rule was all about. In the context of Eastern Europe, industry represented progress and the bigger the industry, the better it was; to be condemned as an agricultural country would entail remaining the backyard of Europe. Romania's dissention within the Warsaw Pact could be partly traced to Khruschev's harebrained scheme. Thirdly, the plans of the 1950s and their further refinement under Brezhnev did not draw the conclusions from the fact that the East European allies were themselves operating command economies in which export priorities were not dictated by market prices but by administrative fiat. Thus, from Gorbachev's viewpoint, it was not enough to agree with the governments concerned on a new programme of cooperation since this was likely to remain a dead letter as long as the East Europeans still controlled all the manpower, the allocation of resources, the production facilities and decided the export markets. For genuine coopera- tion, which would be beneficial to the Soviet Union, it was necessary to break this monopoly of individual governments over their key industries. That in itself was a bigger task that Gorbachev could handle and involved a whole host of social and political problems, among which the role of the party as a guiding force in society is the most obvious one. A middle way had to be found between hurting the East Europeans' sense of pride in their own achievements and attachment to nationalism, the continued operation of a command economy thoroughout the bloc and the desirability of greater economic cooperation and specialisation. This may appear a tall order but Gorbachev met it by looking at East European economies as a mirror of his own. The East European leaders could be compared to the bureaucratic elements in the Kremlin; their economic troubles were similar to those of the Soviet Union. The solution, in the short run, could not be found in dismantling the command economy, but rather in improving on its most glaring failures. The Soviet leader decided to insist on cooperation between his country and his allies at the enterprise rather than state level. Once contracts were signed, once targets had been set, it would be more difficult for any government to frustrate them. It was precisely this insistence which aroused the greatest apprehensions in East Europe.

From the East European viewpoint, this technique implied integration not with a stronger economy, but with a weaker one. Secondly, the East Europeans suspected that such cooperation would be dictated by Moscow's economic priorities and would not address their own needs. In this respect, the fact that the Kremlin was insisting on collaboration in the electronics and machine-building industries was particularly worrying, since the USSR required help in precisely these fields.[13] As Soviet commentators asserted flatly, there are five main priorities, all in the area of research and all closely related to machine-building industries. These are: conversion to electronic components, integrated automation of production lines which would include the introduction of more flexible manufacturing systems, nuclear power engineering, development of new materials and biotechnology.[14]

Finally, cooperation at the enterprise level is inherently more difficult to control nationally, which is precisely why Gorbachev preferred it. In the absence of real prices, convertible currencies or any other accurate economic indicators, the performance of such joint enterprises would always be open to manipulation. For all these reasons, the East Europeans put up stiff resistance at the Bucharest CMEA meeting but the indications that Gorbachev was serious about his intention came soon enough: within days of the Bucharest meeting, all the leaders of the Warsaw Pact were summoned to Moscow[15] for what were officially termed as "frank" discussions.[16]

We still do not know what was decided at that meeting, but the 1987 Soviet State Yearly Plan did state that there would be a "considerable development" in cooperation within the CMEA, especially in the field of "direct production links".[17] The opposition of the East Europeans nevertheless continued, for the GDR's 1987 plan spoke pointedly of cooperation on the "mutual advantage" principle between *state* economies.[18] Gorbachev expected a united front of opposition and was well prepared for it. In late November 1986 therefore, the Soviet Union shifted from a general emphasis on cooperation to attention on specific branches of industry in particular East European states. The GDR came in for most of the attention during the regular meeting of the USSR-GDR economic commission in December.[19]

Honecker was nevertheless confident that he could withstand this well. He was wrong, for Gorbachev played his second
card.

THE SECOND SHOCK

In January 1987, the Soviet leader introduced another
element to the argument: that of political reform. The Soviet
leader's speech to his party's Central Committee Plenum on 27
January 1987 was indeed remarkable. He admitted the failure
of previous Soviet economic policies; he berated the bureaucracy and blamed high party officials for the situation; he
attacked aspects of central planning for stifling initiative and
called for greater investment in consumer industries. Ultimately, he blamed nepotism and promotions within the party
on account of personal relations, rather than merit. The
speech sent a seismic shock throughout Eastern Europe.
Gorbachev could afford to criticise Brezhnev's inheritance for
it was not of his making; the East Europeans represented the
system which the new Soviet leader was now attacking.
Hungary and Poland, both already embarking on efforts which
sought to mobilise forces outside the party to the task of
economic reconstruction enthusiastically embraced the
speech. Romania's Stalinist leader, Nicolae Ceauşescu, reacted
violently to Gorbachev's criticisms and vowed that they were
inapplicable in his country. The GDR and Czechoslovakia were
stunned by the criticisms of policies which they had faithfully
applied for years. The argument between Moscow and the East
European capitals was no longer one about economic reforms
and cooperation but about the survival of the regimes
themselves. Gustav Husak's *raison d'etre* was the application
of Brezhnev-style government and that was now rejected by
the mentors themselves. Gorbachev may yet regret the
implications of his reforms in Eastern Europe for there is little
doubt that they have tended to destabilise local regimes. At the
same time the immediate effect, from the Soviet Union's
viewpoint, was rather beneficial. The spectre of an East
European united opposition front to Soviet economic demands evaporated within weeks of the speech. If the real
question was one of the leaders' actual survival, economic
cooperation was the least common denominator on which they
could agree, provided Moscow could be persuaded not to insist
on the application of social and political reforms in the region.

The breakdown was amply illustrated by the debates which followed. Romania's Ceauşescu desperately tried to resurrect a united opposition front with the support of Czechoslovakia and the GDR; unfortunately for him, his country was already in such an economic mess that no-one was particularly interested in what he had to say. As both Berlin and Prague realised, supporting wayward Romania could be politically unwise and the benefits which could be derived from such support would always be derisory. The debate in the GDR and Czechoslovakia, however, changed from economics to political reforms. Czechoslovakia's tame trade unions were encouraged to continue with "democratic centralism", a euphemism for total political control over all walks of life[20] and Honecker told his people that he had "not the slightest reason to conceal the course" which his country had followed in the past.[21] Brave words indeed but not the sort of pronouncements which would push the Soviet leader off his course. Within weeks of these events, Soviet Foreign Minister Shevardnadze visited both Berlin and Prague and pointedly told the press that he came on "the orders of Comrade Mikhail Gorbachev" to discuss economic matters. The Hungarians sat on the sidelines, shedding crocodile tears about the fact that Eastern Europe "did not evaluate the changes in the USSR in an identical way". That, for the moment, suited the Soviets.

In April and May 1987, Gorbachev paid visits to all his difficult allies: the GDR, Czechoslovakia and Romania. In all, he was careful to emphasise that each communist state would be allowed to implement its own policies; in each one, however, he insisted that his country would no longer accept shoddy goods in return for energy exports and repeated his view that only cooperation at the enterprise level would succeed. The main aim, for the moment, remained economic cooperation and Soviet officials were quite happy to point this out. Professor Albakin, the Director of the USSR Academy's Institute of Economics and one of the main architects of the Soviet reforms, repeatedly followed Gorbachev's travels with assurances that the economic mechanism which the Soviet leader was attempting to replace is one which has outlived its usefulness but which nevertheless was essential in its time. The implication was that the past need not be denigrated as long as the present economic mechanism is reformed. The fact that economic cooperation took priority over political considerations was starkly obvious for all to see. Throughout the post-

war period, tiny Bulgaria remained the Soviet Union's most
reliable ally. Its leader, Todor Zhivkov, pursued, chameleon–
like, every twist of Soviet policy and applied it immediately in
his own country. It must have been particularly galling to him
and interesting for the other CMEA members that in May
1987, it was Bulgaria which was openly criticised by the Soviet
press for engaging in merely "superficial" contacts with the
USSR and for being "unadventurous" in the goods it wanted
to produce.[22] The gloves were quite clearly off; Moscow was
this time determined to achieve what it wanted from Eastern
Europe. Political reforms were not always the most important
and certainly not the most immediate priority. When the
conflict between hardliners and reformers within communist
parties became especially acute, as was the case in Czechoslovakia, the Soviet leadership supported a compromise rather
than a radical political solution: Milos Jakes was appointed to
succeed Husak. With hindsight, it is clear that this was part of
Gorbachev's greater strategy: economic reforms would bind
the region together and ever closer. When these are applied,
and when cooperation is in full swing, there would come the
time for political reforms as well which the Soviet leader still
believes are absolutely essential. However, as long as the
political situation is fluid, it would be better to uphold the
leaders already in power, provided that they can deliver the
goods and economic cooperation required. Deliver they will,
for they understood that Moscow was now aiming at higher
stakes and that their survival depended on accepting the Soviet
demands. Within two years, Gorbachev has managed to
achieve what no other Soviet leader has done before him:
impress on the East Europeans that the USSR would not
support bankrupt economies forever; limiting their room for
manoeuvre and at the same time appearing to the world as a
more enlightened, liberal and lively personality. The East
Europeans were told that they are free to run, but back into
the camp rather than outside it. They were assured that their
independence would henceforth be respected, but that closer
economic cooperation was a necessity. Ultimately, most fell
into line.

DEVELOPMENT OF COOPERATION

The decree of the Soviet Council of Ministers on 19 August
1986, paved the way for 21 government ministries and

departments and 76 of the largest enterprises to have direct access to foreign markets. The Kremlin claimed that in 1987, such firms accounted for 20 per cent of the Soviet foreign trade turnover, and more than 65 per cent of the country's engineering exports in particular.[23] These companies were given very broad powers to trade with their East European counterparts. With Bulgaria, it appears that cooperation was concentrated on the production of transport and manufacturing machinery, such as motor trucks and electric hoists which comprise almost 25 per cent of all Bulgarian machinery exports to the Soviet Union. A joint venture, Avtoelektronika, works on electrical motor equipment and combines Soviet production with factories at Plovdiv. With the GDR, the majority of cooperative ventures centre around robotics and chemical processing plants. East German planners have admitted that their country's exports of microelectronics to the USSR are set to reach a record 4 billion marks by 1990 alone, conducted through 120 joint production agreements.[24] From Hungary, the main area of expansion was in the manufacturing of motor cars, of which the Raba plant in Gyor supplies rear axles for Soviet buses while the Soviet Union delivers power units, front axles and pumps. In addition, 6,600 buses would be bought from Hungary in 1988.[25] A joint Soviet-Hungarian venture, Micromed, develops and sells medical equipment in the city of Esztergom.[26] In Czechoslovakia, the main areas of cooperation are in specialised steels, chemical industries and pig iron and, again, trade is planned to rise rapidly.[27] Polish/Soviet cooperative ventures account for 25 per cent of all joint trade and agricultural tools, ships and navigation equipment as well as manufacturing tools are scheduled for special attention.[28] With Romania, a different technique was applied which nevertheless still underlines Gorbachev's ability to draw his allies together through economic cooperation while refusing to be drawn into long political arguments. Ceauşescu's attraction to the West was based on his claim to be independent from Moscow and making himself useful as a negotiator. He mediated in the Middle East and between the two Superpowers; he acted as a diplomatic channel between the USSR and China. With the appearance of Gorbachev, all these venues disappeared. The Soviet Union conducts summit meetings with the United States, talks to Israel with a view to establishing an international conference and negotiates with China. President Ceauşescu was simply marginalised and exposed for what he always was: a ruthless dictator, ruling a marginally useful country. With his charm gone, his economy in tatters and his population on the brink of starvation, Ceauşescu would ultimately have to come back into the Soviet fold. Gorbachev stipulated his conditions, which entailed greater cooperation in oil extraction in Siberia as well as Romanian deliveries of what were always hard currency goods, such as food and clothing. By being persistent and cool headed, Gorbachev solved a problem which bedevilled all his predecessors. Romania's "independence" is a thing of the past and the country is kept alive by Soviet support. By refusing to rise to Ceauşescu's bait, Gorbachev has exposed the dangers inherent in moving away from the socialist camp. As long as Ceauşescu remains in power, he will be the best advertisement for what happens to wayward allies. Romania is kept afloat by the Soviet Union mainly because the Kremlin would not wish to see a political breakdown there; at the same time, even in this particular case, Gorbachev insisted on exacting his price, and goods which do not meet the specified quality controls are returned to Romania. The Soviet leader has displayed equal flexibility, in other areas of concern to his CMEA partners.

PRICES FOR CMEA ECONOMIC COOPERATION

The Soviet Union has claimed that prices for CMEA cooperation are based on the "world market". We have little indication whether this is so, but we do know that the main problem does not reside in pricing the finished product, but rather in the price decided for labour, expertise and spare parts utilised in this joint production. On this score, the Soviet Union admitted that "the situation is less favourable", since no comparable world prices are easily obtainable.[29] By mid-1987, over 400 Soviet enterprises had direct ties with East Europe and Moscow wants this expanded to 700 "in the near future". This is particularly the case in the machine-building sectors, in scientific R&D and the design of new manufacturing processes. Most of the cooperation involves items on which the pricing mechanisms are by no means clear. The Soviet leadership had one opportunity on 1 January 1987 to reform the pricing merchanism in the CMEA trade but this was not seized upon when new legislation came into force. On the contrary, the USSR insisted on establishing individual prices for particular items supplied through direct production links although it allowed one concession to the East Europeans by accepting that one price would not be taken as a precedent in the formation of future prices for similar products.[30] In effect, the debate on price formation was postponed to the future, while cooperation is enforced in the present. Moreover, in the absence of firm rules on prices, Soviet enterprises were allowed to negotiate their own contract prices. Since much of the cooperation at this stage is politically motivated, the Soviet enterprises probably have the clout to dictate their own prices. There is, however, evidence that the haggling continues; Alexei Antonov, the minister responsible for most of the CMEA negotiations had indicated that his country insists on taking into account the "larger scale of production" which the availability of the Soviet market entails.[31] East Europeans have indicated their concern on this point many times but their options are limited.[32] If anything, the Soviet pressure has intensified, especially in the area of production quality. At least in Bulgaria, quality control is assured by the physical presence of Soviet inspectors who make sure that what is shipped to the USSR is at the specified standard.[33] It is quite probable that the additional costs implicit in this stringent inspection may be passed on to Bulgaria.

RESEARCH AND DEVELOPMENT

The Soviet Union has insisted on cooperating in R&D, claiming that duplication of research within the socialist camp has resulted in "losing" between five and seven billion roubles a year.[34] Losing on what? Presumably, Soviet economists were referring to money spent on researching the development of similar products in many East European countries. Research and development in most countries is not coordinated. However, because of the presence of proper price mechanisms, workable patents protection legislation and flexible application of technological discoveries in market economies duplication is reduced and different branches of Western economies can rely on each other. The one area where duplication in R&D exists is in the defence field, precisely where the forces of market economy in the West apply less. In East European command economies, the duplication is massive, but this is less because of nationalist policies and more a systemic factor. By insisting on direct cooperation, the Soviet Union is attempting, for the moment, to square the circle: to retain command economies while reducing their disadvantage in R&D. The fact that Moscow insisted on cooperation in an area in which it was weak could hardly be reassuring from the East Europeans' point of view.

CONVERTIBLE CURRENCIES

The discussion on convertible currencies displayed similar trends: the Soviet Union promised greater flexibility in the future while pushing for greater integration in the present. Despite the fact that the East Europeans regularly demanded the establishment of a convertible currency, or at least concrete steps towards this goal, at the 43rd session of the CMEA in October 1987, the decision on the full convertability of the transferable rouble was again postponed, although the member states agreed to a Soviet proposal that national currencies could be made convertible on a bilateral basis and they also accepted the Soviet-inspired need for a "division of labour among them in the period of 1991–2005.[35]

A programme aiming at full convertibility of all currencies would be phased in over a ten-year period beginning in 1991.[36] Although joint ventures would be the first to implement the new system,[37] the rouble is still over-valued by about 400 per cent against the US$ and thus continues to be a most unreliable unit of accounting between states.[38] The debate on convertible currency revealed another important aspect of Gorbachev's policy: his impatience with formal rules of conduct within his alliance. Important decisions of this nature were usually taken on a unanimous basis. Nevertheless, and despite the fact that the East Germans and Romanians abstained, the decisions were still carried out and the plan is certainly implemented: Czechoslovakia and the USSR agreed on the convertibility of their currencies in February 1988.[39] This merely means that cooperating enterprises would be able to conduct their accounting in their own national currencies and does not solve the problem. Commercial credits to these enterprises would still be accorded on the basis of what Prague termed an "appropriate" exchange rate, without elaborating [40] and it was subsequently admitted that much more liberalisation of currencies would be required to make the agreement meaningful.[41] Thus, not only was the debate on currency convertibility postponed while the East Europeans' criticism was deflected, but the ability of one country to block changes was arbitrarily removed and the rules of the game changed. Gorbachev was in fact making a virtue out of the East Europeans' dissention. Unlike previous Soviet leaders, he abandoned the chimeric hope of the "socialist camp" although he persists in paying his perfunctory eulogies to the idea. The practice, however, is of a Soviet Union that deals directly with each East European country by ascertaining the local problems and arranging a mutually satisfactory trade. In the words of the Soviet leader:

> The only important thing is that any country's lack of desire or interest to participate in a project should not serve as a restraint on others. Anyone who wants to participate is welcome to do so; if not, one can wait and see how the others are doing. Every country is free to decide if it is prepared for such cooperation and how far it is going to be involved.[42]

He could have added that any country which does not want to join projects dear to the Soviet heart, would have to see for itself how it could manage without Soviet help. Integration does, therefore, take two forms. On the one hand, there is bilateral cooperation in technological and R&D fields. On the other hand, there is multilateral cooperation in areas such as raw materials pricing and marketing. And finally there is Soviet economic help on a case-by-case basis, such as Romania's. This mixture proved irresistible for all the region's states.

FORMALISING THE NEW INTEGRATION

Gorbachev was not content merely to implement his new integration policies, but he is also attempting to have them recognised as such by the West. This is exemplified by Soviet efforts to regulate relations with the European Community which was previously unrecognised as a separate entity by the CMEA. Despite the fact that Gorbachev persists in denying any attempt at transforming the CMEA into a supernational organisation,[43] he has hurried along negotiations with the European Community, which he hopes would lead to the recognition of the two bodies at the same time.[44]

Should this agreement herald closer cooperation between the two organisations, it is bound to have an effect on the GDR in particular; East Germany already had free access to the European Community through unhindered inter-German trade but, according to the conditions attached to the mutual recognition of the two trade blocs, its status may be changed. And, of course, the recognition of the CMEA as a separate entity also entails, by implication, the acceptance of the Soviet Union as the leader of its bloc, the acceptance of the division of Europe according to different economic systems and more, rather than less integration within the CMEA. Although the Kremlin hopes to improve on its US$22.6bn exports with the European Community through this accord the most significant aspect would be symbolic, in reinforcing Moscow's position above that of its East European allies.[45] Significantly, the Soviets always refer to an economic community and foreign affairs specialists in Moscow are at pains to point out that their country would not recognise a united European political entity, but merely an economic organisation which could be compared to the CMEA.[46]

It is precisely for this reason that some West European countries were prevaricating, although Chancellor Kohl was reputedly eager to conclude an agreement by the time his country's presidency of the community expired in the summer of 1988.[47] His wish was fulfilled and the Soviet Union had external as well as internal options to direct the CMEA's further integration.

CONCLUSIONS

The CMEA has undergone major changes in the last few years. This is not, in itself, novel, for most Soviet leaders have attempted to tackle the seemingly intractable problem of Eastern Europe. What makes Gorbachev's efforts stand apart is their persistence and their durability. Should they be successful, they would remove, at a stroke, the ability of many East European states to conduct an independent economic policy and, with economic integration, closer political coordi-nation would come too. Some East European governments are at the moment engaged in industrial collaboration in the hope that Gorbachev's insistence on political reform will subside. It will not; the demands for political reforms have been post-poned until economic integration succeeds. Others, such as Hungary, embrace political reforms while hoping to contain closer economic cooperation. For Gorbachev, all this ulti-mately matters little; by appearing receptive to East European complaints—while assuring his allies of his best attention to their needs and by recognising their diversity and national requirements—Gorbachev may be forging, for the first time since 1945, a true alliance based on at least true economic requirements. The Soviet Union certainly shows no intention of relaxing its demands on cooperation, as Talyzin, the Chairman of the USSR State Planning Committee stated when unveiling the 1988 Soviet plan.[48]

At the same time, Moscow is also more open about its assistance to the region, especially in the field of oil deliveries. During the signing of the trade protocol with Czechoslovakia in November 1987, which heralded a great increase in bilateral trade, Moscow pointedly announced that Prague would benefit next year from deliveries of oil which would be "11 per cent to 13 per cent cheaper" than in the previous year.[49] The Soviet Union has also encouraged cooperation between the East Europeans. In 1987, for instance, a full 35 per cent of all Bulgarian-GDR trade was in goods jointly produced; in 1988, this is scheduled to rise to no less than 85 per cent.[50] Cooperation is also proceeding apace in nuclear energy programmes and sharing of electricity.[51] The CMEA's own administrative structure has been revamped. Reports speak of a 25 per cent staff reduction at the organisation's headquarters in Moscow with responsibilities passed over to certain joint enterprises.[52]

The intention is clear: soon, the most plausible way for cooperation will be, of necessity, through joint production ventures. To be sure, most of the problems still remain unsolved. The problems associated with currency convertibility and price formation are still outstanding. Nevertheless, the Soviet pressure will continue, if only because joint ventures with Western companies are slow in materialising despite strenuous efforts.[53] Some CMEA members with most to lose out of this joint cooperation, such as Hungary, are currently putting up some resistance. In April 1988, an unusually great number of meetings at high level between the Hungarians and Soviets were reported. Prime Minister Ryzhkov visited Budapest at the same time to meet the apprehensive Hungarians,[54] and officially mentioned "difficulties" in bilateral trade.[55] Despite the admitted sharp exchanges of views between the two countries, and the Hungarian refusal to cooperate in the production of a new Soviet motor car the fact still is that Gorbachev's "charm offensive" has left Eastern Europe with few alternatives.[56] Hungary, more dependent on export than any other country in the region has suffered badly from its temporary inability to conclude a separate agreement with the European Community[57] as the Community was still deliberating a wider agreement with the Soviet Union first.[58] All in all, it would be difficult to escape Gorbachev's embrace.

By not allowing such difficulties to prevent the application of his ultimate aims, Gorbachev is forging an alliance which may, one day, look upon the Soviet Union as the genuine leader rather than an insensitive bullying power. He could offer them markets which the West could not; he has shown them that economic expansion westward has not worked in the past and he could offer them guaranteed supplies of raw materials, provided they participate in its extraction. The policy pursued by Gorbachev is certainly sophisticated. He appeared to listen but did not hear; he appeared moderate but in fact played for high stakes. A social explosion and a major challenge to the Soviet Union's position may still come from the area in the future. However, if the present integration efforts are implemented, they could make any "defection" from the socialist camp much more difficult. The Soviet leader implemented his reforms in the CMEA with great skill, but the reforms he enforced were still dictated by Soviet priorities. These remain unchanged: how to rule an empire, while reducing the liabilities and deriving all the benefits? This is a much wider question, for which the answer will depend on the power struggle in the Kremlin and Gorbachev's performance as a leader. For once, the Soviet leader represents progress rather than orthodoxy. The least that could be said about CMEA reforms is that they have prepared the ground for the political battle between the Soviet Union and Eastern Europe, a battle which will probably still have to be fought.

NOTES

[1] Bond & Klein, "Impact of Changes in the global environment on the Soviet and East European economies" in US Congress, Joint Economic Committee, *East European Economies: Slow Growth in the 1990s*, Washington, DC, USGPO, 1985.
[2] *Neues Deutschland*, 26 October 1986.
[3] *Pravda*, 25 October 1986.
[4] See Y. V. Andropov, *Speeches and Writings*, Oxford, Pergamon Press, 1983, p. 9.
[5] See, for instance, *Territorialnyye struktury natsionalnykh khozyaystv stran SEV*, Moscow, Nauka, 1979 and *Otraslevyye strukturyi promyshlennostyi stran SEV*, Moscow, Nauka, 1985; compare with Y. S. Shiryayev, *Mezhdunarodnyye proizvodstvennyye sistemy*, Moscow, Vysshaya Shkola, 1981.
[6] For further details, see Bryson and Melzer, *Planning Refinements and Combine Formation in East German Economic "Intensification"*, Carl Beck Paper No. 508, University of Pittsburgh, Center for Russian and East European Studies, 1987.

[7] *Vneshnaya Torgovlya SSSR*, Moscow, Finansy i Statistika, various years.
[8] Z. Medvedev, *Gorbachev*, Oxford, Blackwell, second edition, 1987, p. 204.
[9] For its official manifestations, see *Soviet Weekly*, 3 October 1987, p. 5.
[10] B. Donovan, "Is the East German Economy Running Into Trouble?", *Radio Free Europe Research*, Background Report No. 64, 14 April 1988.
[11] "The GDR's Economy at the End of 1987", in *Wochenbericht*, The German Institute for Economic Research, No. 5, 1988.
[12] *Pravda*, 4 November 1986.
[13] See for instance *Pravda*, 4 November 1986.
[14] G. Mirov of the International Institute of Economic Problems of World Socialist Systems of CMEA, in *Problemy Mashinostroyeniya I Avtomatizatsii*, No. 9, September 1986, p. 3.
[15] *Financial Times*, 30 October 1986.
[16] *Guardian*, 13 November 1986.
[17] *Pravda*, 19 November 1986.
[18] *Neues Deutschland*, 27 November 1986.
[19] *Neues Deutschland*, 17 December 1986.
[20] *Rude Pravo*, 10 February 1987.
[21] *Neues Deutschland*, 7 February 1987.
[22] *Trud*, Moscow, 21 May 1987.
[23] N. Baturing & V. Demchuk, "USSR-CMEA Member-Countries: Further Progress in Production Cooperation" in *Foreign Trade*, October 1987, p. 14.
[24] ADN, 13 January 1988; G. Beil, "The GDR—Largest Trade Partner of the USSR", *Foreign Trade*, November 1987, pp. 22–3; for the Soviet interests in the GDR economy, see the results of bilateral conversations as reported by ADN, 11 January 1988 and *Neues Deutschland*, 14 January 1988.
[25] *MTI*, 16 October 1987.
[26] Other joint ventures with Hungary are listed in "Joint Ventures on Soviet Territory", *Foreign Trade*, January 1988, pp. 45–47.
[27] V. Monakhov, "USSR-Czechoslovakia: Economic and Technical Cooperation", in *Foreign Trade*, January 1988, pp. 2–4.
[28] Y. Voinov, "USSR-Poland: Cooperation Forms are Improving", in *Foreign Trade*, January 1988, pp. 5–7.
[29] Baturin & Demchuk, op. cit., p. 15.
[30] For previous pointers to the price formation debate, see E. I. Punin, *Scientific and Technical Revolution and World Prices*, Moscow, Mezhdunarodniye Otnosheniya, 1977.
[31] A. Antonov, "New Mechanisms of Cooperation", in *New Times*, No. 47, November 1987, p. 4.
[32] See, for example, Hungary's Mihaly Simai, "There is Much to be Changed by the Third Millennium", in *New Times* (Moscow), No. 38, 1987, pp. 10–11 and "CMEA Reform: Complexity and Gradual Progress, in *Figyelo*, Budapest, 10 December 1987.
[33] See *Sofia News*, 2 March 1988.
[34] *The CMEA Member-Countries in the International Exchange of Technologies*, Moscow, Mezhdunarodniye Otnosheniya, 1986, p. 107 and ff.
[35] *Pravda*, 13 October 1987; Y. Shiryaev, "CMEA: Restructuring the Cooperation Mechanism", in *International Affairs*, January 1988, pp. 20–32.
[36] J. Diehl, "Soviet Rewriting East Bloc Economic Rules, in *International Herald Tribune*, 14 October 1987.
[37] See *International Herald Tribune*, 15 October 1987.
[38] G. Merritt, "Rouble: For a Grand Slam, Free It Up", in *International Herald Tribune*, 22 January 1988; see also V Loshak, "The Rouble and Perestroika", in *Moscow News*, 7 February 1988.
[39] L. Colitt, "Moscow, Prague agree currency convertibility", *The Financial Times*, 29 February 1988.
[40] CTK, 2 March 1988.
[41] *Hospodarske Noviny*, No. 7, 19 February 1988.
[42] M. Gorbachev, *Perestroika*, London, Collins, 1987, p. 168.

[43] Ibid.

[44] G. Dadyants, "CMEA and EEC—Moving Towards Each Other", in *Moscow News,* 6 March 1988.

[45] "Comecon trick", in *The Economist,* 17 October 1987, p. 47.

[46] For a clear statement of this fundamental distinction, see Y. Rubinsky, "European Community: 'Political Dimensions', in *International Affairs,* Moscow, February 1988, pp. 41–49 and compare with *Kommunist,* No. 15, 1987, especially p. 34.

[47] J. M. Markham, "COMECON Edging Toward Trade Accord with EC", *International Herald Tribune,* 3 December 1987.

[48] *Pravda,* 20 October 1987; see also V. Krivosheyev, "CMEA: Facing Perestroika", in *Moscow News,* 29 October 1987.

[49] *Tass,* 27 November 1987.

[50] Bulgaria Home Service, 28 November 1987, 1830 GMT, in *BBC, Summary of World Broadcasts,* EE/W0004 A/1, 10 December 1987.

[51] J. M. Kramer, "Chernobyl and Eastern Europe" in *Problems of Communism,* November-December 1986, p. 40 and ff; for a list of existing projects, see V. Sobell, "The CMEA's Post-Chernobyl Nuclear Energy Program" *Radio Free Euorpe Research,* Background Report No. 19, 15 February 1988.

[52] V. Sobell, "Reform of the CMEA Makes Cautious Progress", *Radio Free Europe Research,* Background Report No. 37, 8 March 1988, p. 2.

[53] "Soviet call for moves to boost joint ventures", in *The Financial Times,* 21 April 1988.

[54] For the official Hungarian position on the eve of the discussions, see *Magyar Hirlap's* editorial, 18 April 1988.

[55] *Daily News,* Budapest, 19 April 1988.

[56] *The Financial Times,* 26 April 1988.

[57] K. Okolicsanyi, "The 1987 Economic Results: Stragnation Continues" in *Radio Free Europe Research,* Hungarian Situation Report No. 4, 30 March 1988, p. 28.

[58] *Daily News,* 22 April 1988.

Hungary: a Model for the Soviet Bloc?

GABRIEL PARTOS

The author is a talks writer for the BBC External Services and an expert on Hungary.

THE DEPARTURE of 76-year-old Janos Kadar from the post of General Secretary of the Hungarian Socialist Workers' Party (HSWP) in May 1988 marked the end of an era in his country's recent history which has become associated with his name. He was installed in power by Nikita Khruschev in November 1956 as Soviet troops crushed the short-lived Hungarian Revolution. Somewhat ironically it was Moscow Radio that was first in announcing Kadar's replacement at the top by Karoly Grosz nearly 32 years later.[1] Naturally, the extent of Moscow's involvement in Hungarian affairs in the late 1980s cannot be compared with that of over three decades ago. Nonetheless, as Soviet pronouncements have shown, Moscow was neither surprised nor disappointed by the changing of the guard in Budapest.[2]

During his long years in power, Kadar, like other East European Communist Party leaders, always had to pay considerable attention to ensuring that his policies received Moscow's approval. But unlike others, he succeeded in introducing a range of economic and some political reforms which provided his régime from the mid-1960s to the early 1980s with a degree of legitimacy and even popular support unparalleled elsewhere in the Soviet bloc with the notable exception of Czechoslovakia during the Prague Spring of 1968.

Briefly, what became known as "Kadarism" amounted to a relaxation of Party and State control, primarily over the economy and over many of the everyday non-political activities of the population.[3] The New Economic Mechanism, introduced in 1968, began the lengthy process of dismantling the command economy by strengthening the independence of enterprise managers at the expense of central planners. Private

201

enterprise was given greater scope within the economy. Trade with the West underwent a considerable expansion. The freer play given to market forces brought to an end the shortages of goods that have remained endemic to this day in several East European countries.

Outside the economy the authorities also reduced their earlier intrusive presence in people's lives. Travel to the West was made easier than in the rest of the Soviet bloc. Greater tolerance was shown towards dissent and by the mid-1970s political imprisonment (though not that of conscientious objectors) had been brought to an end. On the whole, the relaxation of controls was not accompanied by measures designed to introduce a degree of democracy, although Hungary did pioneer contested elections for Parliament, a system made compulsory in 1983.

These piecemeal concessions, granted by the Kadar regime over a period of nearly three decades, were not the result of pressure from below. Indeed, the Kadar regime could afford to hand down these concessions because it had succeeded in de-politicising the nation. The crushing of the Hungarian Revolution and the retribution that followed against those involved in it, created, initially, an atmosphere of fear and later one of widespread apathy and cynicism. Such a background provided the Kadar leadership with a relatively free hand at home to proceed as it wished. If anything, concern over what Moscow thought about the nature of the reforms gradually introduced in Hungary was of far greater importance.

This consideration accounted for the Hungarian leadership's persistent denials that anything like a Hungarian model of socialist development actually existed.[4] In pre-Gorbachev days any such admission would have amounted to questioning the pre-eminence of Moscow in the socialist world. So the official Hungarian attitude was to share the Soviet view that the basic laws of socialist development were identical everywhere and Hungary merely adopted some mechanisms and practices which happened to suit the country's own particular conditions.

In one sense it was, indeed, correct to assert that a Hungarian model did not exist; before Gorbachev's accession to power there was no question of any East European country adopting the Hungarian experiments on any scale. On the contrary, Moscow and almost all of its allies, with the exception of Poland, viewed Hungary's market-orientated policies with profound suspicion and certainly did not regard them as a model to follow. However, there were occasional exceptions to that underlying suspicion, especially with regard to those Hungarian policies which appeared to be working particularly well. At the 26th Congress of the Soviet Communist Party in 1981 Leonid Brezhnev paid fulsome praise to Hungary's agricultural practices, based on close links between the cooperatives and private producers.[5] Attempts were even made to transplant these methods into Soviet agriculture. Yuri Andropov was generally believed to be favourable to the Hungarian reforms and had a good relationship with Kadar whom he first got to know as Soviet ambassador in Budapest during the fateful events of the years 1953–57. However, Andropov's rule was too short to allow him to translate his intentions into practice.

It was not until after Gorbachev took over as Soviet Party leader and launched his twin policies of *perestroika* and *glasnost* that Soviet assessments of Hungary's reforms underwent a fundamental shift in a positive direction. This change became particularly obvious from early 1987 onwards as Gorbachev launched his more radical policies for political and economic reform. Yet, it was somewhat paradoxical that just as the Hungarian model was finally gaining acceptance in Soviet eyes Hungary itself should have entered a period of crisis from which it was not expected to be able to extricate itself—even on the most optimistic assumptions—before the mid-1990s.[6]

The crisis first manifested itself in the economy. From 1985 onwards almost all major economic indicators began to show mounting problems. Huge subsidies to loss-making companies, particularly in the heavy industry sector, produced a record budget deficit of Forints 47,000 million, the equivalent of US$ 1,000 million, in 1986. Since then, in spite of cuts in spending, the deficit has remained high. The reduction of some price subsidies and the introduction of new taxation led to a higher rate of inflation which by 1988 was edging towards 20 per cent. Real wages began to decline in the late 1970s and by 1987 they were 10 per cent below their 1977 level. The problems were equally bad in Hungary's foreign trade relations. The mid-1980s marked the beginning of a serious deterioration in the country's balance of payments with its Western trading partners. This made it necessary for Hungary to take on further loans and by 1988 the country's gross hard currency debt had reached US$ 18,000 million—the highest per capita in Eastern Europe.[7]

To what extent did these economic problems mark the failure of the Hungarian model? Can the market and a ruling Communist Party coexist happily in one country? The supporters of more comprehensive reforms argue that the economic crisis has little to do with the market-orientated changes introduced in Hungary over the past 20 years. Rather, it is closely linked to the failure to implement these policies with any degree of consistency. This is certainly borne out by Kadar's "stop-go" policy on reforms over the years. In the early 1970s many of the decentralising measures were reversed under pressure from Moscow and from domestic hardliners. The removal from the Politburo in 1975 of Rezsoe Nyers, architect of the 1968 New Economic Mechanism, appeared to be a symbolic end of the reform process. Yet as economic growth, productivity and the rise in incomes started to slow down, the Party leadership realised that the cost of returning to the methods of the command economy was simply too high. So, the reforms were relaunched at the end of 1978, albeit initially in a rather modest form.[8]

In the early 1980s new innovations were added to Hungary's economic experiments. Private enterprise and individual initiative, particularly in the form of franchise operations or through companies sub-contracting jobs to small groups of their employees, were given much greater scope. Companies began to issue bonds to members of the public to increase their investment funds.

Whatever the merits of these practices, many of them unique to Hungary in Eastern Europe, they hardly touched the bulk of the country's state-controlled economy. Huge subsidies continued to be handed out to loss-making plants to prevent bankruptcies and the consequent spectre of unemployment which was regarded as impermissible in a Soviet bloc country. Large Western loans were squandered on maintaining living standards, importing popular consumer goods and propping up inefficient enterprises instead of being invested in a radical overhaul of industry which has continued to produce many poor-quality and uncompetitive goods to this day.

Besides the lack of consistency in pursuing genuine market-orientated policies and the failure of deeds to match words when it came to the implementation of the harsher aspects of economic reforms, there was almost a complete absence of change in Hungary's political institutions. The paternalistic style introduced by Kadar in the early 1960s worked reasonably well at the time, combining as it did, less intrusion by the authorities into people's lives with an absence of any accountability on the part of the government to the electorate. By the mid-1980s this had outlived its usefulness. It was nonsensical to expect people to take more initiatives in economic activities and to be accountable for their failures while at the same time carrying on with the political institutions largely inherited from the Stalinist era, such as a parliament almost invariably rubber-stamping any government proposals without genuine debate. Even the introduction of compulsory multi-candidate elections to Parliament, which was first put into practice during the elections of 1985, arranged for a so-called national list of 35 members, consisting of leading Party officials and prominent public figures, who were all elected without a contest.[9]

Radical reformers inside the Party, led by Imre Pozsgay, who in 1982 became General Secretary of the Patriotic People's Front, the umbrella-body of mass organisations, began to push for genuine political changes to introduce a degree of accountability on the part of the authorities and to make the country's institutions more democratic. From 1985 onwards Pozsgay began to voice his views with greater frequency as he got more access to the media. Since then he has been calling for a comprehensive reform of the system of political institutions by making the State authorities more accountable to the electorate, by turning the National Assembly into a more genuine debating chamber and by holding referenda on issues of national importance. Pozsgay and other reformers have been arguing that these changes should be complemented by greater constitutional safeguards for individual and collective rights, by replacing arbitrary bureaucratic controls with government based on legislation and supported by a more independent judiciary. Perhaps most important of all, the time has arrived for the dismantling of the "Party-State", the traditional communist control over the day-to-day affairs of the government and the supposedly autonomous official bodies.[10]

Until Gorbachev began to advocate similar ideas in public, such proposals were considered beyond the pale even in Hungary, perhaps the most liberal of the Communist-ruled countries. They certainly did not fit in well with the paternalism of "Kadarism" as practised for over two decades. Not surprisingly, therefore, as the crisis in Hungary deepened in 1985–86, the initial reaction was to move only in the direction of further market-based economic reforms and a policy of austerity rather than contemplate the introduction of far-reaching political changes.

A new Bankruptcy Law was introduced in September 1986 which, uniquely for an East European country, took bankruptcy proceedings out of the hands of the state and left them to the courts to deal with. Fear of creating too much unemployment, social dislocation and potential discontent has, however, led to a less than vigorous application of this law over the past two years.[11] However, in the financial world a more competitive atmosphere was created by the introduction of a two-tier banking system at the beginning of 1987. The National Bank was turned into an issuing bank while five newly-formed commercial banks took over the business of lending money for enterprises.[12]

Nonetheless, these piecemeal economic reforms, useful as they were, could not hide the fact that there was a continuing drift in policy-making throughout 1986 and the first half of 1987. It was becoming increasingly clear that the Hungarian model, based as it was on half-hearted measures, compromises, and a failure to implement tough economic austerity measures, had outlived its usefulness. Likewise, there was an urgent need for genuine political reforms. Above all, Kadar, the advocate of cautious reforms since the early 1960s, was seen more and more as the main obstacle to change. He had become too wedded to the system he had established and maintained to contemplate a move in a new direction. At the same time, though already in his mid-70s, he was reluctant to relinquish his post. When the new post of Party Deputy General Secretary was created at the 13th HSWP Congress in 1985 with the purpose of easing the burden of the Party leader's workload, Kadar made sure that the job would go to the ageing Karoly Nemeth, one of his closest associates. Nemeth was replaced two years later by the outgoing Prime Minister Gyorgy Lazar, who was not in the best of health, and like his predecessor could not be considered a successor to Kadar.[13]

In fact, the veteran Party leader did his best to avoid designating a successor who could undermine his position and might force him to vacate his post earlier than he wished. Yet, following the 1985 Party Congress two leading contenders for Kadar's job did emerge and the struggle for power increasingly began to dominate Hungarian political life over the next three years, leading to Kadar's removal in May 1988. The two rivals in the contest to replace Kadar were Karoly Grosz, who joined the Politburo in 1985, and Janos Berecz, who took over as Central Committee Secretary in charge of agitation-propaganda in the same year.[14] Quite by coincidence, they were both almost exact contemporaries of Gorbachev.

Neither of the two contenders was a supporter of genuine radical reforms of the kind advocated by Pozsgay or, indeed, Nyers who, having largely disappeared from public view following his sacking by Kadar in 1975, began to stage a vigorous pro-reform publicity campaign in the official media in the mid-1980s. If anything, Grosz had a hardline reputation. Nor was Berecz any different; as late as the end of 1986 he had a bruising confrontation with the Hungarian Writers' Union when its membership proved too independent for the Party's liking.[15]

In spite of the fact that neither Grosz nor Berecz could claim to be in the vanguard of reform, they both realised that "Kadarism", as a system of self-imposed limitations, was on the way out now that Gorbachev was pushing for wholesale restructuring in the Soviet Union. So the struggle for power did not remain merely a matter of personalities but it also involved a growing debate on the viability of the Hungarian model as practised under Kadar and about the need to go beyond it both in the economy and in politics. Pozsgay and the reformers did their utmost to keep the pressure on the leadership to steer the discussion in the direction of their radical proposals. In the autumn of 1986 Pozsgay went on a six-week study tour of the Soviet Union which provided him with a useful opportunity for making wide-ranging contacts with Gorbachev's most fervent supporters. By March 1987—less than two months after Gorbachev's major pro-reform speech—he was publicly calling on Hungarians to look to Moscow for change, in effect, to take the Soviet Union as their example to follow.[16] This was all the more remarkable because the official Hungarian view at the time was that changes then being contemplated in Moscow had already been implemented in Hungary. This was, of course, true, after all many of Gorbachev's proposed experiments, whether giving enterprise managers greater independence or having multi-candidate elections, were all part of the long-standing practice of the Hungarian model.

Nevertheless, Grosz were perceptive enough to realise that it was precisely through attacking that kind of complacency about past achievements which—mixed with doubts about how to proceed—now pervaded the Kadar circle, that he could most expediently launch his bid for power. By May 1987 he was asserting that Hungary may have been ahead of the Soviet Union in many of its reforms, but it should still adopt the "consistency" and "dynamism" with which new ideas were being acted upon by the Soviet leadership.[17] Above all else, he was calling for much needed action to put Hungary's economy on a sound footing again even if that involved unpopular measures.

Kadar remained reluctant to act and unwilling to go. During a visit to Sweden in April 1987 he declared that the Hungarian authorities would be "spared the expense of paying his pension" for some time yet.[18] But the pressure for change was building up, caused in part by a highly critical report on the state of the Hungarian economy, drawn up by a group of leading economists. Entitled *Turning-point and Reform*, the document called for bold market-orientated reforms of a kind not seen in any East European country, perhaps best described as the Hungarian Model—Mark II. Although information about the report was suppressed in the media for six months, it was the most talked about document in official circles and among intellectuals in the first half of 1987. Its authors also received considerable support from Pozsgay's Patriotic People's Front and even the Party's top economic officials eventually gave their grudging and heavily qualified approval for substantial parts of the document.[19]

The clamour for an end to the procrastination and delays, which became more and more vocal after an inconclusive Central Committee session in November 1986, finally brought enough pressure to bear on Kadar for the veteran leader to sanction some changes at the top. Grosz replaced the ailing Lazar as Prime Minister and Berecz joined the Politburo in June 1987. Clearly both of them would have preferred to have been made the Party's Deputy General Secretary and, thereby, de facto, successor-designate to Kadar, but the job was given instead to Lazar who represented no threat to Kadar's position.

Taking over as Prime Minister was something of a mixed

blessing for Grosz. He would now get a chance to act and introduce some of the more rational economic policies he had been calling for. At the same time he would get the blame for the unpopular austerity measures that could no longer be avoided. Moreover, as no short-term improvement in the economic situation could be expected, he would also be saddled with responsibility for any failures.[20] It was not illogical to view Grosz's transfer to the post of Prime Minister as an attempt by Kadar to discredit a contender for his job.

Whatever the personal politics behind Grosz's transfer to the government, this move marked the beginning of the end of the Kadar era. It was the first major Hungarian response to the Gorbachev challenge to restructure society and, in the Hungarian context, to work out a new Hungarian model. Grosz acted with a sense of urgency to implement much delayed economic reforms, to allow a much wider scope for public debate in the spirit of *glasnost* and to hold out some hope of institutional reform in political life.

The changes in the economy began with a three-year stabilisation programme, announced by the government in September 1987, which sought to restore the country's financial standing by 1990.[21] By then the budget should be largely balanced, the accumulation of foreign debt should be brought to a halt and subsidies to unprofitable industries should be substantially cut. The first major step towards more rational economic policies was the introduction of a Western-style taxation system on 1 January 1988. Hungary became the first Soviet bloc country to have a comprehensive personal income tax and value added tax.[22] The purpose of the former was to make people more aware of exactly how much they were contributing to State spending and, thereby, to have a vested interest in wanting to cut down on waste and bureaucracy. VAT was designed to replace a host of taxes on producers, most of which penalised successful enterprises, which could afford to pay the taxes, and subsidised the loss-makers which could not.

Other economic reforms also point towards a more market-based economic system. A limited form of employee share-holding was introduced at the end of 1987. In early 1988 leading banks and financial institutions joined together to establish a coordinating body for dealing in shares and bonds which began to act as an embryonic stock exchange. But the measure which is expected to have, perhaps, the greatest impact is the new Company Law, due to come into force on 1 January 1989. It legalises individual share ownership after a break of 40 years, encourages savings to be put directly into successful enterprises and removes a large measure of State control from the newly-formed public limited companies which are to be run by their boards of directors answerable only to their shareholders.[23] Together with the expansion of private enterprise, it could lead to the development of an economy based on different forms of ownership.

Grosz has also inaugurated a policy of greater openness in public discussion and more tolerance towards those expressing different points of view within and outside the Party. Many Hungarians regard *glasnost* and the political concessions that have recently been granted as the price the authorities have had to pay to ward off social discontent at a time when they cannot hold out the prospect of any short or medium-term improvements in the economy. In a sense this trade-off allows people greater freedom to criticise the authorities and to set up their own independent organisations while they have to accept that there is no alternative to further belt-tightening. Perhaps the most popular of the concessions granted so far has been the liberalisation of passport regulations.[24] Since the beginning of 1988 Hungarians have been almost completely free to travel as they wish anywhere in the world, including the West. Such unrestricted opportunities are not open to the citizens of any other Soviet bloc country.

As Prime Minister, Grosz was already, in effect, running the country and he showed that he had the will and energy to implement many long-delayed projects. However, Kadar and his half-a-dozen long-standing associates in the Politburo continued to represent a serious obstacle to further radical change. Grosz began to organise for a takeover of power and made full use of the media at a time when greater openness was providing more and more opportunities for such a course of action. He also built up the status and importance of the Prime Minister to utilise it in his bid for power. He kept the initiative by always being the first to make important announcements, such as the decision to hold a special Party Conference in May 1988, which was the first of its kind in over three decades.[25]

In the months prior to the Conference Grosz began to build up a grand coalition of supporters to unseat the Kadar leadership. He had for a long time been courting the technocrats; he had a natural base of support from the middle-aged, middle-ranking officials who considered the time was ripe for them to replace their senior colleagues; and he began to build up an alliance with the radical reformers inside the Party. As the time of the Conference approached he started to drop broad hints about the need for elderly, infirm politicians to retire.[26]

Faced by this mounting pressure, Kadar was forced to step down as General Secretary and take the largely honorary post of Party Chairman at the HSWP Conference held on 20–22 May 1988. He and seven of his colleagues were also removed from the Politburo. Their replacements included some of Grosz's closest associates as well as the two ultra-radicals, Pozsgay and Nyers.[27] Their promotion to the Politburo appeared to be a clear indication of Grosz's readiness to go beyond the confines of "Kadarism" and to launch a fresh version of the Hungarian model. His task was to put the economy back on its feed again and to lay the foundations for a new national consensus to make people accept the financial sacrifices needed for an economic recovery.

Grosz immediately set to work and there have been a bewildering number of changes since the Party Conference. Some have been symbolic in character, such as the replacement in June of the old Party *apparatchik*, Karoly Nemeth in his role as head of State by the respected academician, Professor Bruno F. Straub, who is not even a member of the HSWP.[28] This gesture was intended to demonstrate that the Party's monopoly over important posts was being relaxed.

There have also been more substantive innovations. The July session of the Central Committee was the first of its kind to be open to journalists and its debates were covered in considerable depth in the media. It was a major advance in informing the public how decisions are made at the highest levels of the Party. That meeting was also a milestone in offering Central Committee members a choice between two alternative economic programmes in place of a policy already worked out and approved by the Politburo. In fact, the Central Committee opted for the tougher of the two programmes.[29] Finally, the Central Committee gave its blessing to proposals for legislation on safeguards for freedom of association and assembly. The radical proponents of these laws viewed them as providing a much-needed legal framework for independent political, social and cultural activities. Whether the new provisions are applied in a liberal or a restrictive way remains to be seen. After all, the courts can ban any organisations whose aims they consider unconstitutional. Nevertheless, the first steps have been taken towards replacing arbitrary police action by the rule of law.

The reforms now being pushed through in Hungary point in the direction of a new Hungarian model. That is based on a system in which the Party takes a less direct role in the management of political and economic life. In the political sphere public discussions are encouraged, minority views can be expressed and independent organisations are tolerated. In the economy the market is given much greater scope and various forms of private ownership are re-established.

These are far-reaching innovations in a country where the main political and social structures have seen little change since the days of Stalin. Grosz has an advantage over his predecessor, Kadar, in that he can rely on Gorbachev's wholehearted support for these reforms, as long as they do not endanger the stability of the Hungarian regime. Gorbachev may well regard Hungary as something of a laboratory where economic and political experiments can be tried out to see whether they work or not. During his meeting with Grosz in Moscow in July he assured the Hungarian leader of the similarity in their aims and endeavours.[30]

For the other Soviet bloc countries, apart from Poland, these reforms appear too radical. Czechoslovakia and Bulgaria are prepared to countenance some Hungarian-style economic restructuring and, along with Poland, they have, for example, adopted Hungary's more commercially orientated banking system. Under their present leaderships, East Germany and Romania are not even contemplating any move away from their highly centralised economic systems.

A change at the top in any of these countries, which Gorbachev may well encourage, could make the Hungarian example more attractive to them. For his part, Grosz has the difficult task of persuading Hungarians that he can revive the economy and introduce greater democracy even though he remains firmly opposed to the return of a multi-party system.[31] His experiment has no guarantee of success; indeed, it could collapse just as easily as "Kadarism" did in the mid-1980s.

NOTES

[1] Moscow Radio carried Gorbachev's greetings to Grosz at 1800 gmt on 22 May 1988, 30 minutes before Hungarian TV and radio reported that Grosz was the new General

Secretary of the HSWP. See file EE/Pol/Rels/SU/H at the archives of the BBC *Current Affairs Research & Information Section* (hereafter CARIS).

[2] On 30 April 1988 Soviet TV broadcast a Hungarian TV interview with Grosz, originally shown on 1 January 1988, as reported in *BBC Summary of World Broadcasts* (hereafter SWB), Third Series, SU/0161/A2/3. For Gorbachev's greetings to Grosz and Kadar, see *Pravda*, 23 May 1988.

[3] For a discussion of "Kadarism", see Charles Gati, *Hungary between East and West*, Duke University Press, Durham, USA, 1986.

[4] HSWP Central Committee Secretary Matyas Szueroes, reported in SWB, Second Series, EE/8364/A1/4.

[5] *Pravda*, 24 February 1981.

[6] HSWP Central Committee Secretary Miklos Nemeth's speech, *Magyar Hirlap*, 14 July 1988.

[7] For extensive statistics, see Economist Intelligence Unit, quarterly *Country Reports* on Hungary issued in 1987 and 1988.

[8] Rezsoe Nyers, "Economic reform policy in Hungary between 1957–1987", in *Il Politico*, Vol. 52, No. 4 1987, pp. 695–703.

[9] Barnabas Racz, "Political participation and developed socialism: the Hungarian elections of 1985", in *Soviet Studies*, Vol. 39, No. 1, 1987, pp. 40–62.

[10] Among Pozsgay's many interviews and statements on reform, see, for example that of 14 April 1986, SWB, Second Series, EE/8232/B/8, and *Magyar Nemzet*, 14 November 1987.

[11] *Magyar Hirlap*, 17 October 1987; Karoly Okolicsanyi, "The heat is on in the large unprofitable enterprises", *Radio Free Europe, Hungarian Situation Report* No. 1, 11 January 1988.

[12] *Financial Times*, 14 January 1988.

[13] Gabriel Partos, "Hungary's leadership reshuffle", BBC, *CARIS*, Talk No. 73, 25 June 1987.

[14] Alfred Reisch, "The Kadar question: to go or not to go?" *Radio Free Europe, Hungarian Situation Report* No. 13, 23 December 1986.

[15] Judith Pataki, "Writers and regime clash at congress", *Radio Free Europe, Hungarian Situation Report* No. 13, 23 December 1983.

[16] *Magyar Nemzet*, 16 March 1987.

[17] In a Hungarian TV interview, 21 May 1987, as reported in *SWB*, Second Series, EE/8577/B/4.

[18] *Associated Press*, Stockholm, 22 April 1987.

[19] *Koezgazdasagi Szemle*, No. 6, June 1987.

[20] Gabriel Partos, "Hungary's leadership reshuffle", BBC, *CARIS* Talk No. 73, 25 June 1987; Alfred Reisch, "Mission impossible for new Prime Minister", *Radio Free Europe, Hungarian Situation Report* No. 6, 30 June 1987.

[21] *Magyar Hirlap*, 17 September 1987.

[22] "Hungary: fiscal furore", in *Eastern Europe Newsletter*, Vol. 1 No. 6, 12 August 1987, *Magyar Hirlap supplement*, 25 September 1987.

[23] "Towards a capital market", in *Eastern Europe Newsletter*, Vol. 2, No. 11, 1 June 1988; Heti Vilaggazdasag, 7 May 1988.

[24] *Magyar Hirlap*, 28 November 1987; but for continuing restrictions see article by Ferenc Koeszeg in *Wall Street Journal*, 26 February 1988.

[25] In an interview on Budapest Radio on 5th December 1987, see *SWB*, Third Series, EE/0020/B/5.

[26] *Magyar Hirlap*, 28 April 1988.

[27] *Nepszabadsag*, 23 May 1988.

[28] *Magyar Hirlap*, 30 June 1988.

[29] For radio & TV coverage see SWB, Third Series, EE/0204, Supplement C 1 and EE/0205, Supplement C; for newspaper coverage, see *Nepszabadsag*, 14 & 15 July 1988.

[30] *New York Times*, 10 July 1988.

[31] *Magyar Hirlap*, 28 April 1988.

Mikhail Gorbachev and the Western Media: How Open is Openness?

EDWARD FOSTER

Having studied at Cambridge, the author was subsequently engaged in journalism and completed the research on this work at the RUSI. He is now a researcher on the RUSI Western European Security Programme.

IN THE forty-odd years since the end of the last war in Europe, the continent has been divided by an ideological struggle for the hearts and minds of its inhabitants. Citizens on either side of the divide have been subjected to incessant barrages of propaganda and counter-propaganda, delivered in tones of greater or lesser stridency according to the prevailing political climate. In Eastern Europe and the Soviet Union the limited circulation of the Western press has created a situation in which the radio stations of the West, operating free of the strictures imposed by the governments of these countries, play a vital role and they are for many people living behind the Iron Curtain the only corrective to the official manipulation of news and events. Meanwhile the accession to power in the Soviet Union of the younger and more dynamic generation of leaders has reduced surface tension between the Superpowers and suggested to many in the West that the oppressive social constraints—which must remain an obstacle to real harmony for as long as the West continues to proclaim its moral superiority—may be lifted. It should be remembered that the aims of Gorbachev and the new ruling élite are to make the USSR more efficient, not to satisfy the wishes of Western liberals: the question of whether the Soviets and their allies allow these radio stations free access to their considerable audiences, of whether they admit a genuine plurality of ideas, consequently goes to the very heart of our interpretation of *glasnost*.

THE RIGHT TO IMPART AND RECEIVE INFORMATION

This need not be taken to mean that life for everyday people in these countries will not improve; moves that encourage

215

individual initiative should be welcome in any society. Gorbachev's behaviour in this respect, however, leads to the conclusion that he considers this aspect of human rights not as an end in itself, but as an expedient lever to use in negotiations with the West. For human rights are precisely the issue at stake in the information debate. Article 19 of the Universal Declaration of Human Rights—the chief banner of all crusades of this sort—provides for the upholding of a crucial freedom:

> Everyone has the right of freedom of opinion and expression: this right includes the freedom to hold opinions without interference and to seek, receive and impart information and ideas through any media and regardless of frontiers.

In 1975 this principle was further elaborated at the Conference on Security and Cooperation in Europe (CSCE) in Helsinki. Clause 2 of the "Third Basket" of its Final Act records the agreement of the participating states:

> to make it their aim to facilitate the freer and wider dissemination of information of all kinds, to encourage cooperation in the field of information and exchange of information with other countries, and to improve the conditions under which journalists from one participating State exercise their profession in another particular State.[1]

To date, these noble declarations have failed to be put into practice by those in authority in the USSR and Eastern Europe.[2]

RADIO STATIONS BROADCASTING TO THE EAST

It is fortunate, given the unique importance of their function, that there is such a healthy diversity of radio stations in existence and in a position to supplant the state-controlled media in the languages of these countries. Famous among these are the two long-established Anglo-Saxon networks, Voice of America (VOA) and the BBC External Services. Funded by higher bodies but operating free of their control (VOA's operating grant comes from Congress' Board of International Broadcasting; the BBC External Services is financed by the Foreign Office), these two are similar in that they report and comment on events as seen from their respective capitals. Fulfilling a similar role are Deutsche Welle and Deutschlandfunk[3], the foreign-service operators of the West German Radio Council; Radio France Internationale; and Radio Vatican (which has a strong following among the Catholic populations of Poland and Slovakia).

Alongside these and complementing them are the Soviet and East European "surrogate" stations Radio Liberty and Radio Free Europe, whose positions are those of independent radios-in-exile. The former broadcasts to Eastern Europe, the latter to the nations of the USSR; and the autonomy enjoyed by each of their language services—to shape programming to meet the interests of each target nationality—has led to it being said that RFE/RL (the two stations were merged into a single administrative body in 1972) is not one organisation but 21.[4] Concern in Washington in the early 1970s led to them being removed from CIA control, and like VOA they now answer to (and are funded by) the Board of International Broadcasting. Their administrative centre is in Munich; their transmitters are situated in West Germany and also—for technical reasons—in Spain and Portugal.

Despite the elaborate jamming measures, which will be examined in detail, there is no doubt that these stations draw considerable audiences. According to market research under-taken in recent years, VOA has been the Western station most listened to in the USSR, with a weekly audience of around 15 per cent of the radio-listening public.[5] RFE leads in popularity among the audiences of Eastern Europe to which it transmits with a weekly figure of 48 per cent.[6] A more graphic demonstration of radio penetration comes from another survey which indicated that 36 per cent of a polled group of Soviet citizens first heard of the Chernobyl disaster on Western radio and a further 13 per cent subsequently tuned in to hear of developments.[7] This may also apply to some of the Kremlin olympians: information is supplied on a need-to-know basis in the USSR and it is said that both Khruschev and Andropov were regular listeners. It goes almost without saying that these figures are dismissed with scorn by the Communist media.

Although the ceaseless official campaign to vilify these stations has seldom matched the epic tones recorded in 1968 by the Novosti reporter over beleaguered Prague who styled them "the real masters of the black heavens",[8] the disinforma-tion agencies have taken every opportunity to portray them as viciously anti-socialist distorters of reality and willing catalysts of rebellion. Former VOA director Charles Thayer concedes that in the mid-1950s, at the height of the Cold War, the station "fell victim to broadcasting bitterly sarcastic anti-Stalinist attacks",[9] but the negative value of offering a mirror-image of Soviet propaganda was eventually understood and a policy of common sense objectivity re-established. Notwith-standing, Marshal Ustinov still chose in 1980 to refer to the Voice as "the transoceanic falsifiers of history". The sometimes painful detachment of the BBC has led Soviet analysts of Western media to urge particular vigilance: despite its "halo of objectively and justice, it strives . . . to achieve its reactionary aims".[10] Slander aimed at Deutsche Welle has attempted to play on fears of German revanchism, even calling its commen-tators a "Goebbels breed". The particular position of RFE/RL has made them special targets to be discredited, and this will be examined more closely.

In another comment from the Czech uprising, East Berlin's predictable *Neues Deutschland* remarked that the trouble "did not begin on the streets. It all started with unlimited press freedom". This notion, that seditious broadcasts detonate or orchestrate social unrest has been repeated many times since. In August 1987, VOA, the BBC, RFE and Radio Vatican were alleged to have broadcast precise details of a demonstration to be held in Vilnius in the USSR,[11] and TASS accused VOA and RFE/RL of inciting a similar disturbance in Riga.[12] The charge reappeared in *Pravda's* assessment of Western coverage of the riots in Nagorny-Karabakh.[13] With their own reporters firmly excluded from the area, the BBC relays of information from local sources (sources methodically marginalised by *Pravda*) were represented as seeking to inflame a volatile situation. This analysis was echoed a month later in the Warsaw *Trybuna Ludu's* piece on the impact of RFE on the strikes in Gdansk and Nova Huta.[14] And although Hungary has pursued a comparatively moderate line, its leader Karoly Grosz recently called for research into "the broadcasts of Radio Free Europe and the Voice of America and the way . . . they incited the Hungarians to revolt" in 1956.[15]

JAMMING EFFECTS

The Soviets and their allies have been jamming the radio stations of their ideological opponents on and off since 1948, when Moscow first blocked the Russian-language services of

VOA and the BBC. This treatment was extended to cover Radio Liberty and Radio Free Europe virtually at their inception and has remained in their case ever since. Poland ceased jamming RFE in 1956, Romania stopped all interference with incoming transmissions in 1963 (when the USSR stopped jamming VOA and the BBC) and Hungary followed this example shortly thereafter. Jamming is, after all, an expensive activity: at least two transmitters and an additional coordinating installation are needed to trap each offending signal. By the time Poland gave up jamming in 1956, it was costing the country a sum equivalent to the entire budget of VOA. Large, powerful[16] transmitters intercept the signal over the target area from an equal distance, while to cover the gaps in the screen (notably the "twilight immunity" caused by ionospheric instability), a further network of back up stations is in place to "protect" major population centres. All told, it is estimated that there are between 2 and 3,000 jamming stations spread across the USSR, calling on the services of up to 15,000 people. Costings in the Soviet Union are always difficult to assess, but the annual expenditure is unlikely to leave much change from US $1bn. This operation had reached something like its present level when Moscow stopped jamming VOA and BBC broadcasts in 1963.

With Western protests growing as Soviet tanks delivered fraternal assistance to the Czechs in 1968, jamming of these stations summarily recommenced. This continued until 1973, when the jammers were once again switched off as the second stage of the CSCE conference convened in Helsinki. By 1980 relations were deteriorating once more as the Western media gave increasing coverage to the rise of Solidarity in Poland, and the Kremlin clearly felt that free reporting was endangering the status quo: jamming of all Western stations was resumed on 20 August. Significantly, this sweeping extension was accomplished without any fall-off in the scale of operations against RFE/RL—a concrete indication that the entire system had remained in place during its years of silence.[17]

EAST BLOC ATTEMPTS TO ENSHRINE JAMMING OPERATIONS IN INTERNATIONAL LAW

In spite of the enormity of this jamming effort, Kremlin policy for many years was to ignore its existence altogether. Although it flew in the face of clearly-stated international agreement, enquiries on the subject were rebuffed with charges of "impermissible interference" in the country's internal affairs," "hostile propaganda", even "aggression". Meanwhile the perpetrators sought to formulate an argument that would justify the activity in the abstract. At the International Telecommunications Convention in 1952, the Soviet delegation objected to the adoption of an express prohibition on the grounds that it should allow for cases of "incitement to aggression, to war, to the use of force", while voicing the high moral view that "no nation would think of blocking factual information". Any possible loophole left open by the absence of an outright prohibition was closed at the 1979 World Administrative Radio Conference, at which it was stipulated that "Administrations shall cooperate in the detection and elimination of harmful interference" and prompt remedial action was urged.[18].

In the same year the Soviet Union sought an amendment to the 1978 Declaration Concerning the Contribution of the Mass Media to Strengthening Peace and International Understanding (which reiterated the principles of the Universal Declaration of Human Rights and urged wider reporting of its violations) to the effect that transmission could take place only with the consent of the receiving country. This nullifying and unworkable proposal was rejected by the United States and its allies as amounting to "censorship at source".[19] Since the USSR itself boasts the world's second largest radio service, it was compelled to accept the Declaration without this catch-all rider.[20] In doing so, it nevertheless took the opportunity to complain that the West was engaged in disseminating "subversive propaganda".

Unfortunately, "propaganda" is a slippery term that has always eluded an internationally accepted interpretation. Yet it is clear that the Soviets were searching for a legal safeguard against the "incitement" referred to by their delegate in 1952. With this aim, Moscow dusted off the 1936 League of Nations International Convention Concerning the Use of Broadcasting in the Cause of Peace during the early 1980s, and this was belatedly ratified by the Supreme Soviet in 1982. Drafted in a simpler age as a barrier against the Nazi propaganda machine, the Convention contained a provision for arbitration in the event of a dispute.[21] The USSR's rejection of this article therefore offered it free rein to jam at will, and after Bulgaria, Hungary, Afghanistan, the GDR, and Czechoslovakia had followed the Soviet example, the Western countries (led by the UK in July 1985) one by one repudiated the Convention. The United States had not been a member of the League when the document was drawn up and had never ratified it.

It is at this point that Mikhail Gorbachev enters the scene. He was elected General Secretary of the Communist Party of the Soviet Union in March 1985, and the arrival of a new and energetic Soviet leader was bound to reactivate the debate. That September the USSR's East European allies (led by Czechoslovakia) revived proposals for a "New Information Order" before the UN Information Committee. Although the idea of such an order is supported by many Third World regimes anxious to maintain a monopoly of information, this was once again blocked by Western members. To Gorbachev, it must now have been clear that efforts to force acceptance of the principle had run out of steam and that another approach was required. The legal justification would continue to be argued on the basis of national sovereignty, but the new Soviet leader needed to regain the initiative. By going over to the offensive he could redirect attention away from the restrictive nature of his system and on to ground of his own choosing. Meanwhile, in keeping with the new image of businesslike candour he wished to project, the USSR slowly broke its silence on its jamming activities. A delegate at the Budapest Cultural Forum that November came close to a public admission that "lies and poison" were in fact being shut out. Finally, on a visit to Moscow in January 1986, US Information Agency Director Charles Wick was told that the need to jam foreign stations "would soon abate". These remarks coincided with a renewed broadside from the Soviets and the Bulgarians at the Budapest forum accusing RFE/RL of conducting "large-scale psychological warfare", and allegations that they violated international law.[22] Taken together, these developments mark the beginning of a new and subtle campaign of selective targeting. News commentator Vladimir Pozner repeated the distinction before an American audience in June of that year: jamming might even occasionally prove "counterproductive" but was still called for in the case of RFE/RL, given their "subversive" nature. This discrimination became more and more apparent. A West German delegation to Bulgaria was told that Deutsche Welle might soon be able to broadcast freely, and in September 1986 Radio Warsaw answered a listener's letter with a denial that any "state-

owned" station was being jammed, but justifying interference with RFE on the grounds that it aimed to "foment
unrest".[23]

GORBACHEV'S NEW APPROACH

Rather than continue to press for the unobtainable—a blanket
acceptance of the legitimacy of its jamming activities—the
Kremlin was seeking to exploit a possible line of cleavage
between the national radio networks of the Western states and
the surrogates. RFE/RL constitute the most direct challenge
to official disinformation. Staffed largely by exiled opponents
of the regimes in question, they can broadcast current affairs
material compiled with an intimate understanding of their
target audiences and drawn from all available sources, both
official and *samizdat*. They also have the resources to give more
airtime to the non-Russian cultures of the USSR, an increasingly sensitive area highlighted by recent disturbances in the
Baltic states and in the Caucasus. In accordance with the spirit
of our times, those at RFE/RL no longer see themselves as
"Cold Warriors" and their editorial guidelines call for the
same even-handed objectivity and avoidance of vituperation as
their more conventional companions. Even so, the surrogates'
particular interest in and feel for their audiences mark them
out as calling for special attention.

At the Reykjavik summit in October 1986, Gorbachev
became the first Soviet leader to declare unequivocally that
jamming was his country's policy. In doing so, he introduced a
new factor designed to open up the most important potential
fault line, that between RFE/RL and its US-backed stablemate
VOA. Pointing at the ring of transmitters beaming VOA into
the USSR, the General Secretary complained of the 'unequal
situation" in which the Soviet Union found itself, and
proposed that the jamming of this station should cease in
return for the opportunity to set up complementary Soviet
stations to broadcast in the USA. The proposal was restated at
a parallel meeting between Charles Wick and Aleksandr
Yakovlev, the newly-appointed Chief of the Central Committee's Propaganda Bureau. Caught wrong-footed, the American
promised to consider this. His opposite number's initial
suggestion was disingenuous—foreign governments are prohibited under US law from owning domestic American stations,
as the Russians must have known—and neither can he have

been unaware that the medium-wave and FM frequencies (as against the short waveband used by Moscow) he requested were all allocated. Later the American party voiced the hope that commercial American stations might make airtime available, but the real Soviet achievement had been to sidestep Western objections to jamming as a whole. Wick found it necessary the following month to counter rumours of a "double-cross" of RFE/RL,[24] rumours restoked in February 1987 by a story that the head of RFE's Polish section was to be sacked as a conciliatory gesture.[25] These rumours revived when Wick went to Moscow for a second meeting with Yakovlev in May 1987—not surprisingly, since VOA had finally been unblocked one week before.[26]

The unjamming of VOA was the second of the salami-slices to be cut to pacify the West. The practical isolation of RFE/RL had started in January 1987, when jamming of the BBC suddenly ceased before Mrs Margaret Thatcher's visit to Moscow. It was presumably hoped that this would sweeten the atmosphere and make the former "Iron Lady" more amenable to Soviet proposals in the arms-control arena. At the time, a Soviet spokesman described this "suspension" as "a gesture of openness", and such language indicates that the possibility remained that the gesture could always be rescinded. The same rationale clearly lay behind VOA's new freedom. To add force to their new strategy of reciprocity—the converse of the tit-for-tat answer to the periodic diplomatic expulsions—the Soviets began medium-wave broadcasts to the USA from Cuba in May despite American protests that these interfered with domestic broadcasting. Interestingly, the same approach was employed in other areas shortly afterwards as part of the bid to sell the new leadership to Western public opinion.[27] At the same time, equipment no longer used to jam Radio Tirana and Radio Beijing was redeployed against minority-language services of Radio Liberty, and at the same time as the Party chief in Estonia publicly accepted that a quarter of his republic's television aerials were regularly tuning in to Finnish TV,[28] his Latvian counterpart laid the blame for disturbances squarely at the feet of VOA and RFE/RL.[29] While the East-West war of words over the airwaves is expected to continue, the Kremlin is prepared for it to be conducted in public with all the stations except for those whose appeal it distrusts the most. Consequently, at the time of writing, RL is still being jammed by the Soviet Union, as are Deutsche Welle and Kol Israel—the radio stations of its host countries.[30]. Czechoslovakia also appears to be dutifully jamming RFE and Deutsche Welle/Deutschland-funk, VOA's relative acceptability has been underlined by its announcement that it expects to open its first Moscow office in the near future. The first Polish government press conference that the Voice was allowed to attend, in July 1987, was used by government spokesman Jerzy Urban to attack the Israeli decision to allow relaying facilities to VOA and RFE/RL,[31] and the Kremlin will undoubtedly still be hoping that the crack will widen. RFE/RL's sponsors and hosts are sure to be offered inducements to drop their support for the troublesome broadcasters; but whatever the pressures that might arise to do away with stations supported by the US taxpayer and held up as "obstacles" to better relations, the absolute legal and moral indefensibility of jamming is a principle upon which the West cannot afford to compromise. So long as the equipment and the justification for its use remain in place, stations currently being received without interference could find themselves jammed just as they did when previous periods of detente came to an end.

1988 has seen moves to play down the severity of the isolation of the outlawed stations. Poland had regained sufficient stability by the beginning of the year for RFE to be re-admitted. Eighty per cent of those polled in a government survey approved of this action,[32], although Jerzy Urban soberly reminded the population of the "necessary discernment of the free flow of information from the methodic abuse of freedom of speech".[33] In January, Wick reported after a meeting with Gorbachev that the Soviet leader "agreed that direct interference could be counter-productive", and incidentally, "to tone down the propaganda war and desist from untruthful statements about the United States".[34] A reader's letter printed in *Izvestiya* took up this point about forbidden fruit, but added that listening to stations formerly jammed was now "uninteresting,...even disagreeable" in the new age of "*glasnost* and democracy".[35] The public relations campaign is currently gathering pace: Jerzy Urban, ever one of RFE's most willing sparring partners, recently declared his readiness—"although this is not a dignified role for an official of the Polish state"—to travel to Munich to discuss the station's activities, and extended a corresponding invitation to an RFE team. [36]

MOSCOW'S MEDIA POLICY IN EASTERN EUROPE

The degree to which coordination of Moscow's allies has been increased since Gorbachev's arrival is clear. The cooperation now evident between Poland and the USSR is a far cry from the relationship apparent in 1980, when the Soviets resumed jamming to block out news of Solidarity from their own population while VOA and BBC broadcasts in Polish were paradoxically left unmolested. It has already been stated that it was the Czechoslavaks who resuscitated (Soviet) proposals for a "New Information Order"—an Orwellian concept of state-controlled media, and it is uncharacteristic of the allies (other than the Romanians, who have in any case maintained the same line on information "sovereignty" from their peculiarly insular corner of the Warsaw Pact) to undertake any initiative without at least a nod from their Soviet elders. Improved ties between news organisations offer a parallel to Gorbachev's drive to create economic links between the members of the Eastern bloc at enterprise level.[37] Warsaw's youth daily published a call in November 1986 for increased cooperation, with particular reference to developments in satellite television, to counter the West's "propaganda attacks".[38] Revealingly, it drew an analogy between President Reagan's SDI project and a supposed "information umbrella", from beneath the safety of which these attacks were launched. In recent years the Socialist countries have been constructing an umbrella of their own, the structural members of which are provided by a programme of Soviet television beaming via satellite to Poland, Czechoslovakia, Hungary, and Bulgaria which started in the early years of this decade. The lateral supports of this shield were put in place more recently: a Czechoslovak-Bulgarian radio and television agreement was made in June 1987, a similar agreement between Polish and East German radio that July, a Polish-Bulgarian agreement on television cooperation in September, a similar East German-Czechoslovak broadcasting agreement in January 1988, and an East German-Polish news agency agreement in February. It should be pointed out that these supports are not themselves conductors—while Moscow makes radio broadcasts to all the countries of Eastern Europe in their native language, this is strictly one-way traffic. There has been no discussion of East European television becoming available to, say, Muscovites, or of direct transmission from one allied state to another.[39] These accords should be considered in the light of these states' unrelenting position on the right to transmit and receive information.

Moscow's growing awareness of the importance of getting the message across is also underlined by its new contacts with Western media organisations. Protocols have been agreed in recent years with Finland, France, Australia, the Netherlands and other countries not within its camp. Soviet TV set up its first foreign affairs desk in April 1988. In an inaugural conference, a panel considered the popularity of tele-bridges—Deputy Chairman Popov of the State Television and Radio Committee remarked that George Shultz had virtually become a regular contributor to the service.[40] A Western review[41] of more than 20 telebridges in 1987 characterised them as crash-courses in ideological immunisation and considered that by lining up two groups of ordinary citizens, sometimes of school age, with a view to a superficial discussion of matters of some complexity, the Soviets' inculcated orthodoxy gave them a public relations advantage. Joint efforts of this kind are positive in that they break down misleading stereotypes, and although the BBC's external broadcasting chief, John Tusa, indicated a radio link-up involving Margaret Thatcher as evidence of a shift in Kremlin policy, [42] they leave unresolved the selective nature of their access.

Soviet television's new role as the torch-bearer of *glasnost* to other nations is in keeping with Gorbachev's reputation as a skilled and telegenic media handler; the Gorbachev who has made such an impression on Western public opinion. It has already been mentioned that Finnish TV is commonly watched in the Baltic states: it is quite possible that the fact that Soviet programmes would be shown in countries where the domestic services have to put up with competition from German, Austrian, and Yugoslav television was an additional spur for it to revise its drab, outdated presentation. TASS described changes aimed at catching the attention of a new television generation with approval: "*Vremya* [Soviet television's news programme]...was notable for information ostentation (*sic*). Now the situation is changing; the news section is shorter and more specific, the general analytical level is better".[43] More recently, a Moscow *New Times* reporter watching Soviet TV from Warsaw nonetheless called for resistance to "hackneyed phrases like 'unbreakable friendship between nations'...Our television is an important vehicle for getting our ideology across to foreign audiences", he went on, consequently it would "largely decide the extent to which the world's millions . . . believe in us".[44] The Sofia *Rabotnichesko Delo* loyally reassured its readers of the popularity of all aspects of the Soviet mass media in Bulgaria and called it "a school for *glasnost*".[45]

A Czech journalist proposed in May 1988 that the growth of Western satellite channels could be offset in his country by re-broadcasting whole Polish and East German programme schedules.[46] Here, it is important to note the complete reversal in Soviet positions, from fear about the spectre of Western television broadcasts beamed to their possessions and allies to a readiness to confront such developments head-on. As late as 1984 Leonid Zamyatin, then Aleksandr Yakovlev's predecessor at the Propaganda Department, was vehemently opposed to "televised propagandistic aggression", and Foreign Minister Gromyko wrote a letter to the same effect to the UN Secretary-General.[47] In June of the following year, Novosti was still warning that Moscow could take countermeasures "not only on its territory but also in outer space", echoing Soviet proposals for Satellite regulation made before the UN in 1972. But the Polish analyst who drew comparisons with SDI caught the shifting mood when he reiterated the view that jamming (of a hypothetical Television Free Europe) would be fraught with technical difficulties and "might have a boomerang effect".[48] He pointed to the frequency with which West German television is watched in the GDR and dismissed the suggested threat. Poland manufactures satellite antenna and is already issuing licences for their private purchase, and Hungary has obtained a licence to rebroadcast Sky Channel, Super Channel and the French TV5 in 1989, in addition to the television channels of adjacent countries already available by terrestrial means.[49] Significantly, Poland is now due to start receiving and re-broadcasting the full complement of Italian TV's—notion-ally Christian Democrat—first channel, in Italian.[50]

Not all East European states take such a sanguine view of future developments. A publication in Czechoslovakia with the alarmist title of *A Trojan Horse from the Air* anticipates similar developments in that country, and calls for preparatory measures to "foster the ideological steadfastness of our citizens". Nevertheless, the pattern of the more relaxed Communist states rising willingly to the challenge from the West, begs the question of what will happen in the Soviet Union itself. Back in 1986, a speaker at the Bulgarian Party

Conference provided an indication by voicing sentiments similar to the Czechoslavak call to ideology.[51] In June 1988 *Novoe Vremya* postulated that events in the USSR would probably follow the Polish model with licences being issued provided neighbours' reception was not prejudiced and that the roof could take the weight of the antenna.

If this is to be the case, the Soviets still have a certain amount of time to prepare for the onslaught of pop videos, cola commercials, and worse. In Poland the least expensive dish and tuner on the market cost $1,200 in 1987,[52] and pressure will not reach critical point until ownership in the USSR's neighbours has become commonplace. In the meantime, new legislation on information—including satellite television—is in preparation.[53]

The Polish conception of the "information war" fought from behind defensive shields is interesting when assessing the Soviet strategy on news flow. Some relaxation in the name of *glasnost* was inevitable, yet anything short of a total abandonment still leaves Gorbachev and his reforms open to the same charges levelled against his predecessors. By admitting that the USSR was jamming Western radio stations, he placed himself in a difficult moral position; in order to extricate himself he is playing for advantage in the same quid-pro-quo manner as characterises his negotiations in the arms-control field, manoeuvring for position, increasing pressure on his opponents, and dividing them to be dealt with piecemeal. His neighbouring allies are in a position to filter the threat and give him time. For time is vital if he is to consolidate his position at home: consider the views of KGB chief Chebrikov, who in 1981 lashed out at "hostile special services" undermining Soviet youth with concepts of political pluralism, and "bourgeois ideologues" who toss to young people the provocative phrase "conservatism of the old ranks".[54] Gorbachev's uncomfortable irony is that an unresisted tidal wave of free ideas could all too easily sweep *glasnost* aside with its instigator; he will therefore continue with a carefully balanced act of "filtering" the information which the Kremlin allows its citizens to receive, while proclaiming his conversion to upholding an "open" society.

NOTES

[1] The Third Basket of the Helsinki Final Act was not couched in legally binding language, but still records agreement reached between the participating nations.

2 The Communist Bloc states maintain that it is the Western governments, not they, who have consistently violated the Helsinki agreement: The mass media should "respect the internal affairs of other countries in the sense that one is supposed to withdraw from interfering in them directly". *The Democratic Journalist*, vol. 32, no. 9, September 1985.

3 Agreement was reached on division of responsibilities between these two stations, taking effect from the beginning of 1977: DW was to broadcast on Short and Medium Wave in Russian and Romanian, DLF in Polish, Czech, Slovak, and Hungarian.

4 W. Buell, "Radio Free Europe/Radio Liberty in the Mid 1980s", in K. R. M. Short, ed., *Western Broadcasting Over the Iron Curtain*, Croom Helm, London, 1986, p. 80.

5 RFE/RL research, April 1984.

6 RFE/RL research, late 1985/86.

7 RFE/RL research AR 4–86, October 1986.

8 B. Kirsch, "Deutsche Welle's Russian Service, 1962–85", in K. R. M. Short, ed., op cit, p. 159.

9 as quoted by A. Heil, "The Voice Past: VOA, The USSR And Communist Europe", in K. R. M. Short, ed., op cit, p. 99.

10 V. Artemov, "The BBC Is Fanning Psychological Warfare", *Tiesa*, 17–18 November 1983.

11 *Pravda*, 22 August 1987.

12 TASS, 23 August 1987, as reported in *BBC Summary of World Broadcasts* (hereafter "SWB"), SU/8655/A1/1.

13 *Pravda*, 4 April 1988.

14 *Trybuna Ludu* (Warsaw), 12 May 1988.

15 *International Herald Tribune*, 11 July 1988.

16 An ironic reflection of the strength of the jamming signal was reported in the *Daily Telegraph*, 11 June 1986. Leonid Zamyatin, ambassador to London and in all probability—as the former head of the Party Propaganda Department—the man responsible for the resumption of jamming in 1980, complained before a Parliamentary Select Committee on Foreign Affairs that BBC jamming was preventing him from listening to Radio Moscow at home. Mystified, the BBC (which does not jam) investigated the ambassador's problem, and was able to report that the interference was being caused by overspill from Soviet jamming of other frequencies. BBC External Services head Austen Kark remarked that it was a "clear case of the messenger mangling his own message".

17 *The Economist*, 26 September 1980.

18 Article N20/15, section 5133.

19 *UNESCO And The Freedom Of Information: Hearing Before The Subcommittee On International Organisations Of The House Committee On Foreign Affairs*, Ninety-sixth Congress, First Session, 1979, pp. 3–5.

20 *The Economist*, 6 June 1987.

21 Article 7 of the 1936 League of Nations International Convention Concerning the Use of Broadcasting in The Cause of Peace.

22 Alov and Viktorov, Aggressive Broadcasting: Psychological Warfare, APN, Moscow, 1985, pp. 141–142.

23 Radio Warsaw, 26 September 1986 SWB EE/8377/A1/1.

24 *Washington Times*, 13 November 1986.

25 Polish news agency PAP, 13 February 1987 SWB EE/8493/i.

26 *Washington Post*, 6 June 1987.

27 On June 17, TASS complained of the difficulties facing Soviet journalists working in the USA, and four days later Soviet television reported that one of its teams had been denied the opportunity to make a film on lingering memories of the Third Reich in the Bundeswehr. 21 June 1987, SWB SU/8601/A1/4.

28 Radio Helsinki, 14 June 1987.

29 *Moscow News*, 28 June 1987; *Cina*, 16 June 1987; *Le Monde*, 19 June 1987.

30 Other than its Hebrew channel. The Israeli government gave permission in June for VOE and RFE/RL to use booster facilities on its soil. *New York Times* 19 June 1987.

[31] PAP, 23 June 1987 SWB EE/8603/B/10.

[32] PAP, 7 March 1988 SWB EE/0095/B/7.

[33] PAP, 16 February 1988 SWB EE/0078/B3.

[34] *Washington Times*, 25 January 1988.

[35] *Izvestiya*, 22 June 1988.

[36] *International Herald Tribune*, 23 June 1988.

[37] See J. Eyal: "Eastern Europe—The Gorbachev Challenge", *RUSI & Brassey's Defence Yearbook 1988*, p. 49.

[38] *Sztandar Mlodych* (Warsaw), 7–9 November 1986.

[39] RFE research, RAD Background Report/90 (Eastern Europe), 26 May 1988. The exception that will allow Hungary to transmit television pictures via satellite to Hungarian communities in the USSR, Slovakia, and Romania is the result of Gromyko's concession to Magyar sentiment during his visit to Budapest in Spring 1988. In this way Moscow means to make felt its disapproval of Ceauşescu's policy of enforced Romanisation.

[40] Soviet TV, 6 April 1988 SWB SU/0129/A1/3.

[41] Radio Liberty research, RL 37/88, 28 January 1988.

[42] *The Times*, 11 July 1988.

[43] TASS, 24 December 1987 SWB SU/0037/B/60.

[44] R. Borecky, "The Message And The Impact Of The Media", *New Times* (Moscow), No. 19, 1988.

[45] *Rabotnichesko Delo* (Sofia), reported by Sofia Radio 5 May 1988 (US Foreign Broadcast Information Service EEU-88-088).

[46] *Tvorba* (Prague), 11 May 1988.

[47] AP (Moscow), 28 November 1984.

[48] *Sztandar Mlodych*, op cit.

[49] Hungarian news agency MTI, 4 February 1988 SWB EE/W0020/B/1.

[50] *International Herald Tribune*, 5 April 1988.

[51] Lalyn Dimitrov at the 13th BCP congress, reported in *Rabotnichesko Delo*, 4 April 1986.

[52] *Trybuna Ludu*, 15 November 1987.

[53] Radio Liberty research: RL 208/87 "Soviet Jurists Discuss Draft Press Law", 1 June 1987; RL 151/88 "Law On *Glasnost*". In Preparation, 13 April 1988.

[54] *Molodoi Kommunist* (Moscow), No. 4, 1981, pp. 38–44.

Hungarian Security Policy and Role in the Warsaw Treaty Organisation

BRUCE GEORGE MP and MARK STENHOUSE

Bruce George has been MP for Walsall South since 1974. He is a member of the House of Commons Defence Committee and Chairman of the Political Committee of the North Atlantic Assembly. The Assembly's sub-committee on Eastern Europe visited Hungary in 1988. Mark Stenhouse MA studied in the Department of Government at Sussex University and is a researcher specialising in Southern European security.

CURRENT HUNGARIAN threat perceptions are partly the legacy of a troubled history, plagued by military defeats, failed rebellions and the impositions of major powers. Hungary has experienced periods of Ottoman, Hapsburg and Soviet domination. Nevertheless, the Hungarians have never surrendered their autonomy easily and they continue to display a considerable degree of independence within the WTO. In recent years the Hungarians have actively encouraged East-West detente, mindful of the dangers created by unstable international relations.

Hungary's military significance within the WTO is limited but its potential geostrategic value as a base for operations is considerable. However, it is instructive to examine the realiablity, capabilities and composition of the Hungarian armed forces. The Soviet-Hungarian military relationship provides several insights into political, social and economic relations in Eastern Europe.

Between 1945 and 1947 the Hungarian army ceased to function as a structured fighting force due to military defeat and a lack of resources. The Hungarian People's Army (HPA) was established in 1948, as the communists were consolidating power. For the next five years the Hungarians endured a highly repressive Stalinist regime. In a savage purge Laszlo Rajk, leader of the "home" communists, was executed as General Secretary Rakosi enforced a system of centralised planning backed by a terroristic State Security Authority. Political oppression was coupled with economic hardship as a consequence of disadvantageous trade links with the Soviet

231

Union. In 1953 Rakosi lost the leadership to Imre Nagy
following the death of Stalin. Nagy's New Course sought to
abandon forced collectivisation and industrialisation but he
was faced with almost insurmountable problems. The unre-
mitting hostility of Rakosi and party apparatchiks towards
greater liberalisation led to the removal of Nagy and his
replacement by the hardline Erno Gero in April 1955. This
sudden reversal of direction placed an intolerable strain on the
Hungarian political system but Nagy's New Course had already
acted as an outlet for political and social frustration which
undermined party authority. The New Course had a

> profound psychological impact in Hungary... and led to a great
> sense of indignation and moral outrage against past abuses. The
> party's authority and cohesion were further weakened by Khrus-
> chev's famous 'secret speech'... which accentuated ferment within
> the ranks of the party and among the intellectuals. By October
> 1956 the party was so weakened and discredited that in the early
> days of the uprising it virtually disintegrated.[1]

The HPA proved incapable of defending communism in 1956
and following the unsuccessful popular uprising, the HPA was
demobilised and reformed.

THE SOVIET MILITARY PRESENCE IN HUNGARY

Russophobia, historically engendered by the intervention of
Tsarist troops to crush the abortive national revolution of
1848, was intensified by Soviet actions during the liberation
and post-war period. Mutual distrust remains an integral
element of Soviet–Hungarian relations. Clearly the Soviets still
regard the reliability of the Hungarian military with great
suspicion. Undoubtedly, the continuing Soviet military presence acts as a safeguard against the recurrence of political
protest.

The thirtieth anniversary of the Hungarian Revolution was
greeted cautiously by the authorities. Low-key commmemora-
tive ceremonies were held in Budapest and Szolnok. The
Hungarian media presented the official version of events,
acknowledging the abuses of the Rakosi regime and the
justifiable indignation this aroused among the proletariat.
They repeated the allegation that the situation was exacerbated
by "counter revolutionaries" working with Western propaganda agencies, claiming that only the timely intervention of a

reformed party under Kadar, together with Soviet assistance, restored order. Imre Nagy's proclamation of Hungarian neutrality and withdrawal from the WTO on 1 November 1956 was short-lived. Janos Kadar was installed as a leader more acceptable to the Soviets and Nagy was eventually executed. In May 1957, Hungary and the Soviet Union signed an accord legalising the stationing of Soviet troops in Hungary. Soviet units have been "temporarily" stationed in Hungary since 1945.

The Southern Group of Forces (SGF) is the smallest Soviet contingent in a WTO state, with the exception of Poland. Nevertheless, the SGF could easily be strengthened by deployments from neighbouring Soviet military districts. The SGF numbers between 65,000 to 80,000 and is divided into three corps. They have two tank divisions and two motorised rifle divisions, stationed in north-western Hungary, in the Lake Balaton, Szombathely, Esztergom, and Kecskemet areas. With 1,250 main battle tanks, 30 nuclear missiles and 420 front-line aircraft these forces are at a high level of preparedness, probably category two.[2]

HUNGARIAN MILITARY ORGANISATION, CAPABILITIES AND REFORMS

Soviet influence is obviously not restricted to the imposition of the SGF. The Hungarian Army is strictly subordinated within the WTO system of command and control. In 1968, the Hungarians sent 20,000 soldiers, albeit reluctantly, to participate together with other WTO members in the invasion of Czechoslovakia. Subordination has been assisted by the relatively high number of Hungarian army commanders who have attended Soviet military academies, as a prerequisite for career advancement. Hungarian claims to possess a sovereign army appear rather threadbare when set against their evident lack of military autonomy.

In domestic political terms the Hungarian military probably enjoys the least prestige and importance among Eastern bloc societies.[3] The HPA is controlled by the Central Committee and Politbureau of the governing Hungarian Socialist Workers' Party (HSWP). After 1956, applicants for top ranking military positions were subject to a meticulous selection process by party officials.

Membership of the HSWP is essential for an ambitious

officer, although the percentage of party members in the HPA is probably the lowest among southern tier armies.[4] Even the highest ranking military leaders struggle to exert great political influence. A notable example was Defence Minister Czinege who never became a Politbureau member, despite holding the position for over 20 years.

The Hungarian army of 84,000, including 50,000 conscripts, is the smallest in the WTO. In addition, the 1986 defence budget of 40 billion forints was one of the lowest among WTO members. A 34,900 billion forint defence budget is envisaged for 1988 with approximately 4 per cent of national resources devoted to defence.

With severe economic problems the Hungarians cannot afford the latest in military technology. Serious economic difficulties have already affected the military with a 9 per cent decline in the number of army reservists drafted this year and a 30 per cent cut in leading military bodies. Furthermore, the restructuring of the HPA resulted in a much smaller force than the 11 divisions which faced the Soviets in 1956. Hungarian divisions only reappeared in the order of battle of the WTO during the mid-1960s. The Soviets have not encouraged a substantial Hungarian military build-up. From the Hungarian perspective, a continuing Soviet presence allows resources to be diverted towards domestic concerns. Hungarian manufacture of their own defence equipment is extremely limited but the state arsenals produce the FUG (or OT-65) Amphibious Scout Car, the PSZH-IV Armoured Personnel Carrier and the AKM-63 assault rifle.

Despite this limited defence commitment, military reforms have been implemented with some alacrity. In a nine month reorganisation the Hungarians eliminated regiments and divisions, substituting them with brigades. Ostensibly this reform was introduced as a cost effective measure, drastically reducing the number of field officers. It may have been carried out at the behest of the USSR as a role model for their military reforms but at present this is mere conjecture.

Hungarian military modernisation involves a very lengthy planning cycle. Obsolete equipment is replaced but only within tight financial constraints. They possess more than 100 T-72s and some 1,200 T-54s, and T-55s. The HPA also have around 350 Soviet made BMP-1s. 122 and 152 mm self-propelled artillery pieces form an important part of their defences, together with the Sagger anti-tank guided weapon.[5]

The air force possesses 10 MiG-23s (Flogger) in addition to its 120 MiG-21s (Fishbed). Hungarian air defence is composed of three fighter regiments and nine interceptor squadrons. In terms of fighter ground attack they have one squadron with 15 SU-25s. For reconaissance they enjoy the use of one squadron equipped with 15 SU-22s. Two squadrons of Il-14s form an important transport/lift capability.[6]

The Hungarians have gone to enormous lengths to protect themselves from air attack, including a surface to air ring around Budapest. Three regiments equipped with surface-to-air missiles, in 20 sites across the country, testify to their concern. The impending departure of the 401st Tactical Air Wing from Torrejon in Spain and the possible deployment of US F-16s in northern Italy, has only served to increase Hungarian fears.

The Danube Flotilla is mostly Yugoslavian made but Hungary does possess a small domestic capability. Twenty-six 25-ton minelayers/sweepers and a number of mine counter-measure vessels form an important part of the Flotilla's activities.

Domestic economic interests are likely to determine military procurement and organisation for the foreseeable future. Despite the strategic importance of the Danube basin, the Hungarians have neither the will nor the resources to devote greater expenditure towards defence. The calamitous defeat suffered by the Hungarian Second Army during the Second World War and the events of 1956, substantially eroded previous public esteem of the military. The Hungarian Army is extremely important as a participant in the realisation of high-priority investments, producing an annual value of 5.3 billion forints and contributing 510,000 working days to the autumn harvests.[7]

HUNGARY'S ROLE IN THE WTO

Hungary joined the WTO as a founder member in 1955, offering a wide range of strategic options for Central/Southern Region deployment and even activities further afield. Hungarian bases and airfields proved to be an important staging area during the October 1973 war airlift operations.

Hungary also constitutes a potential base from which the Soviets can exert military/political pressure on recalcitrant allies or troublesome neutrals. Romanian autonomy in foreign policy and Yugoslavian adherence to a policy of non-alignment has proved particularly irksome for the Soviets in the past. During a crisis the Hungarian army might be compelled to help enforce allegiance to Soviet wartime objectives.

In a future conflict Soviet troops might need to traverse Romania and Yugoslavia. Hungarian forces could assist by threatening Transylvania in order to allow the Soviets free passage on their way south. They could also be used to secure Vojvodina in a campaign to violate Yugoslavian neutrality. Despite these theoretical possibilities the military contribution of the HPA is likely to be restricted to a supporting defensive oriented role. Hungarian forces are trained to secure lines of communication and build bridges in support of amphibious operations. They could also provide a protective cordon around advancing Soviet troops.

In the event of general war, Hungary would probably be used as a launch pad for a WTO offensive across Yugoslavia and into the northern Italian plain. It could offer a route for continuous reinforcements, coupled with military support. Such a campaign might serve to strengthen the Soviet presence in the Mediterranean, and severely restrict the effectiveness of the Italian army and navy.

Another scenerio involving Hungary might feature an offensive up the Danube valley into southern Germany. Exercise patterns such as "Shield-82" and "Friendship-88" give a good indication of potential roles. Hungarian troops have participated in Central Region exercises together with those in the southern tier. However, their role is more likely to be directed towards operations in southern Europe.

MILITARY RELIABILITY AND RELATIONS WITH OTHER WTO ALLIES

The reliability of the Hungarian armed forces depends not only on the quality of their equipment but also on morale, motivation and situation. Most analysts concur that the Hungarians would fight fiercely and effectively against any aggressor to defend their territorial integrity.

Hungarian reliability in an offensive context is highly problematical. The repossession of historical Hungarian territories might be a major incentive. The Treaty of Trianon in 1920 decimated the territory and population of pre-1914 Hungary and led directly to Hungary's alliance with Nazi

Germany in the 1940s. Its legacy continues to adversely affect Hungary's relations with neighbouring states.

The Soviets could exploit Hungarian historical grievances against Czechoslovakia, Yugoslavia and Romania. In particular the treatment of some two million Magyars by the Romanian authorities causes grave concern. The Hungarian minority issue is a seemingly intractable problem. Since the beginning of 1988 some 10,000 of Romania's ethnic Hungarians have fled from severe repression to their motherland. Many cross the border illegally without passports to escape further linguistic and employment related discrimination. Not only do they form one of Europe's largest minorities but the deterioration in Hungarian-Romanian relations has resulted in an unprecedented situation in the Eastern bloc. Never before has one WTO country accepted refugees from another "ally". For many years the Hungarian authorities refrained from any criticism of Romania's human rights record but with the advent of Mr Gorbachev and an increasingly intolerant regime in Bucharest they are now expressing their disquiet. The Hungarians signed a Western resolution on minority rights at the Vienna Human Rights Conference and recent events have done nothing to allay their fears. President Ceauşescu plans to level 8,000 of Romania's 13,000 villages and replace them with 500 large agro-industrial complexes. The majority of these villages are in Transylvania which contains most of Romania's Hungarian minority. In the words of Karoly Grosz the loss of the villages would be an "irrecoverable losss not only for Romania and the Hungarian nationality but for the whole of mankind".[8]

The intensity of the crisis clearly threatens to shatter the fragile "unity" of the WTO's southern tier. Mr Gorbachev is loth to intervene as the Soviet Union faces severe nationality problems, especially in Azerbaijan and Armenia. Furthermore, such tensions are common throughout Eastern Europe. The Turkish minority in Bulgaria has suffered persecution and Yugoslavia is subject to a long running dispute between Serbs and Albanians over the future of the Kosovo province. Eastern Europe's potentially volatile and antagonistic nationalities make it extremely difficult to predict the performance of non-Soviet WTO forces in a war against NATO. Hungarian room for maneouvre may be limited by the presence of Soviet forces but Moscow can only be reasonably confident of their reliability in a defensive context. "The Hungarian army— which has gone through a comprehensive renovation process since 1956—has clearly been unable to solve its most pressing problems of providing for an effective professional cadre, efficiency and morale consistent with required precepts." [9] The nationalist fervour so evident in Hungarian political culture could be exploited by the Soviets during a conflict. Nevertheless, Hungarian nationalism is very difficult to reconcile with an alliance system based on the principle of "proletarian internationalism".

The Hungarian attitude towards NATO is harder to discern. While Moscow trained high ranking officers would probably remain loyal in a conflict, the lower-level recruits and NCOs might quickly lose their enthusiasm if the WTO suffered a number of military reverses.

HUNGARIAN FOREIGN POLICY

It would be erroneous to consider Hungarian security policy in isolation from the country's foreign policy. Even during the height of East-West tensions in the early 1980s, the Hungarians attempted to encourage dialogue and moderation. The views of Gyula Horn, an influential figure in Hungarian politics, are representative of Budapest's "objective" foreign policy. "Apart from the obvious differences, a community of interests exists between the socialist and the capitalist world, mainly in safeguarding world peace and working out concrete ways and means to considerably lower the level of military tension." [10]   Developments in Hungarian foreign policy have been inextricably linked with the emergence of a distinctive form of communism post-1956. In return for stability, the Soviets allowed Kadar a considerable degree of freedom in domestic policies. The introduction of the New Economic Mechanism (NEM) in 1968, a relatively liberal human rights policy based on the Helsinki Final Act and an openness pre-dating *glasnost* have characterised "goulash communism".

Some autonomy has also been evident in foreign policy despite the fact that good relations with the Soviet Union form the cornerstone of the Hungarian approach. The Kadar regime expressed concern at the Soviet invasion of Afghanistan, anxious that their generally good relations with the West would not be damaged by this issue. Economic interests were only one motivating factor. The Hungarians recognised the dangers of international tensions as a potentially vulnerable state. During the mid-1970s a number of

Hungarian leaders voiced concern at plans to deploy SS-20 missiles in the western military districts of the Soviet Union. The Hungarians were also opposed to Soviet "counter-measures" designed to respond to NATO's INF deployment.[11]

The Hungarians have pursued an active and ideologically flexible foreign policy. Membership of the World Bank, the International Monetary Fund and investment protection agreements with six capitalist countries demonstrates their non-dogmatic approach. Moreover, initiatives in conventional arms control and the organisation of the Budapest Cultural Forum have been important in promoting East–West understanding. Only three weeks after becoming head of the HSWP Karoly Grosz announced that he would visit President Reagan, giving a clear indication that Hungary's foreign policy objectives are likely to remain directed towards the cultivation of close contacts with the West, including the EEC, and the maintenance of detente.[12] However, relations with the Soviet Union could rapidly deteriorate if large scale political protest returns. Mr Gorbachev has given an undertaking to his WTO allies promising greater freedom in the management of their own internal affairs, in an apparent reversal of the old Brezhnev Doctrine. It remains to be seen whether this policy will be maintained if countries such as Hungary continue to experience serious economic and political difficulties, which could eventually threaten the stability and cohesion of the WTO. Conversely, if Mr Grosz's reforms are successful Mr Gorbachev might feel inclined to review the present role of the SGF. As the Soviets have withdrawn from Afghanistan their military presence in Eastern Europe could be re-evaluated. In the light of continuing economic problems the Soviets might regard the SGF as a unaffordable luxury if Hungary manages to solve its own domestic problems. Intense speculation surrounded the possible withdrawal of the SGE, particularly after press reports of "an impassioned plea" by Karoly Grosz during the WTO's annual summit in Warsaw in 15 July 1988. The WTO failed to offer a unilateral withdrawal of the SGF and any future agreement seems likely to be conditioned by the course of MBRF talks and other international negotiations.[13]

CONCLUSION

However, Hungary's ever worsening foreign debt may provoke a crisis with wide ranging ramifications. The introduction of an austerity programme, including unemployment, income and value added tax is unprecedented in Eastern Europe. It remains to be seen whether falling living standards will lead to a high level of social and political unrest. The gravity of the situation has already facilitated a series of changes unparalleled in post-war Hungary.

In many respects the Hungarians have earned a justifiable reputation for openness within the Eastern bloc. During December 1987, the Defence Commission of the National Assembly discussed the security implications of the State Budget in open session for the first time since the communist takeover. The removal of Janos Kadar as general secretary of the HSWP on May 22nd 1988, was the first example of an Eastern bloc party leader relinquishing power after an open debate of his leadership. Gustav Husak's departure in Czechoslovakia resulted in relatively minor changes but it seems that Hungary has embarked on one of the most radical reversals of direction in Eastern Europe since the Prague Spring of 1968. Not only did the Central Committee replace Kadar with Prime Minister Karoly Grosz but eight of the Politburo's 13 members were removed in favour of such radical figures as the architect of the NEM Rezso Nyers and Imre Pozsgay. As general secretary of the People's Patriotic Front, Hungary's leading reformist institution, Pozsgay strongly advocated fundamental political reform exemplified by the phrase "I believe in democracy with no qualifying adjectives".[14]

There is a certain irony in the fact that Karoly Grosz was once regarded by the Hungarian intelligentsia as a hardliner, while Janos Kadar was responsible for many important political and economic reforms. Grosz is seen a a technocrat committed to the introduction of more market forces into the economy but less enamoured of calls for far reaching political reforms. However, Mr Grosz has accepted the need for greater "democratisation" and the decision of the HSWP to chose a non-communist scientist as President is extremely significant. Bruno Straub has been an "independent" member of parliament since 1985 and is expected to hold the position until 1990 when a new constitution is planned. The political importance of the post has clearly declined but Straub's nomination strengthened the forces of modernisation and change. In the age of *glasnost* and *perestroika* Hungary's reform programme is likely to be fully supported by Mr Gorbachev. Hungary's geostrategic importance to the Soviet Union ensures that the Kremlin will continue to play close attention to Hungary's domestic travails. The role Hungary plays in the WTO is likely to be radically altered by the accession of Grosz but the future of Soviet-Hungarian relations are not only dependent on the success of Gorbachev, they will also be determined by the economic, social and political progress of the new regime in Budapest. In a sense the Hungarian political elite are faced with similar problems to those besetting the Soviet Union, namely how to achieve economic progress and "socialist pluralism" without destroying the leading role of the party.[15] Of all the WTO nations Hungary has gone the furthest in matching or even exceeding the political and economic reforms espoused by Gorbachev.

NOTES

[1] Stephen Larrabee, "Soviet Crisis Management in Eastern Europe". In *The Warsaw Pact: Alliance in Transition*, David Holloway and Jane Sharp (eds). Ithaca, New York: Cornell University Press, 1984, p. 115.

[2] *Jane's Defence Weekly*, 28 March 1987.

[3] Joni Lovenduski and Jean Woodall, *Politics and Society in Eastern Europe*, Macmillan 1987, p. 306 and George Schopflin, "Hungary". In *Leadership and Succession in the Soviet Union, Eastern Europe and China*, Martin McCauley and Stephen Carter (eds) Macmillan 1986, p. 110.

[4] Ivan Volgyes, *The Political Reliability of the Warsaw Pact Armies. The Southern Tier*, Durham N.C.; Duke University Press 1982, p. 73.

[5] *The Military Balance 1987–1988*, IISS p. 51.

[6] Ibid.

[7] Hans-Georg Heinrich, *Hungary, Politics, Economics and Society*, Frances Pinter 1986, p. 173.

[8] *The Guardian*, 28 June 1988.

[9] Zoltan D. Barany and Ivan Sylvain, "Hungary". In *Warsaw Pact: The Question of Cohesion Phase II—Volume 3 Union of Soviet Socialist Republics, Bulgaria, Czechoslovakia and Hungary*. Teresa Rakowska-Harmstone et al. ORAE Extra Mural Paper No. 59, Ottowa, March 1986, p. 454.

[10] Gyula Horn, "Hungary in the System of International Relations—Conclusions and Perspectives", *Kulpolitika, Hungary and the World*, Budapest 1987, pp. 12–13.

[11] The failure of the WTO to fully endorse the Soviet position on NATO's INF deployment was partly due to Hungarian resistance, although the Romanians were the most forthright in the opposition to "countermeasures".

[12] According to *Magyar Nemzet*, 30 January 1988, in 1987 some 19 million foreigners crossed Hungarian borders and almost 12 million came as tourists. 23 per cent came from non-socialist countries, a 30 per cent increase on 1986. Given Hungary's economic plight, the need to encourage tourism is an urgent priority which is another reason why bonds with the West are likely to be strengthened.

[13] Grosz reportedly stated that the withdrawal of Soviet troops is now linked to the non-deployment of US F-16s in Italy. During his visit to the US, Grosz confirmed that he had discussed the presence of the SGF with Gorbachev but any withdrawal would have to be associated with similar agreements in other European areas.

[14] *The Guardian*, 24 May 1988.

[15] On the eve of a visit to Moscow, Karoly Grosz remarked that the process taking place in the Soviet Union was in many respects identical with events in Hungary.

Space and Verification of Conventional Arms Reductions

The author was a Senior Research Fellow at the Stockholm International Peace Research Institute heading the research programme on the military use of outer space. He was co-editor of the ABM Treaty—to defend or not to defend? *published by Oxford University Press in conjunction with SIPRI in 1987 and is currently a research fellow on the Space and International Security Programme at the RUSI.*

ONE OF the most important applications of observations from space is in the field of international security, particularly through the arms control process. In their Summit statement of December 1987, President Ronald Reagan and General Secretary Mikhail Gorbachev declared that they had "discussed the importance of the task of reducing the level of military confrontation in Europe in the area of armed forces and conventional armament".[1] It is, therefore, now important to work out a verification procedure for a possible agreement involving any reduction in and elimination of some conventional forces in Europe.

While a number of other sources of information have to be used for effective verification, observation from space could form an important element of the verification procedure because of its non-intrusive character. The importance of the recently concluded INF Agreement is in the fact that it laid down the principle of on-site inspection which the Superpowers had not accepted previously. In a conventional forces agreement, the number of inspections are likely to be fixed so that it may not always be easy to know when to ask for additional on-site inspections. Here lies another advantage of observations from space. It can help raise questions regarding compliance with the terms of an agreement and facilitate further on-site inspections. Surveillance from satellites could, therefore, enhance the implementation of the provisions of the Stockholm Conference on Confidence and Security—Building Measures and Disarmament in Europe (CCSBMDE).

On-site inspection does not only mean the presence of inspectors on the territory of the inspected country. There could be ground-based sensors which could, for example, be monitoring the movements of forces in and out of agreed areas. Information generated from such remote sensors could be transmitted to the monitoring country via communications satellites. Thus, space has an additional role to play in the verification process.

WHAT IS TO BE VERIFIED?

A treaty on conventional arms reduction may involve reducing equipment and manpower. The former may be achieved, for example, by destroying them, or by withdrawing them to specified agreed areas and storing them in secured places. This section does not contain an exhaustive list of every possible object and activity to be verified but only some examples because this will depend on the nature of any agreement finally concluded. Under the CCSBMDE, for example, major military manoeuvres are to be monitored by on-site observers to reduce the fear of surprise attack. However, should there be a conventional arms control agreement in Europe, it would be essential to monitor such objects as armoured vehicles, deployment of missiles and aircraft and troop concentrations.

The type of targets likely to be monitored are listed in Table 1 which also lists the capabilities of sensors required for different interpretation tasks in verification. It is clear that verification involves determining the type and number of vehicles, tanks, and aircraft, and their locations, as well as the battle-organisation of forces. The existence, size and development of manoeuvres, even of the back-up forces will need to be determined. For such tasks, various kinds of sensors are essential. These are discussed below.

IMAGING SENSORS ON BOARD SATELLITES

A sensor is an essential element of a surveillance system. It can record images of objects on the earth's surface by detecting electromagnetic radiation emitted or reflected by them. From orbit, it is possible to photograph objects in considerable detail. Weather is no longer a contraint because radar can penetrate heavy clouds and heat-sensing infrared sensors can see in the dark. Such devices can also indicate whether camouflage is being used. The various types of sensors

TABLE 1—RESOLUTION (IN METRES) REQUIRED FOR INTERPRETATION TASKS

Target	Detection[a]	General identification[b]	Precise identification[c]	Description[d]	Analysis
Bridges	6	4.6	1.5	0.9	0.3
Communications					
Radar	3	0.9	0.3	0.15	0.04
Radio	3	1.5	0.3	0.15	0.15
Supply dump	1.5	0.6	0.3	0.03	0.03
Troop units	6	2	1.2	0.3	0.08
Airfield facilities	6	4.6	3	0.3	0.15
Rockets and artillery	0.9	0.6	0.15	0.05	0.01
Aircraft	4.6	1.5	0.9	0.15	0.03
Command and control					
headquarters	3	1.5	0.9	0.15	0.03
Missile sites					
(SSM/SAM)[e]	3	1.5	0.6	0.3	0.08
Vehicles	1.5	0.6	0.3	0.05	0.03
Land minefields	9	6	0.9	0.03	—
Railway yards and					
shops	30.5	15	6	1.5	0.6
Roads	9	6	1.8	0.6	0.15
Urban areas	61	30.5	3	3	0.3

[a] Requires location of a class of units, object or activity of military interest.
[b] Requires determination of general target type.
[c] Requires discrimination within target types of known types.
[d] Requires size/dimension, configuration/layout, components construction, count of equipment, etc.
[e] SSM and SAM refer to surface-to-surface missiles (i.e., intercontinental or intermediate range missiles) and surface-to-air (i.e.,anti-aircraft) missiles respectively.
Source: Reconnaissance Handy Book (McDonnell Douglas Corp., USA), p. 125.

used on board satellites to observe the earth's surface are summarised in Table 2. Considerable progress has been made in their capabilities.

IMAGING DEVICES IN VISIBLE SPECTRUM

Notable advances have been made in infrared sensors, particularly in the focal-plane surveillance systems. One of the important elements of such a system, for example, is micro chips containing mosaic of detectors on one side and a corresponding number of small signal processing elements called charge-coupled devices (CCDs) on the other. These are high density information storage devices constructed from gallium arsenide or mercury-cadmium telluride. Information in the form of images of a scene is stored as electrical charges

TABLE 2—TYPES OF SENSOR ON BOARD SATELLITES

Sensor	Functions	Type of satellite
Photographic	Detection of: ABM systems; ICBM; military facilities and troop movements	US photographic reconnaissance satellites such as the KH series and many of the Soviet Cosmos satellites
Infrared	Detection of: missiles; aircraft and cruise missiles (potential)	US and Soviet early-warning satellites
Ultraviolet	Detection of fluorescence from gases surrounding booster or nose cones during ballistic flight	US and Soviet early-warning satellites
Electronic signal detectors	Radio and microwave telephonic transmission interception; radar signal detection; missile telemetry interception	Soviet Cosmos and US electronic reconnaissance satellites
Radar	Detection of: naval surface ships; many other ground-based military objects	US Seasat I and Soviet Cosmos ocean surveillance satellites
Thermal infrared scanners or radiometers	Night-time reconnaissance; detection of targets such as buried structure or underground construction	US Seasat I

on a linear or two-dimensional array of closely spaced electrodes or the CCDs. The stored charges can be read in a sequential manner. Each picture element ("pixel") (see Figure 1) is about one 30 millionth of a square metre. In the USA, 12,000 detectors using mercury-cadmium telluride have been produced in a mosaic for a focal plane sensor.[2] An advantage of this material over a silicon-based sensor is that the detector can spot objects in both the near- and far-infrared region of the electromagnetic spectrum. Recently, a platinum sillicide hybride chip containing a mosaic of 65,536 detectors has been developed.[3] Such a focal plane detector called the TEAL RUBY is planned to be orbited aboard the AFP 888 satellite. The purpose of the TEAL RUBY is to detect aircraft in flight from space.

Performance of such devices is measured in instantaneous field of view (IFOV) rather than ground resolution.[4] The former is defined as the size of the spot on the ground "seen" by one particular point in the image or seen by a scanning sensor at the instant of observation (see Figure 1). Thus, better definition or resolution of an image is obtained if the pixel size is small. It should be recalled that the quality of an image does not depend solely on the IFOV or resolution of a system but

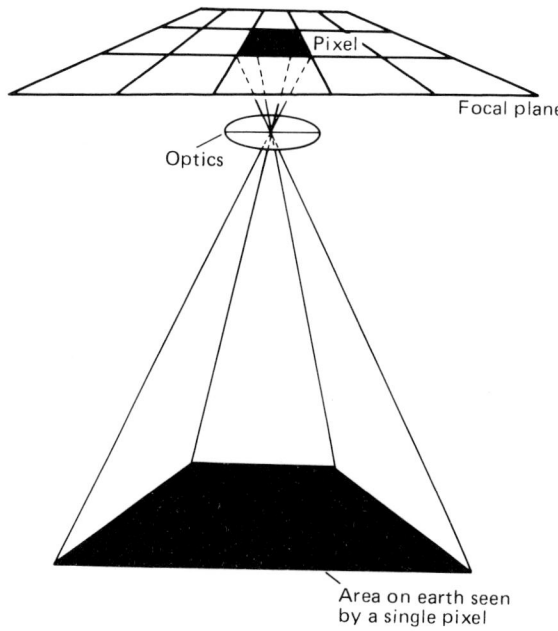

FIG. 1—Concept of a pixel.

also, for example, on the contrast of the image. The latter is defined as the ratio of the brightness of an object to that of the background. In a standard scene, the contrast ratio is 2.5:1. In this case the IFOV is about half the resolution indicated in Table 1.

The concepts of IFOV and resolution are best illustrated by Figures 2 and 3. In Figure 2, the small shaded area (Figure 2a) represents an area on earth of 25m × 25m which a single pixel can see. In this case the sensor, which consists of an array of pixels, is said to have an IFOV of 25m. In the same figure the larger shaded area is made up of eight 10m × 10m areas and it is difficult to see what sort of object is being viewed. In fact the sensor is looking at two Soviet T–54/55 tanks standing in a row one metre apart. Details of the dark areas are not resolved unless the IFOV of the sensor is improved to about one metre. Even then, the two tanks are only just separated. The IFOV of the sensor has to be about half a metre before the tanks could be recognised as tanks and seen clearly as two tanks.

To recognise a tank, a sensor with a resolution of about one metre is needed (see Figure 4c). However, a fighter aircraft such as a US F–15 could be recognised as an aircraft with a

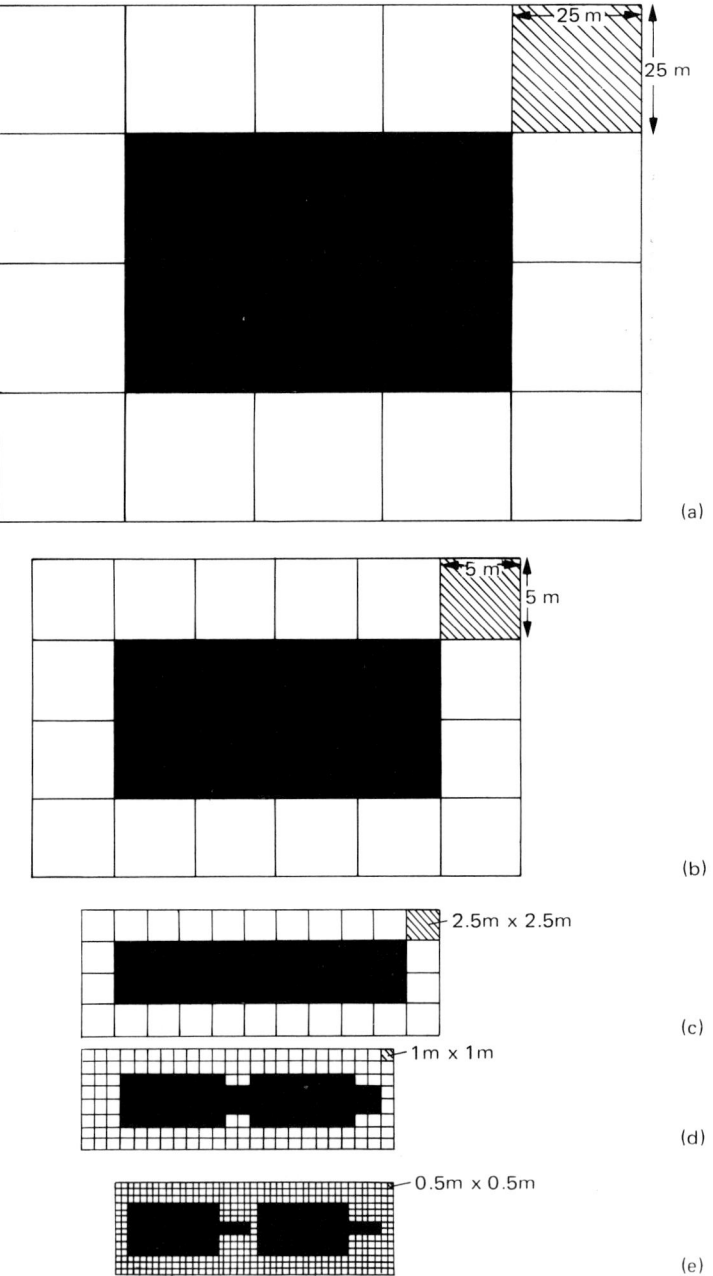

FIG. 2—The effect of resolution on satellite image of two Soviet T–54/55 tanks deployed in a row.

sensor of only five metre resolution. The aircraft could be described in general terms if a sensor has a resolution of between five and two and a half metres (see Figure 3).

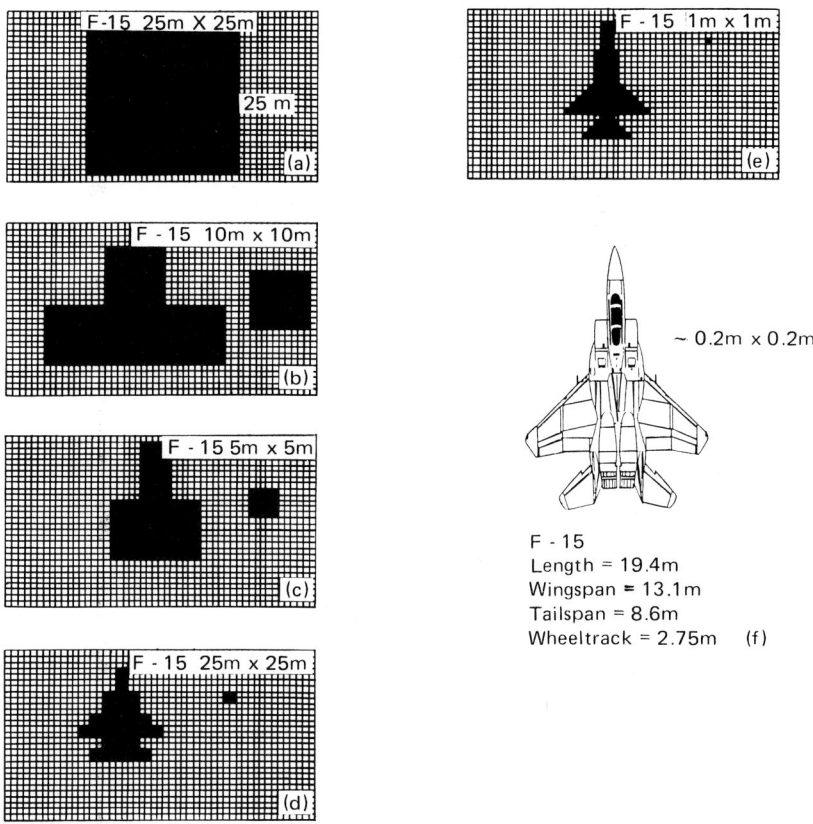

FIG. 3—The effect of resolution on the satellite image of an F–15 aircraft which may be carrying conventional weapons. The small shaded area in (a) represents an area of 25m × 25m which a single pixel can seen on the earth. The sensor thus has a resolution of 25m IFOV. As the resolution is progressively improved, as in b, c, d, and e, to an IFOV of 1m, the F–15 is recognisable as an F–15 aircraft. For a complete description a resolution of .02m is necessary (as in f).

US and Soviet military reconnaissance satellites have resolutions better than half a metre. However, such high resolution systems may not always be necessary because data available from civil remote sensing spacecraft with IFOV of 10m could be useful. For example, from Figure 2a, it can be seen that for one tank, there are four pixels being used covering an area of 400 sq m on the ground. For two tanks, there would be six pixels used giving the ground area of 600 sq m and for three tanks the latter would be 800 sq m. If the number of tanks in a row is increased, it is found that there is a linear relationship between the number of tanks in a row and the area on the ground recorded on the sensor.

Thus, if it is known that the sensor is recording the presence of tanks, then the number of these could be determined from the area recorded by the sensor. Theoretically it is possible to detect the removal of a single tank, as it represents two pixels or an area of 400 sq m. In reality, however, a tank or any other object has to be observed against a background consisting of a multitude of objects. Since the CCD data is in digital form, it is possible to remove the effects of the background to a large extent if the scene were recorded prior to the deployment of the tanks. It may also be possible to apply this technique to observations of troop movements and other targets.

In Figure 4, a comparison is drawn between a tank and an anti-personnel vehicle. It can be seen that with an IFOV of 2.5 m, the two vehicles are resolved as two separate objects but they cannot be described. A tank can be recognised at an IFOV of 1.0m, (Figure 4c). An additional technique to distinguish between a truck and a tank or between other types of vehicles is to measure their speed. For example, the French earth resources satellite, SPOT 1, carries two cameras (one panchromatic and one multispectral) pointing slightly off nadir in different directions along-track. The result is a short time-lag between the two recorded images of the same scene. To detect motion, one image is subtracted from the other.[5]

These illustrate that the resolution of a sensor system could be improved by reducing the size of pixels. The best resolution from civil satellites so far is 10m (IFOV) obtained from the French SPOT spacecraft which orbits at about 800 km. If, however, this altitude is lowered, considerable improvement in the resolution would be achieved. The effect of this is to reduce the area observed by the sensor on the earth's surface (see Figure 1). In any case, the Soviet Union is now selling images from its Cosmos earth resources satellites with resolution of between five and 10m.

Imaging radars

Another important sensor is the synthetic-aperture radar (SAR) which has a relatively short antenna which is made to

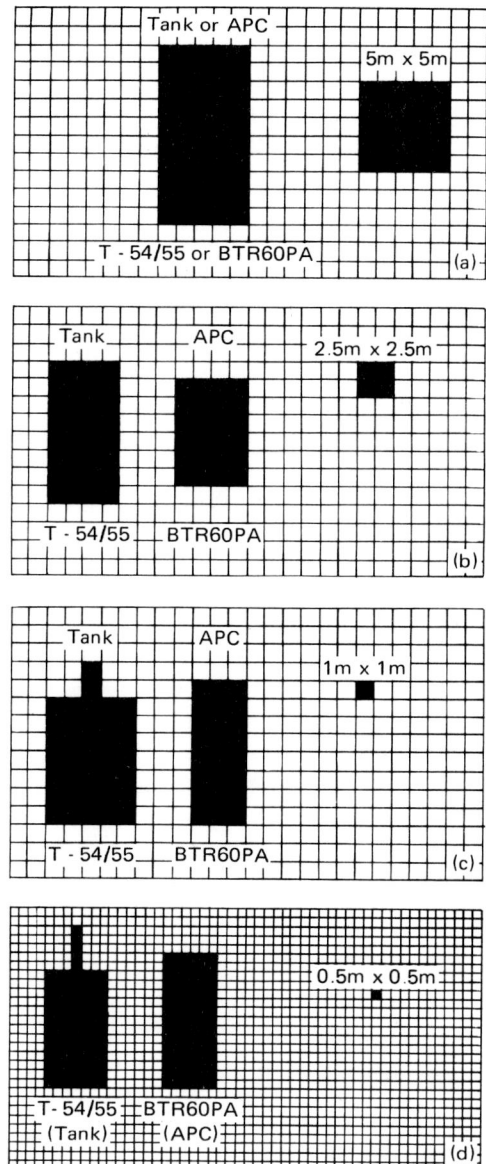

FIG. 4—Comparison between a Soviet T–54/55 tank and an armoured personnel carrier (APC) at various resolutions.

behave like a long one with a narrow beam. The resolution of a radar depends on, among other factors, the length of its antenna. A short antenna can be simulated to represent a very long one by taking advantage of the motion of the satellite. Signals from a short antenna are added electronically and synthesised to give the effect of a long antenna. The latter is a requirement for high resolution.

Electronic sensors on board spacecraft

Much can be learnt about military activities on earth by monitoring the electronic signals they generate. An example of this is that of the signals originating from military communications between bases during military manoeuvres. One of the most important types of sensors carried on board electronic reconnaissance satellites are electronic devices consisting of sensitive antennas to monitor signals originating from missiles during their test flights, for example. The telemetric measurements from missile tests would yield data on booster size, weight and range of the missiles. Moreover, such sensors also measure the characteristics of signals to enable assessment of qualitative changes in the performances of various types of systems emitting the signals. Electronic reconnaissance satellites could also locate and pinpoint the position of objects producing signals.

Moreover, under a future conventional arms reduction treaty, there may be a limit established on the movement of permissible equipment such as tanks, helicopters and fuel trucks. These could be tagged with tamper-resistant electronic signal generating devices. The signals being transmitted continuously could then be monitored by electronic reconnaissance satellites.

THE ROLE OF COMMUNICATIONS SATELLITES

Space and ground-based surveillance systems generate a considerable amount of data. This could be from the ground-based sensors placed around, for example, areas where treaty limited tanks may be deployed. Any movements of these could be detected by seismic or pressure devices. Signals from tags on military vehicles detected by electronic reconnaissance satellites would also need secure transmission to the monitoring country.

Transmission of such information in a reliable and secure way could be achieved using communications satellites. Besides the USA and the Soviet Union, France, the UK and NATO also use military communications satellites. Therefore, this element of the space segment is already widely available.

CONCLUSIONS

Considerable progress has been made in the sensors deployed onboard civil satellites for remote sensing. Not only this but image processing and interpretation systems are also becoming widespread. Nations other than the two big powers, are now able, to a certain extent, to use observations made from space to monitor compliance with some existing arms control treaties and possibly some of the ones currently being discussed. Moreover, nations other than the two big powers have the capability to use and to orbit communications satellites which may be essential for secure transmission of data collected over inaccessible places.

If there is to be a wider participation in multilateral arms control agreements, international or multinational verification is important. Should there be an agreement on reducing conventional weapons in Europe, the region would need to get involved in the verification process. As Europe, at present, has no space-based reconnaissance capabilities, it could use, to some extent, data obtained from civil earth observation satellites. For example, observations from SPOT 1 type satellites could be used for verification. Europe also has significant capability to launch and operate communications satellites, an important element for verification tasks. Moreover, Europe also has significant launcher capability. Thus, a European regional satellite monitoring agency (RSMA) has been proposed.[6].

However, many political and institutional problems must be solved before multinational remote sensing from space gains acceptance for arms control verification.

Notes

[1] "Soviet-US Summit Statement", *Soviet News*, No. 6405, 16 December 1987, p. 450; and "Soviet Union–United States Summit in Washington—Joint Statement, December 10, 1987", *Weekly Compilation of Presidential Documents*, Vol. 23, No. 49, 14 December 1987, p. 1497.
[2] Calif, A., "Aerojet ElectroSystems' investment in sensor technology yields $500 million in contracts", *Aviation Week & Space Technology*, Vol. 129, No. 3, 18 January 1988, p. 66.

[3] Welling, W., "Focal plane arrays: brightening the picture for ground, air and space defence", *Military Technology*, No. 5, 1988, pp. 24–29.

[4] This is defined as the minimum distance between two identical small objects when they can still be distinguished as two separate objects.

[5] Stern, M., "A new tool: SPOT image for studying rapid movements", paper presented at the *Twenty-First International Symposium on Remote Sensing of the Environment*, Ann Arbor, Michigan, 26–30 October 1987.

[6] Jasani, B., "A regional satellite monitoring agency", *Environmental Conservation* Vol. 10, No. 3, 1983, pp. 225–6.

Military Space: A View from France

DR ALAIN DUPAS

*Dr. Alain Dupas is conducting advanced studies for the French space agency,
CNES, and is head of the Department of Science, Technology and Space in the CREST
(Centre d'étude des relations entre technologies et stratégies). He is the European editor of the journal
Space policy. The views expressed in this article are those of the author alone and do not reflect the policy
or opinions of CNES or CREST.*

FRANCE WAS the third country in the world to launch an artificial satellite with her own carrier rocket: A-1 ("Asterix") was orbited in November 1965 by the first Diamant launcher. As has also been the case in the USSR and the USA (and later in China), the building of this space carrier rocket originated in the development programme of ballistic missiles. The French Ministry of Defence was thus very much involved in the birth of the French space programme. At that time, military circles in France were already interested in applications of space systems for the support of military forces.

However, contrary to what happened in the two major space powers, this interest remained mainly theoretical, and the French space programme was totally civilian for more than 15 years. Things have only really been changing since the beginning of the 1980s, with the Syracuse military space communication system, and the future Helios reconnaissance satellite. This very late start was mainly caused by budgetary constraints; the new military space projects had to compete against more traditional defence programmes for limited resources. Such financial problems caused, for example, the cancellation of a precursor to the Helios project: the SAMRO programme in 1982.

Budgetary constraints are much more difficult to manage for the French military space budget than for the civilian one: costs of most civilian projects are shared between countries of the European Space Agency (ESA), while military space programmes are purely, or at least mainly, national. Cancellation of the SAMRO project came after the failure of cooperation talks between France and West Germany (FRG).

Efforts to combine the only two European military space communication systems in some way, the French Syracuse and the British Skynet projects, have been unsuccessful up to now. So far, the only success in cooperation at the European level in a military space programme has been the Helios project: Italy and Spain have agreed to engage in the development of the Helios system, for 15 per cent and 5 per cent of the cost respectively. This cooperation, is however, limited to the building of hardware: reconnaissance data from the Helios satellites will be used on a purely national basis. Observation time will be allocated between the three international partners according to their share in the development programme.

Western European countries are used to cooperating extensively in civilian space activities. The successes of the Ariane family of launchers and of the scientific satellites and probes (for example, the Giotto spacecraft sent towards the famous Halley comet in 1985) are a tribute to this very fruitful cooperation organised and managed by the ESA. Cooperation in the field of military space is much more difficult to start and implement. The ESA charter states that its activities should be confined to "peaceful" uses of outer space and conducted in accordance with international treaties. The American and Soviet experiences show that these constraints do not prevent the use of space systems for reconnaissance such as electronic, photographic and ocean reconnaissance, telecommunications, and early warning.

However, ESA involvement in such military space activities is probably impossible for political reasons: some of the thirteen ESA members are neutral countries (Austria, Sweden and Switzerland), or NATO countries which would object to any military programme engaged within ESA (Denmark, Norway and Ireland, for instance). This limitation does not apply to the commercial use of Ariane rockets to launch "peaceful" military satellites developed outside the ESA framework (at least by ESA member countries): Telecom-1 satellites, with the Syracuse military payloads, have been orbited by Ariane rockets, and Helios spacecrafts are also to be launched by Ariane.

A real European military space programme would need other institutional arrangements than ESA. Some thoughts have been given to the possible role of the Western European Union (WEU) organisation. The WEU's charter enables the cooperative development of military systems by its seven member states. Another option is a dedicated European military space agency. There is not, however, any strong drive towards such developments. Most European NATO member countries have simply no recognised needs for military space assets: they have no independent deterrenT capability, and rely indirectly through NATO on information from, or use of, American military satellites. This is, for example, the case for intelligence information on Soviet weaponry or military manoeuvres.

The FRG is *de facto* prohibited from becoming a nuclear power and is a member of NATO's integrated command. However, she plays such an important role in central Europe's defence that she would certainly benefit from space intelligence data collected by a national or European space system. She has considered a joint venture with France for the development of an optical reconnaissance satellite (the SAMRO system quoted earlier). Cloudy skies are, however, prevalent in central Europe, and an optical imaging system was not considered optimal for operational observations in the region. The FRG walked out of the project, but could very well be a partner for the development of a Helios follow-on using a synthetic aperture radar (SAR) providing an all-weather imaging capability in the 1990s. Political considerations could, however, deter FRG from embarking on military space activities: such developments could have a negative impact on relations between the FRG and the Soviet Union.

The UK is a nuclear power but is a member of the integrated NATO command, and has a special relationship with the United States. She is, however, the only Western country apart from the USA which has developed dedicated military telecommunication satellites: the Skynet spacecrafts (the French Syracuse system is only a guest payload on a civilian satellite). Information has also circulated about an British project for a ferreting satellite. The UK is thus a potential partner for France, or other European countries, for some kind of military space activity. Up to now, exchanges of views between France and the UK on military space communication systems have not been fruitful, as already noted.

France is not a member of NATO's integrated command, and has an independent nuclear deterrence capability with a triad of nuclear vectors: land based ballistic missiles, submarines carrying nuclear missiles and aircraft with nuclear bombs and cruise missiles. She has a fleet operating far from her bases in Europe, distant territories in various parts of the world (Caribbean Sea, South America, Indian Ocean, Pacific Ocean), and special relationships with many African countries. France is thus a special medium nuclear power; a member of an alliance (the North Atlantic Treaty), but with largely independent military forces, and some world-wide responsibilities. It may be difficult for France to live up to this ambition, with the share of her military budget staying under 4.5 per cent of GNP. However, all French governments since the 1960s have tried to stick to this policy, which is almost a non partisan issue in the country.

The interest of the French Ministry of Defence for some military space capabilities comes from this special international posture. Space communications systems are needed to connect the metropolitan territory with overseas bases, forces operating far away (in Chad for instance), distant ships, and so on. Reconnaissance data is useful for targetting purposes, independent appraisal of adverse military capabilities or of international crises in various regions of the world, and the evaluation of situations in which French forces are involved (again in Chad for instance). Last year's operations in Chad, when the French government was largely dependent on United States data to know the position of Libyan and rebel Chadian troops, have probably had a very positive role in the revival of the French reconnaissance satellite programme.

The main directions of the French military space programme were approved at the highest level (the "Defence Council" chaired by the French President) in 1985. They are summarised in French by a three word sentence: "Communiquer, écouter, voir" (communicate, listen, see), which points to the following military applications—communication, electronic reconnaissance and photo reconnaissance.

The first of these applications was implemented well before 1985; the first Telecom-1 satellite, with a Syracuse military payload, was orbited in 1983. Two others (Telecom-1B and Telecom-1C) have been launched since. Each Syracuse payload involves two 7–8 GHz transponders able to insure secure communications with fixed or mobile terminals with 1.3 to 3 m diameter antennas. At the end of 1987, the Syracuse system was using 27 operational terminals, of which 16 were on land and 11 on surface ships. A more advanced system, Syracuse-2, will be introduced in 1992, with the first second generation Telecom-2 satellite. The total payload (and cost) will be equally split between civilian (Ministry of Postal Services and Telecommunications) and military (Ministry of Defence) users, with five military transponders. About 100 small terminals (with antennas as small as 0.4 m) will be connected, some of them on nuclear submarines and AWACS planes. Three Telecom-2 satellites have been ordered, and a fourth will be ordered if needed. The system will be operational for more than 10 years.

The second application, electronic reconnaissance, will be implemented at some unspecified time in the future, and the status of the corresponding project is unknown. The third application, photo reconnaissance, is the objective of the Helios programme. The Helios satellite will be launched by Ariane in a polar sun-synchronous orbit about 850 km above the surface of the earth. It will use a bus modified from the one used by the French Spot earth resources satellite, the first model of which was orbited at the beginning of 1986. The capability of its optical system is unknown, but is much better than the resolution of 10 metres achieved by Spot in the panchromatic mode of its camera. Resolutions between one and two metres have been quoted by the French press, but are unconfirmed up to now. The Helios programme is conducted under the responsibility of the "Direction générale de l'armement" (DGA) in the Ministry of Defence, but the development of the satellite is managed by the French civilian space agency CNES (Centre National d'Études Spatiales). At least three satellites will probably be ordered, each with a planned lifetime of four years.

Since March 1985, a special committee has been advising the French Minister of Defence on long term military space issues: the "Groupe d'études spatiales" (space studies group) or GES, which is preparing a "plan pluriannuel spatial militaire" (PPSM—multi-year military space plan) covering the next 15 to 20 years. In the framework of this group advanced space applications are considered, such as the one which may result from the research conducted during the future US/French oceanographic mission, Topex/Poseidon.

Between 1987 and 1988 the military space budget has benefited from the largest increase in the whole budget of the Ministry of Defence: a 79.8 per cent growth for appropriations (from 757,7 MF—millions of French francs—to 1,362.2 MF) and a 59 per cent growth for obligations (reaching 2,764 MF). The growth will continue at a slower rate and the military space budget will reach about three billion francs by 1991. This is about half the level of the civilian space budget, and 2 per cent of the total French defence budget. These numbers enable a comparison with the American situation: in the US, the Department of Defence (DoD) space budget is about twice the NASA budget, and is equal to about 6 per cent of the total DoD budget. In absolute terms, the difference between the French and US military space effort is enormous: about $500 million for the French, against about $20 billion for the American. The ratio is thus of the order of 40 to one. It is worth noting that the US budget includes the SDI research programme.

With her autonomy in security affairs deriving from her independent nuclear deterrent, France is very concerned by the US Strategic Defense Initiative (SDI), announced in March 1983, and what could be the Soviet counterpart to this programme. The French President François Mitterand clearly let the US President Ronald Reagan know that he disagreed with SDI during a summit in Bonn on 4 May, 1985. There are two main reasons for the French President's negative attitude: first, France is not interested in a programme in which she would not have any decision-making powers; second, SDI cannot really affect nuclear deterrence for at least 50 years, and it is not worth trying to change an equilibrium which has prevented war for 40 years now.

This position led France to reject the invitation made on 26 March, 1985, by the US Secretary of State Caspar Weinberger to NATO countries, Israel, Japan and Australia, to officially participate in the SDI research programme in the fields in which they were interested. This refusal concerned an official agreement between France and the United States, and did not prohibit French companies, even state-owned ones, from seeking SDI contracts directly (some contracts have indeed been signed by French companies).

President Reagan's SDI has been perceived by the French President as a challenge in different fields, and this perception has led François Mitterand to propose three initiatives of his own. The first one was related to the space challenge of SDI and was quite ambiguous: in February 1984, in The Hague (Netherlands), Mr Mitterand called for a European space station which could be used for military applications. However, this proposal was probably important in shaping the President's ideas about the importance of space, and may have led to the later French drive in favour of an important European manned space programme, and the engagement by ESA of the Hermes (spaceplane) and Columbus (visitable space station) projects.

The second initiative was supposed to be an answer to the technological challenge of SDI. It was the proposal in April 1985 for a "Eureka" European technological research programme. The Eureka programme has since been implemented. It is, however, mainly a civilian programme, with a total budget far less than that for SDI, and a focus on advanced products for the commercial market. It cannot really be compared to the SDI research programme, and must be judged on its own merits which are not inconsiderable. The question of a military equivalent to the Eureka programme is however now opened, and could very well be discussed over the next few years.

The third presidential initiative was probably trying to respond to a perceived political challenge of the SDI against the ability of the Western European nations to organise their own defence. In January 1986, François Mitterand called for an extended European air defence to counter shorter range Soviet ballistic missiles. The prospects for concrete realisation of this proposal are difficult to assess. They are linked to the whole process of disarmament negotiations in Europe, but are a real possibility for the future.

The change of parliamentary majority in March 1986, and the "cohabitation" period of March 1986 to May 1988, with socialist President Mitterand in the Elysée Palace and a centre-right government with Jacques Chirac as Prime Minister, has not changed the French position on SDI in depth. In particular, the French government has continued to support the conventional (restrictive) interpretation of the 1972 ABM Treaty, which is indeed a bilateral treaty, but which is considered by France to be very important for the future credibility of her "force de frappe". However, to protect the French military deterrence against some kind of Soviet SDI breakthrough, and change in the 1972 ABM treaty, the Ministry of Defence has already taken measures to enhance the penetration of French nuclear warheads.

The SDI has also probably led France to define her position on the question of space militarisation. On 12 June, 1984, the French representative at the Geneva conference on disarmament made three main proposals. First, to forbid the development of anti-satellite (ASAT) weapons able to reach high

DY—J

altitudes (this restriction was provided to avoid the problem of low altitude (ASAT either existence or development). Second, to forbid for a renewable five year period the testing and deployment of directed energy weapons. Third, to extend to new "rules of the road" the agreement of 15 June, 1975, on international registration of space objects.

France has also been concerned for a long time by the monopoly of the two large space powers on space reconnaissance. The USA and USSR have agreed not to interfere with "national technical means" of verification of the other party, which means in fact early warning and reconnaissance satellites, but this immunity has not been extended to third party satellites. France would be interested in an international agreement providing immunity to satellites of all countries. She also proposed in 1978 an International Satellite Monitoring Agency (ISMA), which could provide an international means to control disarmament agreements and to help international management of crisis. This proposal was not well received by the two space superpowers at the time. However, on 8 June 1988, the USSR proposed the establishment of an International Monitoring and Verification Agency.

France has again considered a proposal of this kind: on 2 June, the French Minister of Foreign Affairs, Roland Dumas, called for the creation by the United Nations of an agency for processing and interpreting space imagery. This proposal raises a lot of practical and political questions, but it could be the beginning of a new process, with possible new French initiatives in the near future. It comes at a time when France with her civilian programme Spot and military project Helios clearly has the technical capability to provide or help provide images for international treaty monitoring and crisis management.

The French point of view could thus be summarised in the following terms: classical military space applications could be very useful both for the preservation of a national deterrence capability, and for international stability; deployment of weapons in space, even in the framework of defensive systems, could destroy the international equilibrium built on deterrence; and international treaties should be negotiated to favour positive applications of military space technologies and to prevent negative ones.

Military Developments in Space

GENERAL JOHN L PIOTROWSKI, USAF

The author is Commander-in-Chief, North American Aerospace Defence Command and United States Space Command. This contribution originated in a lecture given at the RUSI.

POTENTIAL SCENARIOS of future conflicts suggest that, if military forces are to achieve their desired results, they must operate effectively in four media—on land, at sea, in the atmosphere, and in space.

There is, in my opinion, one overarching military space imperative. The imperative is to step up to the challenge, to more effectively employ space and space-based systems in our strategy of deterrence. To fulfil that goal requires us, first, to coldly and objectively assess our own military space capabilities and the capabilities of our adversaries. Having made that evaluation, we must then have the resolve to take those actions and invest those resources necessary to preserve our ability to deter and, should deterrence fail, to use space and space systems to meet national and Alliance warfighting objectives.

Before I outline some thoughts on relative military space capabilities, let me say a few words on the United States Space Command.

The United States Department of Defense has been actively involved in space-related programmes since the mid-1950s. In 1958, the National Aeronautics and Space Administration was established to administer the civilian aspects of our space programme, while the Department of Defense retained responsibility for the national security aspects. Within the Department of Defense, each of the military departments pursued space programmes tailored largely to the needs of their own particular services. Over time, both the Air Force and the Navy established sizeable acquisition organisations that also conducted space operations.

As our dependence on space systems grew dramatically during the 1960s and 1970s, these systems continued to be operated for the most part by research and development agencies of the three separate military departments. Unlike land, sea, and air forces, our space-related resources were not formally incorporated into the unified and specified command structure—no single organisation or commander was responsible for their peacetime operation or integrated employment in wartime. While that arrangement was awkward, it was manageable in peacetime. However, lessons learned during daily operations and military exercises have indicated that the customary way of doing business in space could be much improved for crisis management, the transition to war, or during wartime. Thus, the Joint Chiefs of Staff, with the approval of the Secretary of Defense, recommended the activation of a unified space command comprised of Army, Navy, and Air Force components under a single Commander-in-Chief. President Reagan approved that recommendation, and the United States Space Command was activated in September 1985, with its headquarters at Peterson Air Force Base, Colorado Springs, Colorado.

The command is comprised of three component organisations: the Air Force Space Command headquartered in Colorado Springs; the Naval Space Command at Dahlgren, Virginia; and the Army Space Command, also located in Colorado Springs. The command is supported by a headquarters staff of approximately 600 personnel. Including the three component commands, there are approximately 15,000 personnel assigned to the United States Space Command.

The Air Force Space Command has the mission of operating and providing resource management for 29 worldwide missile warning and space surveillance and satellite operations facilities. The Naval Space Command is comprised of the Naval Astronautics Group, which operates the Navy Navigation Satellite System and the Naval Space Surveillance System, a ground-based array of sensors that monitors the activity of orbiting spacecraft. The Army Space Command plans the Army's participation in military space programmes, and operates the Department of Defense's Communications Satellite system.

MISSIONS

The missions of the United States Space Command cover two broad areas: space operations and aerospace defence. Space operations include both space control and space support activities. Aerospace defence consists of surveillance and warning as well as the requirements, development and planning necessary to support the mission of defending against a ballistic missile attack should a decision be made to deploy ballistic missile defence systems.

OPERATIONS

Space, like the world's oceans, is an international regime. As a nation, the United States maintains its right to free passage through space and, like ships at sea, our spacecraft are also the sovereign possessions of our country. Space control operations, like sea control operations, are planned and carried out with the objective of guaranteeing our right to unimpeded operations in, and free passage through, space. But again, space, like the sea, is a medium in which our right to free passage is sometimes challenged and in time of war, would have to be enforced. Deterring aggressive behaviour in space and successfully countering it if deterrence were to fail, requires that we are first able to "know" what is happening there, and that requires the ability to effectively survey space.

Today the United States Space Command operates a satellite early warning system which detects the launch of space boosters early in their flight, providing a prompt indication that a new foreign payload is being placed in orbit. A worldwide, ground-based array of mechanical radars, newer phased array radars, and optical sensors track the new spacecraft once it achieves earth orbit. Data from these sensors is continuously transmitted—much of it over communications satellites—to the United States Space Command's Space Surveillance Center located at Cheyenne Mountain Air Force Base, Colorado. The Space Surveillance Center organises, coordinates, and directs the activity at each surveillance sensor to ensure they are used as efficiently as possible.

The people who operate our space surveillance network are faced with a tremendous task. They complete approximately 55,000 space observations each day to keep track of some 6,000 objects in orbit up to geo-synchronous altitudes; more than 300 are operational payloads. In the event of a crisis or conflict, especially involving the interests of the Soviet Union, that number would probably increase significantly.

It is entirely possible that future conflicts could begin in space. It is difficult to imagine the Soviets beginning a conflict and not attacking our space-based systems in the process. For that reason, we closely monitor Soviet activity which could represent a potential threat to friendly space objects. In the event of one of our spacecraft being threatened, personnel in the Space Defense Operations Center would provide warning to the owner and operator of the threatened satellite.

SPACE CONTROL

The space control mission also requires that we have the ability to deny an adversary freedom of action in space during conflict. Soviet space systems and the military capability they provide to their land, sea, and air forces pose a serious threat to our terrestrial forces. Currently the United States does not possess an operational anti-satellite system capable of deterring aggressive Soviet acts in space, responding in kind to Soviet preliminary use of their anti-satellite system, or negating their space systems which place our people and forces at risk. Unfortunately, while the United States has debated, delayed, and deferred for nearly 10 years on this issue, the Soviets have proceeded at a deliberate and alarming pace with the development and deployment of anti-satellite and other threatening space systems.

SPACE SUPPORT

The United States Space Command is also in the support business. In a similar way to the military transportation commands, it provides support for US and Allied forces, commands, and troops who are engaged in combat or who support combat operations. Space support operations include supporting the launch and on-orbit operation of communications, navigation, and surveillance systems assigned to the United States Space Command, and ensuring that the vital information they provide is received when and where it is needed.

The first warning of a strategic attack on the United States or our allies would probably be provided by our missile warning system. First, space-based early warning satellites would detect either submarine or land-based ballistic missile launches early in flight. The Ballistic Missile Early Warning System radars, including the one located at Fylingdales Moor, or phased array radars located on the borders of the United States, would then detect and track the threatening missiles. Data from each of these sophisticated and reliable sensors is transmitted to and processed by personnel in the Missile Warning Center located in Cheyenne Mountain. The combination of space-based systems using infrared sensors and ground-based radars provides "dual phenomenology" detection and tracking of threatening events needed to provide the National Command Authorities with prompt and reliable warning information.

The technology embodied in our missile warning radars also gives these sensors the capability to simultaneously conduct the space surveillance operations discussed earlier. In fact, our missile warning facilities conduct a large percentage of the thousands of space observations made every day.

The growing importance of space as an operational medium and the incredible capability space systems provide to land, sea, and air forces suggest that the United States Space Command faces a future filled with challenges and opportunities. There is no doubt that the major challenge will be posed by a Soviet military space programme, a programme characterised by continuous growth, increasing technological sophistication, and by systems that significantly enhance the warfighting potential of Soviet terrestrial forces. The Soviets fully recognise the strategic importance of space, and it is equally clear that they seek to dominate us in that arena.

SOVIET SPACE OPERATIONS

Since 1957, the Soviets have conducted nearly 2,000 space launches, at least 90 per cent of which have had military or military-related missions: photo-reconnaissance, military communications, surveillance, navigation, and targeting. Today, Soviet space systems and the capabilities they provide are fully integrated into the operation of their land, sea, and air forces. Soviet space systems have greatly enhanced the capabilities of Warsaw Pact forces and, therefore, represent a significant military threat to NATO. The scope of that threat is as worrisome as it is broad.

The Soviets possess the largest and most responsive space infrastructure in the world. Compared to the United States and our Allies, they have twice as many launch pads, annually conduct almost 10 times as many space launches and put five times the weight into orbit. Not only are their space forces numerically larger than those of the collective NATO nations, but the Soviets also have a superior space system replenishment capability. Whereas it takes the United States, for example, months to prepare and launch a spacecraft, the Soviets can launch their space systems in weeks or in some cases, days. Normal space launch operations for the Soviets are characterised by one space launch every four days. In a crisis, however, they can reduce this "normal" interval considerably. During the Falklands Conflict, for example, the Soviets conducted 28 space launches in just 69 days. Not only did this represent launching one-third of their annual total in just over two months, it also exceeded all the successful launches conducted by the United States in 1985 and 1986 combined.

Moreover, the Soviet space arsenal includes an impressive and formidable array of anti-spacecraft weaponry: an operational co-orbital spacecraft interceptor with a conventional warhead that is able to engage many of our militarily significant low-earth orbiting spacecraft; "soft-kill" lasers of Sary Shagan possibly capable of causing degradation or damage to Allied space systems; direct ascent nuclear-armed GALOSH anti-ballistic missiles which also have an inherent anti-spacecraft capability; and electronic warfare capability. This mix of weaponry could be used to control space during a crisis or during the critical early stages of a conflict—to deny freedom of action in that medium to opposing forces.

Other Soviet space systems—such as the Radar Ocean Reconnaissance Satellite (RORSAT) and Electronic Ocean Reconnaissance Satellite (EORSAT)—are designed to track Alliance naval forces and provide terrestrial and maritime weapons platforms with the targeting data required for prompt attack by land, sea, and air-launched stand-off weapons. Soviet space control and space-based force enhancement capabilities, including existing and potential capabilities provided by manned Soviet systems, have added a new and important dimension to the threat arrayed against the Alliance.

THE FUTURE

In the relatively short period of 30 years, space systems have in many ways transformed the manner in which terrestrial forces are employed, and that transformation will continue. Yet, because space forces are unseen, there is often a tendency to underestimate or misunderstand the many critical contributions they make to the warfighting capabilities of the more visible terrestrial and maritime forces. In the case of NATO, all its forces—land, naval, and air—would be less effective in deterrence and war-fighting without the pervasive presence of space systems.

We need, first, to acknowledge that the Soviet Union has its own very ambitious space goals. Towards that end, the Soviets have developed the world's largest and—by a wide margin— most responsive space infrastructure. We must be aware that they have developed space systems and a launch capability which are more than adequate for peacetime, but that appear to be better suited to war-fighting, and war support, than to anything else.

Next, we need to be aware that those Soviet capabilities that are worrisome in peacetime would become threatening in crisis or conflict. Thus, we must structure our own military space capabilities to operate effectively in an environment that would most assuredly be contested in crisis or conflict.

This, in turn, requires a more responsive launch capability to ensure access to space in peace and conflict. It also requires more robust and more survivable satellite constellations. And, finally, we require an authentic and fully operational anti-satellite capability to deter aggression in space—and should deterrence fail, to enforce our right of free passage or disable the space systems that an adversary is using to support hostile action against our terrestrial and maritime forces.

One of my tasks as the Commander-in-Chief of the United States Space Command is to give visibility to space, to allow it to show its contributions, to illuminate its effectiveness, and to reveal its potential. While much can be said, unfortunately much more must remain unsaid.

The role of the United States Space Command is to provide national agencies, United States unified and specified commands, and our Allies with the necessary space-borne assets needed to carry out national and international security objectives. This role will increase in importance in the future. The United States Space Command will continue to work with these users to identify and define requirements which can be accomplished by space-based systems. Through creative applications of existing systems, new technological advances in computer and communications systems, manufacturing procedures, sensing and other space-related systems, the United States Space Command will continue to support future requirements.

The United States' ability to carry out a viable space programme in the future and to meet the demands placed upon it will not be easy. As the Soviet Union continues its major national commitment to the military exploitation of space, this commitment will undoubtedly result in major new space systems in the future. Imaginative leadership and forward looking decision-making will enable the United States Command to continue accomplishing its mission in support of the common defence.

The UK in Space—Implications for Defence and National Security

PROFESSOR ROY GIBSON

The author is a special advisor to the Director General of the International Maritime Satellite Organisation (INMARSAT). This contribution originated in a lecture given at the RUSI.

THE INTERACTION between civil and military space activities has been treated in most of Europe with unnecessary hypocrisy. In many discussions the speaker goes out of his way to make it clear that he speaks only about "civil" space, but it is simply not possible to make a clear distinction between what is purely civilian and what is relevant and helpful to defence and national security interests. There is, and should be, considerable overlapping.

Even the European Space Agency has difficulties in maintaining its virginity. The ESA Convention obliges the Agency to pursue only "peaceful purposes", and several of the non-NATO Member States of ESA are constantly vigilant for possible infringements. However, in 1979–80 when Arianespace was being formed to market and launch the Ariane rocket after qualification by ESA, ESA Member States were unwilling to say in the inter-governmental agreement that Ariane should only be sold to customers with exclusively peaceful purposes in mind. Ariane has, in fact, already launched satellites with defence payloads, and the UK's Skynet 4 is on the Ariane manifest. No criticism is implied; it is difficult to draw a hard and fast line.

There was an important Western European Union symposium held in Munich in late 1985 which did put a lot of fresh air into the system. For the first time there were open discussions about the interrelationship between civil and military space programmes in Europe. Indeed the Director General of ESA expressed the opinion that the ESA Convention would permit the Agency to undertake the procurement of a defence surveillance satellite system, provided always that the ESA Council gave the approval which is necessary for all new programmes.

This corresponds to the attitude many organisations and academic institutions have taken for years in relation to the Advisory Group for Aerospace Research and Development (AGARD). Opposition to, or non-membership of, NATO has not prevented many of them from attending AGARD's technical seminars, or even from accepting contracts.

There are, of course, other reasons besides a wish for plain speaking—for wanting more openness on this subject. At the same WEU seminar, captains of Europe's aerospace industry, supported by an impressive array of Ministers, emphasised the commercial necessity of having the maximum transparency and cooperation between civil and military space programmes. They were, of course, envious of the massive commercial advantages which flow to the US civil space programme and industry by reason of the military space programme. It is common knowledge that the US military space budget has long since overtaken NASA/NOAA (National Oceanographic and Atmospheric Administration) expenditure and, whatever the security restrictions may be, experience and technology gained from the sophisticated defence programmes washes liberally over onto the civilian programme.

FORMATION OF THE BNSC

In the UK there was little formal liaison between civil and military space activities until the formation of the British National Space Centre (BNSC). Indeed, the civilian side itself was fragmented: the Science and Engineering Research Council (SERC) financed and was wholly responsible for the national and international space science programme; the Department of Trade and Industry (DTI) was responsible for most of the remainder, although there were small pockets of independent activity in other department and governmental institutions. The DTI's spending on the national space technology programme (as opposed to the UK's subscriptions to the ESA applications programmes) was effected through the Royal Aircraft Establishment (RAE) of the Ministry of Defence. In addition to the DTI's money, the MoD added some of its own funding and the RAE managed this joint space technology programme, through its Space Department, which was also responsible for the MoD's own—classified—programme. Given the very slender technical staffing available to the DTI's Space Division—and the tremendous burden of monitoring ESA affairs in Council and innumerable programme boards—the real coordination of the joint technology programme took place at RAE Farnborough.

In the months following the formation of the BNSC, there were internal reorganisations, secondment of staff and so on which, without any increase in overall complement, provided an integrated structure with responsibility for the whole of the UK's civil space programme—science and applications, both nationally and internationally, including the MoD/DTI joint technology programme.

Liaison between that newly coordinated civil programme and the classified MoD space programme was effected at several levels: by having a senior MoD official on the BNSC's Management Board; then by having the DG and a senior Director (the Director of Projects and Technology) being made privy to the MoD's own space programme; and by the Head of RAE's Space Division within the BNSC also being responsible for the classified programme. And, of course, finally through a certain commonality of staff between the two programmes.

This new organisation produced, in September 1986, a National Space Plan which, although wholly civil and in no way replacing the MoD's classified programme, did take good account of defence and national security needs. This is not to say that the Plan gave civilian objectives second priority, but merely that the various elements of the Plan were put together in full knowledge of their utility and relevance to defence needs. The MoD member on the BNSC Management Board often needed to be schizophrenic; he wanted on occasions to encourage the other Members of the Board to include this or that element into the recommended programme, whilst being quite firm that these elements were not part of the MoD's operational requirements and therefore no additional MoD funding could be expected. Not an unusual situation and the Plan was probably none the worse for it.

The BNSC had been asked to produce a coherent national plan, that is to say, one looking at national interest overall, and willingness of a Board Member to pay from his existing departmental vote was not a criterion for inclusion in the Plan—otherwise the Plan would have been a mere restatement of the existing situation. The proposals were intended to take action of the very complex set of justifications for Governmental funding of a "civil" space programme—including national security considerations. The aim was to show how good defence value could be obtained from the civilian space programme without deforming it in any way. This helped our general thesis that there is not just one single justification for increased civil space expenditure, but rather a mixture of several different motives, one of which is surely defence and national security.

In the event the UK Government did not find it possible to make any significant increase to the funding of civil space, and this policy was forthrightly announced at the ESA Ministerial Council in The Hague in November, 1987. The UK is not participating in the Ariane 5 launcher and Hermes space plane programmes, but is negotiating a small (one third of GNP) participation in the Columbus—space station related programme. No government funding is to be made available to continue the Rodasat remoting sensing programme with Canada, nor to take the Hotol, next generation launcher, programme to its next stage.

LACK OF A NATIONAL SPACE PLAN

More alarming even than the UK's exclusion from the new generation of programmes—many of which have a real significance for our defence and national security—is the continued lack of a National Space Plan. However modest its expenditure may be, the country needs a space policy not only to guide our relations with ESA but also to enable civil, military, public and private sector investment to be coordinated.

The various programmes are so interlinked that it is foolish to try to deal with them separately. Even the space science programmes have an importance to our technological objectives. Without such a policy, a country reacts to international or national proposals to undertake this or that space programme without a blueprint, without an overall strategy. This has long been realised by the French, and several other European nations are following suit by establishing national space agencies and national space plans. Germany, Italy and Norway have all taken this route, as has Canada on the other side of the Atlantic. In each case the relevance of the civil space programme to national security has been noted.

The UK's abdication comes just at the time when there is increased interest in the use of space technologies for defence and national security—not only at a national level but also on a regional and perhaps even wider basis. The French, with some support from Italy and Spain, are going ahead with the HELIOS surveillance satellite system (based on SPOT), and other systems are being discussed as part of a sharpening world interest in using these new techniques for maintaining peace. It is hard to believe that the UK would not wish to play a part in such systems, say perhaps emanating from WEU which in these past days has been discussing the subject; but the UK's capacity to contribute to their design—let alone their manufacture and operation—will diminish rapidly if our space effort is not increased to something like the proportions prevalent in the other major European countries. Even our ability to advise on the merits of competing systems will be hard to retain.

As is so often the case, it may well be that in time the national security implications will become apparent, and that the UK Defence Budget will be squeezed to find what are after all the relatively small amounts needed. In that way no doubt some of the damage could be avoided and the situation salvaged, but the overall benefit to the UK of this emergency expenditure would be substantially less than a timely investment in the civilian space programme.

Meanwhile there is a real danger that the coordination achieved over the past two years will start to deteriorate, and the parties will want to fend for themselves again. They will individually be the worse for it, and nationally the UK would not be able to assume some of the responsibilities which many people in the UK—and perhaps even more in other countries in Europe—believe to be an appropriate charge on government. Ironically this happens at a time when in the remainder of Europe there is increasing realisation that a planned civil space programme is an essential element in national defence and in underpinning foreign policy.

Part II — Signposts

A Chronology of the Iran–Iraq War

FRANCIS TUSA

The author is Middle East analyst at the RUSI

EVEN AT the stage of the Iranian acceptance of UN Security Council Resolution 598, the Iran–Iraq War ranked as the longest conventional struggle of the twentieth century. The length to 18 July 1988 was a little over seven years and nine months. The cost in terms of money and economic damage might well never be fully known, but estimates range between $250 and $600 billion. The human price is still sickeningly high, yet vague, with estimates of the dead ranging up to and beyond one million military and civilian casualties on the two sides, with at least the same number wounded.

It is too early to say whether even a fully-fledged UN ceasefire will take effect, let alone whether that will be transformed into a long-lasting peace treaty. All the signs as of early August are that Iraq is determined to win as many concessions militarily out of a weakened Iran as they can before returning to the peace table, in the same way that Iran basically ignored the original implementation of Resolution 598 in 1987 because they still thought that they could win on the battlefield. Bearing in mind the mutual antipathy that has built up between the two countries in the course of such a hard-fought war, bets on the length of any ceasefire or peace treaty may turn out to favour only a short duration.

In the following chronology, only the important dates are included, that is to say ones which have a noticeable effect on the course of events of the war, or which are more evident as being crucial now that peace seems to be in the air. It would be possible to name every Iraqi or Iranian offensive by the given name, but only the watershed offensives have been thus treated; all the others are just placed in the time frame in which they occurred.

1980

June– August	Border clashes/artillery duels between Iran and Iraq.

September
4	Iraqi date for the start of the war due to Iranian bombardments.
17	Iraq abrogates 1975 Algiers Treaty.
22	Iraq invades Iran on three axes. Kharg Island bombed by Iraqi Air Force: ineffective.
23	Iran bombs Baghdad.
28	UN Security Council Resolution 479. Iraq accepts with a few reservations, Iran rejects it.

October– November	Iraqis take Khorramshahr and destroy Abadan, but are finally held up at Susangird.

1981

February
4	Gulf Cooperation Council (GCC) established. The members are Bahrain, Kuwait, Oman, Qatar, Saudi Arabia and the United Arab Emirates.

September– December	Iran regains much of the territory lost in late 1980, relieving Abadan. Bahrain uncovers an Iranian-sponsored coup.

1982

March– June	Iran recaptures most of its lost territory.

June
20–30	Iraq announces a unilateral withdrawal from all Iranian territory, Iran says it will continue the war.

July

12   UN Security Council Resolution 514 calling for a ceasefire.

Iran launches "Operation Ramadan", the first invasion of Iraq, in an effort to take Basra.

August

12   Iraq announces a Maritime Exclusion Zone in Iranian waters.

September–   Iran launches more offensives on southern,
October   central and northern fronts, all of which make gains.

1983

February–   Further Iranian offensives in southern battle-
April   front make little headway.

May   Iranian Tudeh (Communist Party) banned.

August–   More Iranian attacks on Basra make little
October   headway for large loss.

October   Iraq receives five French Super Etendard fighters on loan equipped with Exocet.

1984

February   Iraq launches SCUD attacks against several Iranian targets, including civilian ones; Iran replies with artillery. A mutual ceasefire is agreed on 18 February.

Iraq starts concerted strikes on Iranian oil export terminals and tankers; also starts to use primitive forms of chemical warfare.

Iran attacks the Majnoon Islands in the Shatt al-Arab; this sees the start of human wave tactics by the Iranians which result in massive casualties.

March
21   UN inspectors report on chemical weapons use in
                Iran, but Iraq not named as the user.

April–   Tanker War accelerates, with Iran starting attacks
May   on neutral shipping in the southern Gulf.

June
5   Saudi Arabia shoots down two Iranian F–4
                PHANTOMS attacking Saudi shipping.

July–   Iran undertakes reorganisation of armed forces
December   for forthcoming campaigns including intensive
                infantry training.

                Little land activity, but continuing strikes in the
                Tanker War.

1985

March   Large Iranian offensives on Basra front show
                greater Iranian capability, better training and
                fewer casualties.

May   Start of "Irangate" arms supplies to Iran from
                US.

1986

January   Iran completes massive build-up of forces for an
                attempt at a breakthrough in the south.

February
9   Al Dawa offensive takes the Fao Peninsula from
                Iraq, but bogs down elsewhere.
24   UN Security Council Resolution 582 calls for
                ceasefire.

March
14   UN report names Iraq as user of chemical
                weapons for the first time.

April	Iraq counter-attacks in Fao area fail.

May
Iraqi attacks on Iranian refining capacity bear fruit: Iran has to import refined petroleum products.
10–17 Iraq captures the town of Mehran.

July
1–10 Iran retakes Mehran.

August
12 Iraq bombs Iranian oil trans-shipment terminal at Sirri Island in the southern Gulf.

November
4 Iranian internal government rivalry surfaces over the "Irangate" affair after a Beirut magazine publishes a "planted" story.
25 As the culmination of an increasing number of attacks on all forms of Iranian oil exports, Iraq attacks Larak Island in the Straits of Hormuz: no area is now safe from Iraqi airpower.

December
23–24 "Kerbala IV" offensive by Iran against Basra: large-scale, but rushed because of internal political reasons created by the disclosure of the "Irangate" scandal.

1987

January
6 "Kerbala V" offensive by Iran. Breaches Iraqi defences north of Basra. Heavy fighting until 25 February. Large Iranian casualties for little appreciable gain.
13 "Kerbala VI" Iranian Regular Army offensive in Central Region.
26–30 Islamic Conference Organisation summit meets

at Kuwait. Iran boycotts meeting in protest at Kuwait's support for Iraq.

February   Soviet and Kuwaiti officials discuss the chartering of three Russian oil tankers by the Kuwait Petroleum Company. Kuwait approached the United States and the Soviet Union in November 1986 to agree on commercial procedures to give Kuwaiti ships foreign naval protection.

"War of the Cities", which had grown to a peak, is ended by a mutual two week truce.

27–28   "Ya Zahra" offensive around Basra sees the end of "Kerbala V" offensive. Fighting around Basra declines to a lower level, with only sporadic, sharp engagements.

March   More sporadic fighting around Basra. Iraq tends to squeeze Iranian gains, Iran consolidates them. Further fighting in Kurdistan.

April
26   Saddam Hussein of Iraq meets President Assad of Syria in a secret meeting in Jordan.

May
8   Soviet freighter *Ivan Koroteyev* hit by Iranian gunboats in the Gulf. The attack was seen as a protest over Kuwait's chartering of Soviet tankers.

16   The Soviet tanker *Marshal Zhukov*, one of three on charter to Kuwait, hits a mine close to Kuwait. Soviets move four minesweepers to the Gulf. US Navy has little mine clearance capability, and none in the Gulf.

17   USS *Stark* accidently attacked in the Gulf by Iraqi Mirage F–1. 37 crew killed.

21   US Government announces that 11 Kuwaiti ships are to be reflagged by the United States.

Reagan Administration delays $400 million arms package to Saudi Arabia after allegations that the

Saudis did little to prevent the *Stark* incident from occurring.

June

8   Before Venice Summit of seven Western Nations, President Reagan warns of possible pre-emptive attacks on Iranian "Silkworm" anti-ship missile sites.

12   Iran test fires several "Silkworm" missiles.

19   Iran mobilises 20,000 men to man new Revolutionary Guard bases in Gulf.

24   Saudi Arabia agrees to let its minesweepers operate off Kuwait after a fourth ship hits a mine near the al Ahmadi oil terminal. US divers also involved in mine clearance operations.

        USS *Constellation* battle group arrives on station outside Gulf.

July

15   Kuwait requests British flag for its tankers.

20   UN Security Council Resolution 598, calling for ceasefire, passed.

21   Kuwaiti tanker *al Rekkah* reflagged as US *Bridgeton*.

        Mrs Thatcher says in the House of Commons that registration of vessels is a commercial matter, not political.

22   Reflagged tankers, *Bridgeton* and *Gas Prince*, enter Gulf.

23   Kuwait approaches Gibraltar shipping authorities for re-registering.

24   *Bridgeton* hits mine close to Kuwait.

30   Riots at Mecca during the annual pilgrimage. Around 400 pilgrims of all nationalities killed in ensuing chaos.

        Eight US minesweeping helicopters dispatched by plane to Gulf.

        French battle group based around the aircraft carrier *Clemenceau* leaves for the Gulf.

US Government formally asks for British and French help in minesweeping operations.

August

4–8 Iranian Revolutionary Guard Exercises in Straits of Hormuz.

6 US Special Forces team moved to Gulf to combat Iranian patrol boats.

10 Tanker hits mine off Fujayrah (UAE) in Gulf of Oman.

Iraqi aircraft carry out first raids on Iranian land targets since before UN Resolution 598 passed.

11 Britain and France send minesweeper forces to Gulf.

US Navy Air Force F–14 "engages" supposed Iranian fighter.

14 US minesweeping helicopters on USS *Guadalcanal* arrive in Bahrain.

18 Iran recommences attacks on shipping in Gulf after break of five weeks.

21 US Navy establishes Gulf Command.

29 Iraq recommences attacks on Iranian oil exports in the Gulf. This phase of attacks is still under way. To date, Iraq has attacked approximately 45 ships, Iran 40 ships. Although the tally is approximately equal, Iraq has badly damaged or sunk more ships than Iran.

September

4 Italy decides to send three frigates and three minesweepers to the Gulf.

Iran fires "Silkworm" missiles at Kuwait.

7 Netherlands to send two minesweepers to Gulf. British Armilla patrol to provide support.

Kuwait tanker *Al Faiha* reflagged as the British *Tonbridge*.

11 UN Secretary General Perez de Cuellar flies to Iran at start of Gulf peace mission.

13 UN Secretary General Perez de Cuellar flies to Baghdad after failure of his mission in Iran.

14	Belgium decides to send two minesweepers to Gulf.
21	British tanker *Gentle Breeze* attacked by Iranian gunboats north of Bahrain.
	Iranian naval ship *Ajr* caught by US Special Forces helicopters laying mines in international waters. The ship is captured by the US Navy, and is scuttled.
23	Britain closes Iranian Arms Purchasing Office in London.

October

2	Iranian naval manoeuvres in the northern Gulf seem to threaten Saudi oil fields. Saudis and the US Navy move forces to counter Iran.
6	"War of the Cities" gains pace as Iran launches two SCUD B missiles at Baghdad.
9	Three Iranian Revolutionary Guard gunboats sunk by US Special Forces helicopters near Farsi island.
16	US flagged ship *Sea Isle City* hit by "Silkworm" anti-ship missile while lying at anchor in Kuwaiti waters.
19	US escort ships destroy oil platforms in Rostam and Sassan oilfields used by Iranian gunboats with naval gunfire.
22	Kuwait Sea Island oil loading terminal set ablaze by Iranian "Silkworm" missile.

November

8	Amman Arab League summit opens. Iran fires SCUD missiles at Baghdad.
11	Final statement of Amman summit permits individual countries to restore diplomatic relations with Egypt; nine states do so within 10 days.
14	Egyptian military advisors arrive in Kuwait to examine defences. In January, President Mubarak visits GCC countries.

December
2   Pakistani troops in the service of Saudi Arabia
             return home after their contract expires.

             The expected Iranian offensive around Basra
             does not materialise, and there are secondary
             conscription campaigns inside Iran, bringing into
             doubt their ability to conduct campaigns as
             before. The tanker war continues unabated, but
             the theatre moves from the northern Gulf to-
             wards the south. The US fleet finishes their 23rd
             escort operation. Total ships hit in 1987: IRAQ:
             76 of which 68 are tankers; IRAN: 79 of which 67
             are tankers.

1988

January
             Iran extends its conscription period due to
             manpower shortages. United States and Soviet
             Union cooperate over mine-hunting.
20   France announces that she will provide protec-
             tion to any merchant ship that asks for it.

             Western European Union announces mine-
             sweeper force reductions.
25   US Defense Secretary Carlucci announces US
             naval force reductions in Gulf.

February
28   "War of the Cities" recommences.
29   First use of Iraqi long-range missiles.

March
1   "War of the Cities" continues with minor inter-
             ruptions until early April.
16   In response to Iranian successes in Kurdistan,
             Iraq bombs the Iraqi Kurdish town of Halabja
             with chemical weapons.
30   Kuwait accuses Iran of attacking its positions on
             Bubiyan Island.

April
5   Kuwait 747 highjacked by pro-Iranian Kuwait group; lands at Mashad in Iran.
14   US Frigate *Samuel B. Roberts* hits a mine in the Gulf.
17   Iraq launches offensive to retake Fao Peninsula from Iran
18   Iraq recaptures Fao.

           US carries out retaliation for the mining incident of the 14 April, during which Iran losses two oil platform bases, two Frigates, one Attack Craft and two gunboats.
20   Iran fires SCUD at Kuwaiti oilfield.

           Kuwaiti 747 highjack solved at Algiers.
29   US announces that it will protect any shipping under attack.

May
25   Iraq recaptures Salamcheh area from Iran, thus clearing the eastern approaches to Basra of all Iranian troops.

June
2   Hojestalem Rafsanjani appointed acting C-in-C of the Iranian Armed Forces. Attempts an immediate counter-offensive in the Salamcheh area which fails.
25   Iraq recaptures Majnoon Islands from Iran.

July
3   USS *Vincennes,* one of the most advanced anti-aircraft cruisers in the world, shoots down an Iranian Airbus having confused it with a non-existent Iranian F–14. The report into the incident blames inexperienced operators for misreading accurate data.
12   Iraq recaptures Zubeidat area in central front; Iran withdraws from much of Iraqi Kurdistan.
17   Saddam Hussein appeals to Tehran for an honourable peace.
18   Iran accepts without reservation UN Security Council Resolution 598.

Soviet Satellite and Spacecraft Launches 1987

Translated and compiled by

S. DE BANZIE

S. de Banzie is a member of the research staff on the RUSI Soviet and East European Studies Programme

THE FOLLOWING table provides data on the launches of USSR satellites and spacecraft in 1987 and the comments which follow are an official Soviet explanation of their functions. The RUSI comments indicate those satellites which Western experts calculate are intended for military purposes.

Sources:

1. Satellite spacecraft launch data and Soviet Comment reproduced in translation from *Aviatsia i Kosmonavtika* No. 4, 1988, p. 40–41, Voenizdat, Moscow, by kind permission.
2. RUSI Comment by Dr B Jasani, Research Fellow, Royal United Services Institute Space and International Security Programme.

LAUNCHES OF USSR EARTH SATELLITES AND SPACECRAFT IN 1987

Launch Date	Name	Period, in mins	Apogee, (maximum height) in km	Perigee, (minimum height) in km	Orbital inclination	Orbital lifetime in years, (date of cessation of work)
1	2	3	4	5	6	7
5 January	Meteor-2	104.0	973	950	82.5	520
9 January	Kosmos-1811	89.7	367	181	64.9	(13.2.87)
14 January	Kosmos-1812	97.8	677	648	82.5	60
15 January	Kosmos-1813	90.0	387	208	72.8	(29.1.87)
16 January	Progress-27	88.9	280	189	51.6	(25.2.87)
21 January	Kosmos-1814	100.7	815	775	74.0	118
22 January	Kosmos-1815	93.5	558	345	50.7	3

LAUNCHES OF USSR EARTH SATELLITES AND SPACECRAFT IN 1987 *cont.*

| Launch Date | Name | Initial Orbit Parameters | | | | Orbital lifetime in years, (date of cessation of work) |
		Period, in mins	Apogee, (maximum height) in km	Perigee, (minimum height) in km	Orbital inclination	
22 January	Molniya-3	12 h. 16 min.	40800	473	62.8	15
29 January	Komos-1816	104.9	1024	979	82.9	1200
30 January	Kosmos-1817	88.4	224	192	51.6	(13.2.87)
2 February	Kosmos-1818	100.7	810	790	65.0	110
6 February	Soyuz TM-2	88.7	247	197	51.6	(30.7.87)
7 February	Kosmos-1819	88.7	254	197	72.8	(18.2.87)
14 February	Kosmos-1820	88.8	273	186	64.8	(18.2.87)
18 February	Kosmos-1821	105.0	1029	983	82.9	1200
19 February	Kosmos-1822	89.5	332	205	73.0	(5.3.87)
20 February	Kosmos-1823	116.0	1538	1497	73.6	(19.12.87)
26 February	Kosmos-1824	89.7	370	177	67.2	(22.4.87)
3 March	Progress-28	88.8	272	191	51.6	(28.3.87)
3 March	Kosmos-1825	97.7	677	649	82.5	(5.1.88)
11 March	Kosmos-1826	90.3	403	206	72.9	(25.3.87)
13 March	Kosmos-1827	113.8	1434	1387	82.6	10000
13 March	Kosmos-1828	113.8	1435	1354	82.6	10000
13 March	Kosmos-1829	113.9	1437	1402	82.6	10000
13 March	Kosmos-1830	114.0	1437	1404	82.6	10000
13 March	Kosmos-1831	114.0	1440	1412	82.6	10000
13 March	Kosmos-1832	114.1	1443	1414	82.6	10000
18 March	Kosmos-1833	101.9	878	851	71.0	(23.3.87)
19 March	Raduga	24 h. 0.5 min.	36087	35852	1.3	1000000
31 March	Kvant	89.2	320	177	51.6	—
8 April	Kosmos-1834	92.8	443	413	65.0	28
9 April	Kosmos-1835	89.7	367	180	65.0	(3.6.87)
16 April	Kosmos-1836	89.2	313	188	64.8	(2.12.87)
21 April	Progress-29	88.7	257	194	51.6	(11.5.87)
22 April	Kosmos-1837	88.7	260	194	82.3	(28.4.87)
24 April	Kosmos-1838	5 h. 10 min.	17452	211	64.9	(12.11.87)
24 April	Kosmos-1839	5 h. 10 min.	17429	209	64.9	(29.10.87)
24 April	Kosmos-1840	5 h. 10 min.	17530	209	64.9	(29.10.87)
24 April	Kosmos-1841	90.5	403	225	62.8	(8.5.87)
27 April	Kosmos-1842	97.8	678	648	82.5	60
5 May	Kosmos-1843	89.5	312	214	70.4	(19.5.87)
11 May	Gorizont	23 h. 21 mins.	35174	35024	0.9	1000000
13 May	Kosmos-1844	102.0	879	851	71.0	117
13 May	Kosmos-1845	90.4	400	217	70.0	(27.5.87)
19 May	Progress-30	88.8	265	192	51.6	(19.7.87)

LAUNCHES OF USSR EARTH SATELLITES AND SPACECRAFT IN 1987 *cont.*

Launch Date	Name	Period, in mins	Apogee, (maximum height) in km	Perigee, (minimum height) in km	Orbital inclination	Orbital lifetime in years, (date of cessation of work)
21 May	Kosmos-1846	89.2	314	196	82.4	(4.6.87)
26 May	Kosmos-1847	89.7	373	177	67.2	(22.7.87)
28 May	Kosmos-1848	90.2	400	208	72.9	(11.6.87)
4 June	Kosmos-1849	11 h. 49 min.	39342	613	62.9	15
9 June	Kosmos-1850	100.8	825	785	74.0	(21.1.88)
12 June	Kosmos-1851	11 h. 50 min.	39402	592	62.8	14
16 June	Kosmos-1852	114.5	1503	1400	74.0	9550
16 June	Kosmos-1853	114.7	1485	1396	74.0	9550
16 June	Kosmos-1854	114.9	1484	1412	74.0	9550
16 June	Kosmos-1855	115.0	1485	1425	74.0	9550
16 June	Kosmos-1856	115.2	1484	1441	74.0	9550
16 June	Kosmos-1857	115.4	1489	1457	74.0	9550
16 June	Kosmos-1858	115.7	1521	1481	74.0	9550
16 June	Kosmos-1859	115.7	1511	1481	74.0	9550
19 June	Kosmos-1860	89.7	283	255	65.0	0.2
23 June	Kosmos-1861	105.0	1014	995	83.0	1220
1 July	Kosmos-1862	97.7	679	645	82.5	60
4 July	Kosmos-1863	90.8	383	208	72.9	(18.7.87)
7 July	Kosmos-1864	104.8	1019	977	83.0	1215
8 July	Kosmos-1865	89.5	327	204	64.8	(14.8.87)
9 July	Kosmos-1866	89.8	386	177	67.2	(26.7.87)
10 July	Kosmos-1867	100.8	813	797	65.0	108
14 July	Kosmos-1868	94.5	726	279	74.0	1
16 July	Kosmos-1869	97.8	679	647	82.5	60
22 July	Soyuz TM-3	88.6	236	200	51.6	(29.12.87)
25 July	Kosmos-1870	88.7	282	168	71.9	—
1 August	Kosmos-1871	88.3	212	191	97.0	(11.8.87)
4 August	Progress-31	88.8	269	193	51.6	(23.9.87)
18 August	Meteor-2	104.1	974	954	82.5	520
19 August	Kosmos-1872	89.6	333	208	72.9	(30.8.87)
28 August	Kosmos-1873	88.8	274	186	64.8	(1.9.87)
3 September	Kosmos-1874	89.6	333	208	72.9	(17.9.87)
3 September	Ekran	23 h. 43 min.	35619	35459	0.4	1000000
8 September	Kosmos-1875	113.8	1432	1389	82.6	10000
8 September	Kosmos-1876	113.9	1434	1395	82.6	10000
8 September	Kosmos-1877	113.9	1436	1400	82.6	10000
8 September	Kosmos-1878	114.0	1436	1404	82.6	10000
8 September	Kosmos-1879	114.1	1439	1411	82.6	1000
8 September	Kosmos-1880	114.1	1441	1413	82.6	10000
11 September	Kosmos-1881	89.0	297	190	64.8	—
15 September	Kosmos-1882	88.6	253	196	82.3	(6.10.87)
16 September	Kosmos-1883	11 h. 16 min.	19153	19124	64.9	1000000
16 September	Kosmos-1884	11 h. 16 min.	19155	19121	64.9	1000000
16 September	Kosmos-1885	11 h. 16 min.	19153	19122	64.9	1000000
17 September	Kosmos-1886	89.8	384	178	67.2	(2.11.87)

DY—K

LAUNCHES OF USSR EARTH SATELLITES AND SPACECRAFT IN 1987 *cont.*

| Launch Date | Name | Initial Orbit Parameters | | | | Orbital lifetime in years, (date of cessation of work) |
		Period, in mins	Apogee, (maximum height) in km	Perigee, (minimum) height) in km	Orbital inclination	
24 September	Progress-32	88.8	268	193	51.6	(19.11.87)
29 September	Kosmos-1887	90.5	406	224	62.8	(12.10.87)
1 October	Kosmos-1888	24 h. 03 min.	35989	35861	1.4	1000000
9 October	Kosmos-1889	90.4	400	216	70.0	(23.10.87)
11 October	Kosmos-1890	92.9	442	414	65.0	2.8
14 October	Kosmos-1891	104.9	1030	957	82.9	1200
20 October	Kosmos-1892	97.8	678	647	82.5	60
22 October	Kosmos-1893	89.7	374	179	67.2	(15.12.87)
28 October	Kosmos-1894	24 h. 02 min.	35918	35729	1.3	1000000
11 November	Kosmos-1895	90.4	402	217	70.4	(26.11.87)
14 November	Kosmos-1896	89.4	319	203	64.8	(25.12.87)
21 November	Progress-33	88.7	266	193	51.6	(19.12.87)
26 November	Kosmos-1897	23 h. 55 min.	35825	35727	1.4	1000000
1 December	Kosmos-1898	100.8	820	781	74.0	120
7 December	Kosmos-1899	89.3	297	216	70.4	(21.12.87)
10 December	Raduga	23 h. 16 min.	35049	34948	1.3	1000000
12 December	Kosmos-1900	89.8	287	263	65.0	0.5
14 December	Kosmos-1901	89.8	376	181	64.9	(3.2.88)
15 December	Kosmos-1902	92.4	417	373	66.0	3
21 December	Soyuz TM-4	88.6	250	170	51.6	—
22 December	Kosmos-1903	11 h. 49 min.	39342	614	62.8	15
23 December	Kosmos-1904	104.9	1021	989	82.9	1200
25 December	Kosmos-1905	89.3	298	216	70.4	(8.1.88)
26 December	Kosmos-1906	88.8	274	190	82.6	(31.1.88)
27 December	Ekran	23 h. 51 min.	35944	35458	1.5	1000000
29 December	Kosmos-1907	90.2	398	208	72.9	(12.1.88)

In total 116 satellites and spacecraft were put into orbit in 1987.

SOVIET COMMENTS ON TYPES OF EARTH SATELLITES AND SPACE VEHICLES LAUNCHED IN 1987

Meteor-2 is a meteorological earth satellite with onboard equipment for obtaining global images of clouds and underlying atmosphere, using sensors sensitive to the visible and infrared spectra, both for onboard storage and for real time transmission. It continuously monitors radiation penetrating the earth's atmosphere, and obtains global data on vertical temperature distribution. (The first satellite [of this type] was launched on 11 July 1975.) The satellite is equipped with an electromechanical attitude stabilisation system.

Kosmos is the name of a series of artificial earth satellites, which are regularly (since 16 March 1962) launched from Soviet cosmodromes. Their scientific research programme covers:

☐ the study of the concentration of charged particles in the ionosphere, with the aim of researching the distribution of radiowaves, corpuscular flow, low energy particles and the energy composition of the Earth's radiation belts, for the evaluation of radiation danger during prolonged space flights, the process of adaptation to weightlessness, and the primary composition and intensity of various types of cosmic radiation, the Earth's magnetic field, shortwave radiation from the sun and other heavenly bodies, the upper layers of the atmosphere, and the action of meteorites on space craft;

☐ research and experiments on materials studies, the production, under conditions of microgravity, of semiconductors with improved properties and of particularly pure biological preparations, research on the influence of space flight on living organisms and also scientific technical research and experiments in the interests of various branches of the economy and international collaboration, including hydrology, cartography, geology, agriculture and environment studies;

☐ the development of various elements of space navigation systems and equipment for the purpose of ensuring the determination of the positions of civil aircraft and naval vessels and fishing fleets, including those in distress, the development of experimental equipment for relaying telephone and telegraph information, and of equipment, units and elements for the construction of satellites in different flight modes, including docking;

☐ the obtaining of operational information and the continued development of new types of sensors, and new ways of studying, from a long distance, the surface and oceans of the earth and its atmosphere, in the interests of various branches of the economy, science and international collaboration.

Progress-27, -28, -29, -30, -31, -32 and -33 are unmanned cargo ships. The aim of these vehicles is to supply the orbital

space station "Mir" with expendable materials and various
equipment.

Molniya-3 is a series of communications satellites (an
updated version of the Molniya-1 and -2 series of communications satellites) for ensuring the use of long-distance
telephone and telegraph radio communications systems,
transmission of USSR Central Television programmes to
points of the "Orbita" network, and international collaboration. (The launch of the first satellite took place on 21
November 1974.) The satellites' onboard equipment uses
microwaves in the centimetre wavelength range.

Soyuz TM are modernised manned "Soyuz T" spacecraft,
which can carry a larger (up to 250 kg) useful payload. Yu
Romanenko and A Laveykin were taken on the "Soyuz TM-2"
to begin the second main mission of the "Mir" complex. The
Soviet-Syrian team of A Viktorenko, A Aleksandrov and M
Farris travelled on board the "Soyuz TM-3". On completion of
their international programme, the cosmonauts A Viktorenko,
A Laveykin and M Farris returned to earth. Flight times were:

A Laveykin	-174 days, 3 hours, 26 minutes
A Viktorenko & M Farris	-7 days, 23 hours, 5 minutes

"Soyuz TM-4" put on board the "Mir" complex the cosmonauts V Titov, M Manarov and A Levchenko. After the change
over of the teams, the work of the second main mission on the
orbital complex was continued by V Titov and M Manarov. Yu
Romanenko, A Aleksandrov and A Levchenko returned to
earth on the "Soyuz TM-3". Flight times were:

Yu Romanenko	-326 days, 11 hours, 38 minutes
A Aleksandrov	-160 days, 7 hours, 17 minutes
A Levchenko	-7 days, 21 hours, 58 minutes

Raduga is a series of communications satellites with onboard
relay equipment, designated for maintaining telephone and
telegraph communications and the transmission of television
programmes. It is equipped with multichannel communications, using microwaves in the centimetre wavelength range.
(The first satellite was launched on 27 December 1975.)

Kvant is a specialised astrophysics module, designated
for carrying out wide-ranging research in astronomy free
from atmospheric interference, and the solution of a series of
other scientific and economic tasks. It is the first of a series
of specialised modules which make up the multi-purpose,

permanently functioning manned "Mir"-"Kvant"-"Soyuz TM" complex.

Gorizont is a series of communications satellites for ensuring round-the-clock long distance telephone and telegraph radio communications and the transmission of television programmes to stations of the "Orbita" and "Moskva" systems, and also for use in the international "Intersputnik" communications satellite system. (The first "Gorizont" satellite was launched on 19 December 1978). Installed on board is multichannel relay equipment, which operates using microwaves in the centimetre wavelength range.

Ekran is a television broadcast satellite series, with onboard relay apparatus, ensuring, in the decimetre wavelength range, the transmission of Central Television programmes onto the network of collaborative receiving stations. (The first "Ekran" satellite was launched on 26 October 1976).

Kosmos-1827 to Kosmos-1832, Kosmos-1838 to Kosmos-1840, Kosmos-1852 to Kosmos-1859, Kosmos-1875 to Kosmos-1880 and Kosmos-1883 to Kosmos-1885. Each group of satellites was put into orbit by a single launcher.

On 15 May at 21h. 30min. Moscow Time the first launch of the largest launcher "Energiya" took place at the Baykonur cosmodrome, on its maiden test flight, carrying a full scale model of a satellite.

ROYAL UNITED SERVICES INSTITUTE COMMENT ON SOVIET SATELLITES LAUNCHED
IN 1987

Name of satellite	Purpose
Kosmos-1811	Military photographic reconnaissance satellite
Kosmos-1812	Military electronic reconnaissance satellite
Kosmos-1813	Military photographic reconnaissance satellite
Kosmos-1814	Military communications satellite
Kosmos-1817	Military communications satellite. Failed to enter geosynchronous orbit
Kosmos-1818	Military ocean surveillance satellite
Kosmos-1819	Military photographic reconnaissance satellite
Kosmos-1821	Military navigation satellite
Kosmos-1822	Military photographic reconnaissance satellite
Kosmos-1823	Possible military geodetic satellite
Kosmos-1824	Military photographic reconnaissance satellite
Kosmos-1825	Military electronic reconnaissance satellite
Kosmos-1826	Military photographic reconnaissance satellite
Kosmos-1833	Military electronic reconnaissance satellite
Kosmos-1834	Military ocean surveillance satellite
Kosmos-1835	Military photographic reconnaissance satellite
Kosmos-1836	Military photographic reconnaissance satellite

ROYAL UNITED SERVICES INSTITUTE COMMENT ON SOVIET SATELLITES LAUNCHED
IN 1987 *cont.*

Name of satellite	Purpose
Kosmos-1837	Military photographic reconnaissance satellite
Kosmos-1842	Military electronic reconnaissance satellite
Kosmos-1843	Military photographic reconnaissance satellite
Kosmos-1844	Military electronic reconnaissance satellite
Kosmos-1845	Military photographic reconnaissance satellite
Kosmos-1846	Military photographic reconnaissance satellite
Kosmos-1847	Military photographic reconnaissance satellite
Kosmos-1848	Military photographic reconnaissance satellite
Kosmos-1849	Military early-warning satellite
Kosmos-1850	Military communications satellite
Kosmos-1851	Military early-warning satellite
Kosmos-1852	Military tactical communications satellites
Kosmos-1853	Military tactical communications satellites
Kosmos-1854	Military tactical communications satellites
Kosmos-1855	Military tactical communications satellites
Kosmos-1856	Military tactical communications satellites
Kosmos-1857	Military tactical communications satellites
Kosmos-1858	Military tactical communications satellites
Kosmos-1859	Military tactical communications satellites
Kosmos-1860	Military ocean surveillance satellite
Kosmos-1862	Military electronic reconnaissance satellite
Kosmos-1863	Military photographic reconnaissance satellite
Kosmos-1864	Military navigation satellite
Kosmos-1865	Military photographic reconnaissance satellite
Kosmos-1866	Military photographic reconnaissance satellite
Kosmos-1867	Military ocean surveillance satellite
Kosmos-1869	Military ocean surveillance satellite
Kosmos-1870	Military photographic reconnaissance satellite
Kosmos-1870	Military photographic reconnaissance satellite
Kosmos-1872	Military photographic reconnaissance satellite
Kosmos-1874	Military photographic reconnaissance satellite
Kosmos-1881	Military photographic reconnaissance satellite
Kosmos-1882	Military photographic reconnaissance satellite
Kosmos-1886	Military photographic reconnaissance satellite
Kosmos-1888	Military communications satellites
Kosmos-1889	Military photographic reconnaissance satellite
Kosmos-1890	Military ocean surveillance satellite
Kosmos-1891	Military navigation satellite
Kosmos-1892	Military electronic reconnaissance satellite
Kosmos-1893	Military photographic reconnaissance satellite
Kosmos-1894	Military communications satellites
Kosmos-1895	Military photographic reconnaissance satellite
Kosmos-1896	Military photographic reconnaissance satellite
Kosmos-1897	Military communications satellites
Kosmos-1898	Military communications satellites
Kosmos-1899	Military photographic reconnaissance satellite
Kosmos-1900	Military ocean surveillance satellite
Kosmos-1901	Military photographic reconnaissance satellite
Kosmos-1903	Military early-warning satellite
Kosmos-1804	Military navigation satellite
Kosmos-1905	Military ocean surveillance satellite
Kosmos-1906	Military ocean surveillance satellite
Kosmos-1907	Military ocean surveillance satellite
Meteor-2-15	Military meteorological satellite
Meteor-2-16	Military meteorological satellite
Molniya-3-31	Military communications satellite

Afghanistan: Chronology of the Soviet Withdrawal, January–August 1988

FRANCIS TUSA

The author is Middle East analyst at the RUSI

THE FOLLOWING chronology of the Afghan Conflict in 1988 is taken from a number of accessible sources in a variety of media. It is not meant to be exhaustive, especially on the military front where battles around such cities as Kabul and Kandahar rage almost continuously. The points listed are deemed to be important mainly from the point of view of the Soviet withdrawal and the decisions taken on this front. Where the abbreviation "REP" is given, it means that unnamed sources of various sorts have passed information to the West, but its exact veracity cannot be checked. When the Mujahedin Alliance is mentioned, the reference is to the polyglot and increasingly factious grouping of the seven main Mujahedin parties based in Peshawar in Pakistan.

JANUARY

2   First reports from Kabul of supplies reaching Khost.
4   US Under Secretary of State Armacost gives assurances to the Mujahedin that the US will not desert them.
    Khost seige becomes a widespread battle in the area.
    Soviet Foreign Minister Eduard Shervadnadze arrives unannounced in Kabul.
6   Shevardnadze says that Soviet troops can leave Afghanistan within a year, possibly by the end of 1988.
    Soviet spokesman acknowledges ethnic trouble spilling over into Central Asian Republics.
7   US Secretary of State Shultz: aid to Mujahedin may stop once a withdrawal starts, but will recommence if the withdrawal stalls.

Mujahedin build up of forces in Paktia Province.
Soviet spokesman: Soviet withdrawal dependent on a pro-Soviet government in Kabul. Troops will start the withdrawal 60 days after a treaty, possibly in early May.

11 *Pravda*: Withdrawal could start on 1 May; Treaty would have to be 1 March to allow two months for Pakistan to close guerilla bases.
US Report: Soviets making treaties with local Mujahedin commanders on possible evacuation routes.

12 Soviet–Afghan offensive in Mahajalat area to the south of Kandahar.

13 Kabul government announce results of Reconciliation Plan:
40,000 rebels downed arms;
114,000 rebels ceased fire;
116,000 refugees returned home.

14 Jamiat and Hezb-i-Islami turn down Kabul coalition offer.

18 Mujahedin spokesman: Soviets and protegés have been issued with evacuation ID cards.

19 UN Under Secretary General Cordovez starts a pre-Geneva proximity talks 10-day shuttle between Kabul and Islamabad. Mujahedin refuse to meet him.
Khost comes under heavy attack.

20 Pir Gailani, head of the more moderate National Islamic Front of Afghanistan, says that he will meet Cordovez.
Soviet troop build up around Gardez.
Mujahedin step-up attacks on Kandahar and Herat.

25 Afghan government troops withdraw from Gardez-Khost road.
Pakistan demands that a formal interim government is established when Soviet troops withdraw: by this stage, the Soviets have dropped this as one of their own conditions.

26 Cordovez returns from Kabul—little progress.
Mujahedin spokesman: Khost cut off once more.
Battle in Jadran Valley near Khost.

27 Mujahedin Alliance want formal UN recognition before they meet Cordovez.
President Zia of Pakistan says that he will not sign a treaty with President Najibullah of Afghanistan.
Kabul "pardons" six Mujahedin leaders sentenced in absentia.

28 Shevardnadze: a treaty now very close, but the last troops will now only be able to withdraw by early 1989.

31 Mujahedin will set up a provisional government after the Soviet withdrawal.

FEBRUARY

1 US report: No sign of the start of a Soviet withdrawal.
KHAD, the Afghan secret service, believed to have planted a bomb in Peshawar.

7 US report: Soviet Union signing thousands of economic treaties with Afghanistan to retain control of the raw materials even after a withdrawal.
Cordovez meets Yunis Khalis, the Alliance spokesman, in Peshawar. Then leaves for Kabul.

8 Gorbachev speech: Withdrawal to start on 15 May regardless of a treaty; withdrawal to be front-loaded.

9 Yuli Vorontsev (Soviet Foreign Ministry) in Islamabad: guarantees Soviet non-interference in Afghanistan.
Cordovez: Pakistan-Afghanistan agreement on timetable.

10 Mujahedin Alliance agree to form a transitional government.
Gulbadin Hekmatyar (head of the fundamenta list Hezb-i-Islami): All but about 1000 PDPA members will be able to join Mujahedin government.

11 Mujahedin spokesman: fighting to continue until PDPA overthrown.
Pakistan: transitional government to be established before Soviet withdrawal.
Vorontsev: no Treaty, no withdrawal.

15 US Administration starts to consider weapons supply symmetry.
Afghan spokesman: President, Defence, Interior and Foreign portfolios will always be held by PDPA members in any Kabul government.

16 Mujahedin move troops towards Kabul.
Soviet offensive around Shinwar and Dehbala in Nangarhar province.

18 Vorontsev: Pakistan deliberately delaying signing a treaty.

19 US report: further Soviet-Afghan economic agreements in the north.
Pakistan spokesman: bloodbath in Afghanistan unless an interim government is formed.

22 Gorbachev: there is no secret Soviet bridgehead being built in Afghanistan.
Afghan Mujahedin in Iran reject a coalition with PDPA.

23   Mujahedin Alliance announce their Provisional government:
     Executive Council (from the 7 Peshawar groups)
     Government formed from: 14 Mujahedin members;
                             7 refugee representatives;
                             7 "good Muslims living in Kabul".
     Council of 75 to frame provisional laws.
     General Election 6 months after Soviet withdrawal completed.
     Bomb attacks in Kabul on Soviet Embassy and Radio Kabul.
     US spokesman: Interim government to be established after a Soviet withdrawal.
24   Gulbuddin Hekmatyar, says that Pakistan should not sign Geneva Accords.
     Alliance names provisional "Head of State": Ahmed Shah (fundamentalist), Zabulla Mojadidi as Vice President.
25   Zia: no rift with US over Afghanistan policy.
     Armacost: meets the 7 Mujahedin groups in Peshawar.
28   Refugee camps in Pakistan bombed by Afghan air force.
     Mujahedin around Khost bombed heavily.
29   Pakistan has last minute talks on their negotiating position.

MARCH
 1   Najibullah: only guerillas inside Afghanistan have a right to share power.
     Afghan spokeman: Pakistan cannot dictate the shape of the Afghan government.
     US report: fighting around Kabul did not slacken off during the winter.
 2   Cordovez: Pakistan should leave the question of an Afghan government to the Afghans.
     Mujahedin Alliance rejects the Geneva talks.
 3   Afghan Foreign Minister Abdul Wakil: Soviet troops will withdraw over 9 months, the first half to leave within 3 months.
     Cordovez: most of the 4 Geneva Instruments agreed upon.
 4   Pakistani Foreign Minister Noorani flies back to Islamabad to see whether Pakistan should sign.

6 Noorani returns to Geneva via Moscow: insists on an interim government.

7 US start to back Pakistan over the interim government.
US-USSR symmetry issue the main block to a treaty.

8 Cordovez: a settlement may take time.
Mojadidi withdraws his group from the Mujahedin Alliance.

9 Wakil: talks are stalemated by Pakistan.

10 Mojadidi to rejoin Mujahedin Alliance.
Soviet spokesman: if a treaty is signed, they will talk about an interim government.

11 US delegation to Geneva talks reaffirm committment to symmetry.

13 Mujahedin Alliance delegation visits Geneva: Yunis Khalis resigns.
Pakistan reaffirms commitment to interim government in Kabul.
Hekmatyar becomes Mujahedin spokesman.

15 Soviet spokesman: no real deadline for treaty, but a May 15 withdrawal was only feasible will only happen if a treaty is signed on March 15.

16 Afghanistan berates Pakistan for delaying signature of the Accords.
Pakistan: introduces the question of the border between Pakistan and Afghanistan. Kabul angry.

20 Announcement of Afghan elections: to be held for a 10-day period from 5 April.

21 Shultz-Shevardnadze meet in Washington. Hopes of talks resuming.
Khost comes under close seige again.
Zahir Shah, ex-King of Afghanistan, rules out power sharing with PDPA

22 Shevardnadze: Moscow will withdraw some troops even if there is no treaty.

23 Soviet/Afghan troops break the 3 year seige of Urgun in Paktia Province.

24 Shultz: 3 month-1 year on arms shipments to the two sides if a ceasefire starts, to be continued during a withdrawal.

25 Pakistan softens on the issue of an interim government, Soviet Union on the issue of symmetry.

27 Report: US rushing $300 million of arms to Mujahedin to beat a treaty, but ending STINGER sales.

29 Najibullah offers a coalition with the Mujahedin without
    mentioning PDPA participation.
    REP: Soviet troops buying up western consumer goods in
    Kabul in bulk.
    REP: Soviets flying in extra equipment for the Afghan
    forces.
    Hekmatyar: suggestion of a confederation/union of
    Afghanistan and Pakistan.
30 Mujahedin turn down Najibullah's offer of a coalition.
    REP: Soviet hints of a partition of Afghanistan.
31 Soviet spokesman; moratorium of aid to Kabul unacept-
    able.
    US spokesman: US will supply Mujahedin if Moscow
    continues to supply Kabul.

APRIL
 1 Shultz confirms continuation of aid to Mujahedin.
    Soviets reject US continuation of aid.
    Mujahedin spokesman: STINGER ban will harm their
    chances.
    Cordovez: talks could well fail.
 3 Shevardnadze in Kabul for talks with Najibullah: contin-
    gency for Geneva talks failure?
 4 Hekmatyar: war will continue regardless of Geneva.
    Afghan and Pakistani troops clash near Quetta in Balu-
    chistan.
 5 Shevardnadze: Soviet troops will leave regardless of the
    result of the Geneva talks.
    REP: Soviets thinning troops out in Afghanistan.
    REP: creation of an Afghan Prime Minister of the North
    based at Mazr-e-Sharif.
    REP: new province of Sari Pol created in northern
    Afghanistan.
 6 Gorbachev, arrives in Tashkent for talks with Najibullah:
    US fears of a bilateral agreement.
 7 US-Soviet agreement over weapons supplies: "positive
    symmetry", both to supply equal quantities.
    President Zia agrees to sign Geneva Accords.
 8 Cordovez: instruments ready for signature on 14 April.
    Radio Moscow: Soviet troops to begin withdrawal on 15
    April.
    REP: town of Gorband to the north of Kabul taken by
    Mujahedin.

10 Explosions at Pakistani ordnance base at Faisabad: Muja-
    hedin weapons also stored there.
    Gorbachev confirms 15 May withdrawal date.
11 Zia: sabotage at ordnance depot probable.
    President Reagan accepts Geneva Accords.
    REP: 450 Soviet vehicles leave Lowgar Province.
12 Iranian Government rejects Geneva Accords.
    Soviet spokesman: 50,000 troops to leave within 3
    months.
    REP: increasing rocket attacks on Kabul. About 7 per
    week.
    REP: Soviets flying in more equipment for Kabul forces.
    REP: 130 Afghan soldiers desert in Paktia Province.
    REP: Afghan commander of central zone killed.
13 Mujahedin spokesman: Kabul will fall quickly after the
    Soviet withdrawal.
    Soviets claim that Pakistan will dismantle all Mujahedin
    camps.
14 Afghan Accords signed in Geneva.
    Zia: Faisabad explosion was sabotage. Confirms the
    presence of Mujahedin arms there.
    REP: Mujahedin arms dump in Baluchistan (Pakistan)
    blows up.
    REP: 4 Soviet advisors killed by a bomb in Kabul.
15 Mojadidi will not attend a rally to condemn outright the
    Geneva Accords.
17 REP: the KHAD behind Faisabad and other explosions.
19 REP: Increased fighting around Kabul.
    REP: Soviet troops prepare to leave Jalalabad.
    REP: Kabul trying to reach arrangements with Local
    Mujahedin commanders.
24 REP: Soviet–Afghan forces evacuate Marouf, Atagar and
    Darwa Zigag in Kandahar Province and Barikot in Kunar
    Province.
    Najibullah: US prolonging the war.
    Soviets suggest an informal weapons freeze.
25 Najibullah: 20,000 Soviet troops to leave by the end of
    May.
26 Afghan Revolution 10th Anniversary Parade.
    REP: Soviet troops moving into Logar Province to keep
    roads clear.
    REP: Soviet troops evacuating Paktia, Paktika, Nangahar
    and Zabol Provinces.

27   REP: car bombs in Kabul.
     REP: clashes around the Kabul-Peshawar road.
29   Heavy Mujahedin rocketing of Kabul.

MAY
 1   REP: Mujahedin hit large convoy in Paktia Province.
     REP: heavy fighting in Lazha in Paktia Province.
 2   REP: Mujahedin take Arghastan near Kandahar.
 7   REP: Mujahedin take Afghan base of Chamkari in Paktia
     Province.
12   REP: Soviet heavy equipment leaving Jalalabad.
     REP: Soviet troops reinforcing withdrawal routes.
     REP: first Soviet units leaving Jalalabad.
15   REP: Afghan garrisons in Panjshir Valley captured.
16   First Soviet troops leave Kabul for the Soviet Union.
     REP: heavy fighting around Kandahar.
17   Soviet spokesman: rejects the idea of a northern buffer
     zone.
     REP: only five major Afghan garrisons left on Pakistan-
     Afghan border.
     REP: town of Hesarah on Jalalabad-Kabul road taken by
     Mujahedin.
     REP: Soviet withdrawal convoy attacked north of Kabul.
18   First Soviet withdrawal convoy reaches Soviet Union.
     REP: Afghan base of Chowni evacuated.
19   Novosti News Agency announce approximate casualty
     figures: 12–15,000 dead and 35,000 wounded.
     First official complaint to the UN about breaches of the
     Geneva Accords made by Afghanistan.
     REP: Soviet troops withdraw from Nangahar Province.
     REP: Soviets ready to leave Kandahar.
     REP: Alikhel evacuated, Qalat and Asdabad under seige.
20   REP: Soviet dependents being evacuated.
22   Soviet spokesman: 25% of troops to leave by 29 May.
     Jalalabad evacuated by Soviet troops.
23   UN to ask for $1–2 billion of aid for Afghan reconstruc-
     tion.
25   Soviet Defence Ministry figures: 13,310 dead
                              35,478 wounded
                              311 missing in action.
     REP: Mujahedin ambushes in Faryab Province.
26   Soviet Defence Ministry: 100,300 Soviet troops were in
     Afghanistan as of 15 May.

27 REP: Dakha in Nangahar Province under seige.
Najibullah names Prime Minister for new "National Reconciliation Government".
29 Soviet spokesman: the withdrawal may be slowed down or halted unless weapons supply to the Mujahedin stops.
30 REP: Panjshir Valley clear of all Soviet/Afghan troops.

JUNE
1 Gorbachev: accuses Pakistan of breaking the Geneva Accords.
REP: Khalq/Parcham in-fighting in the PDPA delaying announcement of the full government.
2 REP: increased fighting around Kandahar.
6 REP: Chief of Staff of Afghan Air Force killed in Kandahar.
7 Najibullah lists breeches of the Geneva Accords to the UN.
Najibullah announces an unchanged cabinet.
REP: Mujahedin pressure on Kandahar and Jalalabad.
REP: Mujahedin take Ghozni.
9 Shevardnadze accuses Pakistan of treaty violations at UN.
12 Afghan offensive in Kapisa Province.
13 US spokesman: Mujahedin attacks are slowing down Soviet withdrawal.
14 Pakistan and Afghanistan exchange artillery fire.
15 Mujahedin Alliance change spokesman from Gulbadin Hekmatyar to Pir Gailani.
16 REP: heavy fighting in Nangahar, Kunar and Baghdis Provinces.
17 REP: Hekmatyar did not wish to give up spokesman's post.
19 REP: Kalat, capital of Zabol Province, falls to Mujahedin.
21 REP: Nur Gol in Kunar Province falls to Mujahedin.
26 REP: Mujahedin destroy 8 SU-25s at Kabul airport.
REP: heavy fighting in Paghman area near Kabul.

JULY
3 REP: Mujahedin close to capturing Qalat.
Cordovez to launch a UN initiative to form a coalition government.
REP: Soviet withdrawal on schedule; 30,000 troops have left.

4   REP: Mujahedin take town of Moh Agha on Kabul-Gardez road.
    REP: Mujahedin take town of Maidan to SW of Kabul.
6   Mujahedin call for resignation of Cordovez because of his attempts to form a coalition government with PDPA members.
10 Cordovez Plan: Najibullah and government to step down.
                    Ceasefire on 1 September 1988.
                    Loya Jirga (traditional Parliament on 1 March 1989).
12 REP: Soviet–Afghan counter-attacks at Kandahar, Maidan and around Kabul.
14 Disagreements in Mujahedin Alliance over the acceptability of a coalition government.
23 Zia: Soviets have halted withdrawal, and actually sent 10,000 new troops into Afghanistan.
28 Embassies start to evacuate dependents and non-essential staff.
    REP: during July, over 200 rockets per week hit Kabul.
31 Rabbani critisises attacks on withdrawing troops.

AUGUST
2   REP: Soviet troops reinforce Afghan positions around Kabul to stop rocket attacks.
    REP: Soviet offensive in Paghman area near Kabul.
5   Soviets complete withdrawal from Kandahar, but suffer casualties.
    Shevardnadze in Kabul for talks with Najibullah on security situation.
    Claims that there are no Soviet troops in southern Afghanistan. Links Pakistan with international "terrorism against Afghanistan".
    REP: Soviet troops leaving Kunduz in the north.
6   Shevardnadze: withdrawal will be completed on schedule.
    Soviet troops no longer in 25 Afghan provinces.
7   Mujahedin commanders inside Afghanistan accuse Hekmatyar of stopping arms supplies to all but his own group.
8   Shevardnadze: Soviet–Afghan economic aid treaty to last until the year 2000 to be signed in autumn.
    Pakistan shoots down an Afghan jet intruding into local airspace; in a separate incident, two Afghan pilots defect to Pakistan.
    REP: first set of Kabul-based Soviet troops begin to leave.

Half of the 50,000 troops left in Afghanistan believed to be based around Kabul.
REP: Pakistan believed to be sending troops to help Mujahedin. Soviet statement: "If violations continue, we [The Soviet Union] may prolong our presence".

9 REP: heavy fighting around Kandahar.
REP: Pakistan denies that it has troops in Afghanistan.
Kabul accuses Pakistan of over 100 land and air violations of the Geneva Accords since its signing.

10 REP: Mujahedin take the northern provincial capital of Kunduz.
REP: Soviet convoy from Kandahar attacked.
REP: Mujahedin have 11–15,000 troops around Kandahar.

11 REP: heavy fighting near Herat.
REP: Mujahedin cut all Kabul-to-Kandahar roads.

14 Soviets announce that half of their troops have left Afghanistan.
REP: Soviet troops only left in the provinces of Baglan, Balkh, Herat, Kabul, Parwan and Samangan. Confirms that Kunduz captured.
Soviet spokesman: Pakistani troops aiding the Mujahedin.
REP: heavy fighting around Gardez and Khost as well as in Kunduz.
REP: Soviet diplomatic dependents evacuated.
REP: Paghman area near Kabul totally under Mujahedin control.

15 Afghan forces regain most of Kunduz.
REP: Shakandar district near Kabul seized by Mujahedin.
Soviet spokesman: Pakistani interference "cannot be tolerated further".

16 Soviet spokesman: withdrawal will not end before 15 February 1989.
REP: arms and fuel base at Kilagay destroyed by Mujahedin. Up to 2 years supply was kept there.

17 President Zia killed in plane crash.

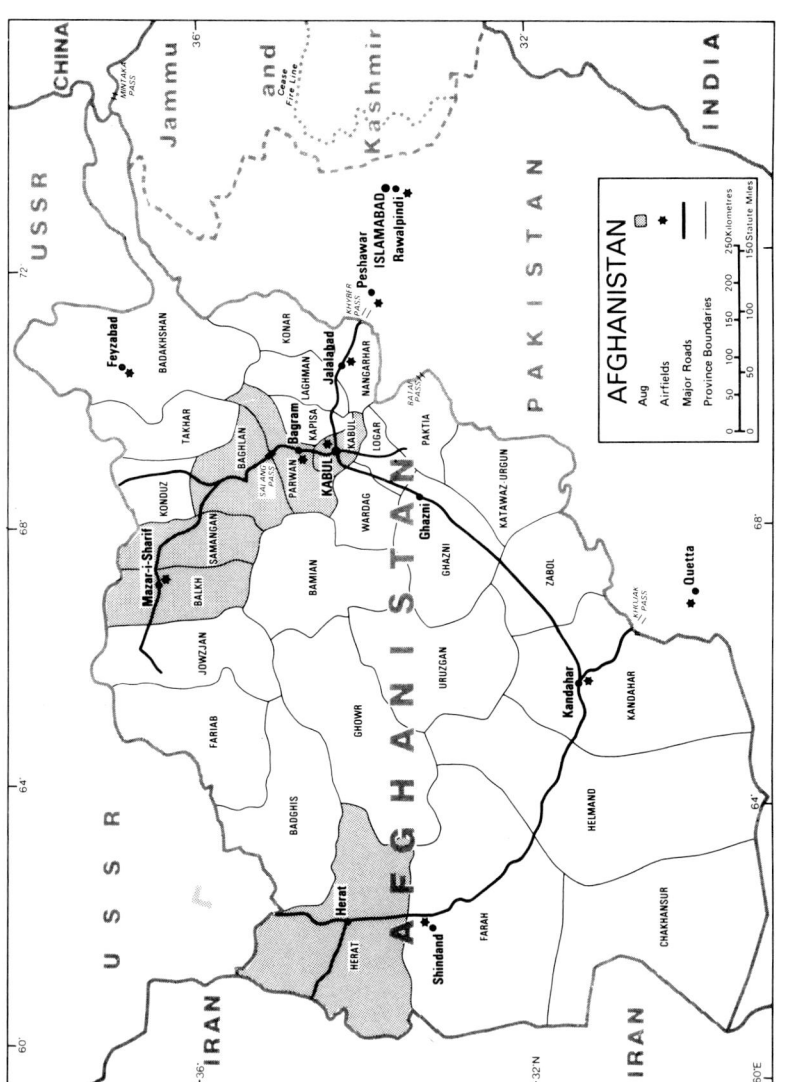

Map 1—Official Soviet troop dispositions as of August 1988.

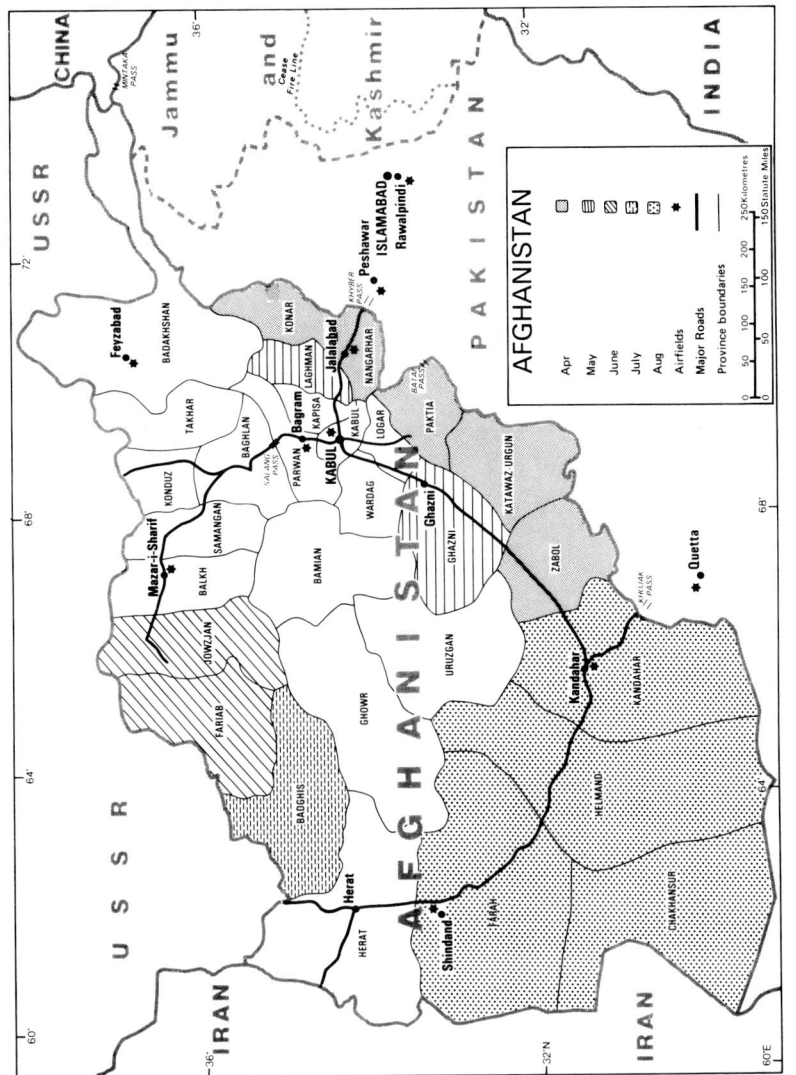

Map 2—Announced dates of Soviet Troop evacuations from Afghan provinces.

THE GENEVA ACCORDS

DECLARATION ON INTERNATIONAL GUARANTEES

FOLLOWS THE full text of the declaration on international guarantees:

The government of the Union of Soviet Socialist Republics and of the United States of America,

Expressing support that the Republic of Afghanistan and the Islamic Republic of Pakistan have concluded a negotiated political settlement designed to normalize relations and promote good-neighbourliness between the two countries as well as to strengthen international peace and security in the region;

Wishing in turn to contribute to the achievement of the objectives that the Republic of Afghanistan and the Islamic Republic of Pakistan have set themselves, and with a view to ensuring respect for their sovereignty, independence, territorial integrity and non-alignment;

Undertake to invariably refrain from any form of interference and intervention in the internal affairs of the Republic of Afghanistan and the Islamic Republic of Pakistan and to respect the commitments contained in the bilateral agreement between the Republic of Afghanistan and the Islamic Republic of Pakistan on the principles of mutual relations, in particular on non-interference and non-intervention;

Urge all states to act likewise.

The present declaration shall enter into force on 15 May 1988.

Done at Geneva, this fourteenth day of April 1988 in five original copies, each in the English and Russian languages, both texts being equally authentic.

(Signed by the USSR and the USA)

AGREEMENT ON PRINCIPLES OF MUTUAL RELATIONS

Follows the full text of the bilateral agreement between the Republic of Afghanistan and the Islamic Republic of Pakistan on the principles of mutual relations, in particular on non-interference and non-intervention:

The Republic of Afghanistan and the Islamic Republic of Pakistan, hereinafter referred to as the high contracting parties,

Desiring to normalize relations and promote good-neighbourliness and co-operation as well as to strengthen international peace and security in the region,

Considering that full observance of the principle of non-interference and non-intervention in the internal and external affairs of states is of the greatest importance for the maintenance of international peace and security and for the fulfilment of the purposes and principles of the Charter of the United Nations,

Reaffirming the inalienable right of states freely to determine their own political, economic, cultural and social systems in accordance with the will of their peoples, without outside intervention, interference, subversion, coercion or threat in any form whatsoever,

Mindful of the provisions of the Charter of the United Nations as well as the resolutions adopted by the United Nations on the principle of non-interference and non-intervention, in particular the declaration on principles of International Law concerning friendly relations and co-operation among states in accordance with the Charter of the United Nations, of 24 October 1970, as well as the declaration on the inadmissibility of intervention and interference in the internal affairs of states, of 9 December 1981,

Have agreed as follows:

Article I

Relations between the high contracting parties shall be conducted in strict compliance with the principle of non-interference and non-intervention by states in the affairs of other states.

Article II

For the purpose of implementing the principle of non-interference and non-intervention each high contracting party undertake to comply with the following obligations:

1. To respect the sovereignty, political independence, territorial integrity, national unity, security and non-alignment of the other high contracting party, as well as the national identity and cultural heritage of its people;

2. To respect the sovereign and inalienable right of the other high contracting party freely to determine its own political, economic, cultural and social systems, to develop its

international relations and to exercise permanent sovereignty over its natural resources, in accordance with the will of its people, and without intervention, interference, subversion, coercion or threat in any form whatsoever;

3. To refrain from the threat or use of force in any form whatsoever so as not to violate the boundaries of each other, to disrupt the political, social or economic order of the other high contracting party, to overthrow or change the political system of the other high contracting party or its government, or to cause tension between the high contracting parties;

4. To ensure that its territory is not used in any manner which would violate the sovereignty, political independence, territorial integrity and national unity or disrupt the political, economic and social stability of the other high contracting party;

5. To refrain from armed intervention, subversion, military occupation or any other form of intervention and interference, overt or covert, directed at the other high contracting party, or any act of military, political or economic interference in the internal affairs of the other high contracting party, including acts of reprisal involving the use of force;

6. To refrain from any action or attempt in whatever form or under whatever pretext to destabilize or to undermine the stability of the other high contracting party or any of its institutions;

7. To refrain from the promotion, encouragement or support, direct or indirect, of rebellions or secessionist activities against the other high contracting party, under any pretext whatsoever, or from any other action which seeks to disrupt the unity or to undermine or subvert the political order of the other high contracting party;

8. To prevent within its territory the training, equipping, financing and recruitment of mercenaries from whatever origin for the purpose of hostile activities against the other high contracting party, or the sending of such mercenaries into the territory of the other high contracting party and accordingly to deny facilities, including financing for the training, equipping and transit of such mercenaries;

9. To refrain from making any agreements or arrangements with other states designed to intervene or interfere in the internal and external affairs of the other high contracting party;

10. To abstain from any defamatory campaign, vilification

or hostile propaganda for the purpose of intervening or interfering in the affairs of the other high contracting party;

11. To prevent any assistance to or use of or tolerance of terrorist groups, saboteurs or subversive agents against the other high contracting party;

12. To prevent within its territory the presence, harbouring, in camps and bases or otherwise, organizing, training, financing, equipping and arming of individuals and political, ethnic and any other groups for the purpose of creating subversion, disorder or unrest in the territory of the other high contracting party and accordingly also to prevent the use of mass media and the transportation of arms, ammunition and equipment by such individuals and groups;

13. Not to resort to or to allow any other action that could be considered as interference or intervention.

Article III

The present agreement shall enter into force on 15 May 1988.

Article IV

Any steps that may be required in order to enable the high contracting parties to comply with the provisions of Article II of this agreement shall be completed by the date on which this agreement enters into force.

Article V

This agreement is drawn up in the English, Pashtu and Urdu languages, all texts being equally authentic. In case of any divergence of interpretation, the English text shall prevail.

Done in five original copies at Geneva this fourteenth day of April 1988.

(Signed by Afghanistan and Pakistan)

Agreement on Interrelationships

Follows the full text of the agreement on the interrelationships for the settlement of the situation relating to Afghanistan:

1. The Diplomatic process initiated by the Secretary-Gen-

eral of the United Nations with the support of all governments concerned and aimed at achieving, through negotiations, a political settlement of the situation relating to Afghanistan has been successfully brought to an end.

2. Having agreed to work towards a comprehensive settlement designed to resolve the various issues involved and to establish a framework for good-neighbourliness and cooperation, the Government of the Republic of Afghanistan and the Government of the Islamic Republic of Pakistan entered into negotiations through the intermediary of the personal representative of the Secretary-General at Geneva from 16 to 24 June 1982. Following consultations held by the personal representative in Islamabad, Kabul and Teheran from 21 January to 7 February 1983, the negotiations continued at Geneva from 11 to 22 April and from 12 to 24 June 1983. The personal representative again visited the area for high level discussions from 3 to 15 April 1984. It was then agreed to change the format of the negotiations and, in pursuance thereof, proximity talks through the intermediary of the personal representative were held at Geneva from 24 to 30 August 1984. Another visit to the area by the personal representative from 25 to 31 May 1985 preceded further rounds of proximity talks held at Geneva from 20 to 25 June, from 27 to 30 August and from 16 to 19 December 1985. The personal representative paid an additional visit to the area from 8 to 18 March 1986 for consultations. The final round of negotiations began as proximity talks at Geneva on May 5 1986, was suspended on 23 May 1986, and was resumed from 31 July to 8 August 1986. The personal representative visited the area from 20 November to 3 December 1986 for further consultations and the talks at Geneva were resumed again from 25 February to 9 March 1987, and from 7 to 11 September 1987. The personal representative again visited the area from 18 January to 9 February 1988 and the talks resumed at Geneva from 2 March to 8 April 1988. The format of the negotiations was changed on 14 April 1988, when the instruments comprising the settlement were finalized, and, accordingly, direct talks were held at that stage. The Government of the Islamic Republic of Iran was kept informed of the progress of the negotiations throughout the diplomatic process.

3. The Government of the Republic of Afghanistan and the Government of the Islamic Republic of Pakistan took part in

the negotiations with the expressed conviction that they were acting in accordance with their rights and obligations under the Charter of the United Nations and agreed that the political settlement should be based on the following principles of International Law:

- The principle that states shall refrain in their international relations from the threat or use of force against the territorial integrity or political independence of any state, or in any other manner inconsistent with the purposes of the United Nations;
- The principle that states shall settle their international disputes by peaceful means in such a manner that international peace and security and justice are not endangered;
- The duty not to intervene in matters within the domestic jurisdiction of any state, in accordance with the Charter of the United Nations;
- The Duty of states to cooperate with one another in accordance with the Charter of the United Nations;
- The principle of equal rights and self-determination of peoples;
- The principle of sovereign equality of states;
- The principle that states shall fulfil in good faith the obligations assumed by them in accordance with the Charter of the United Nations.

The two governments further affirmed the right of the Afghan refugees to return to their homeland in a voluntary and unimpeded manner.

4. The following instruments were concluded on this date as component parts of the political settlement:

A bilateral agreement between the Republic of Afghanistan and the Islamic Republic of Pakistan on the principles of mutual relations, in particular on non-interference and non-intervention;

A declaration on international guarantees by the Union of Soviet Socialist Republics and the United States of America;

A bilateral agreement between the Republic of Afghanistan and the Islamic Republic of Pakistan on the voluntary return of refugees;

The present agreement on the interrelationships for the settlement of the situation relating to Afghanistan.

5. The bilateral agreement on the principles of mutual relations, in particular on non-interference and non-interven-

tion, the declaration on international guarantees, the bilateral agreement on the voluntary return of refugees, and the present agreement on the interrelationships for the settlement of the situation relating to Afghanistan will enter into force on 15 May 1988. In accordance with the timeframe agreed upon between the Union of Soviet Socialist Republics and the Republic of Afghanistan there will be a phased withdrawal of the foreign troops which will start on the date of entry into force mentioned above. One half of the troops will be withdrawn by 15 August 1988 and the withdrawal of all troops will be completed within nine months.

6. The interrelationships in paragraph 5 above have been agreed upon in order to achieve effectively the purpose of the political settlement, namely that as from 15 May 1988: there will be no interference and intervention in any form in the affairs of the parties; the international guarantees will be in operation; the voluntary return of the refugees to their homeland will start and be completed within the timeframe specified in the agreement on the voluntary return of the refugees; and the phased withdrawal of the foreign troops will start and be completed within the timeframe envisaged in paragraph 5. It is therefore essential that all the obligations deriving from the instruments concluded as component parts of the settlement be strictly fulfilled and that all the steps required to ensure full complicance with all the provisions of the instruments be completed in good faith.

7. To consider alleged violations and to work out prompt and mutually satisfactory solutions to questions that may arise in the implementation of the instruments comprising the settlement representatives of the Republic of Afghanistan and the Islamic Republic of Pakistan shall meet whenever required.

A representative of the Secretary-General of the United Nations shall lend his good offices to the parties and in that context he will assist in the organization of the meetings and participate in them. He may submit to the parties for their consideration and approval suggestions and recommendations for prompt, faithful and complete observance of the provisions of the instruments.

In order to enable him to fulfil his tasks, the representative shall be assisted by such personnel under his authority as required. On his own initiative, or at the request of any of the parties, the personnel shall investigate any possible violations of any of the provisions of the instruments and prepare a report thereon. For that purpose, the representative and his personnel shall receive all the necessary cooperation from the parties, including all freedom of movement within their respective territories required for effective investigation. Any report submitted by the representative to the two governments shall be considered in a meeting of the parties no later than forty-eight hours after it has been submitted.

The modalities and logistical arrangements for the work of the representative and the personnel under his authority as agreed upon with the parties are set out in the memorandum of understanding which is annexed to and is part of this agreement.

8. The present instrument will be registered with the Secretary-General of the United Nations. It has been examined by the representatives of the parties to the bilateral agreements and of the States-Guarantors, who have signified their consent with its provisions. The representatives of the parties, being duly authorized thereto by their respective governments, have affixed their signatures hereunder. The Secretary-General of the United Nations was present.

Done, at Geneva, this fourteenth day of April 1988, in five original copies each in the English, Pashtu, Russian and Urdu languages, all being equally authentic. In case of any dispute regarding the interpretation the English text shall prevail.

(Signed by Afghanistan and Pakistan)

In witness thereof, the representatives of the states-guarantors affixed their signatures hereunder:

(Signed by the USSR and USA)

AGREEMENT ON VOLUNTARY RETURN OF REFUGEES

Follows the full text of the bilateral agreement between the Republic of Afghanistan and the Islamic Republic of Pakistan on the voluntary return of refugees:

The Republic of Afghanistan and the Islamic Republic of Pakistan, hereinafter referred to as the high contracting parties,

Desiring to normalize relations and promote good-neighbourliness and co-operation as well as to strengthen international peace and security in the region,

Convinced that voluntary and unimpeded repatriation constitutes the most appropriate solution for the problem of Afghan refugees present in the Islamic Republic of Pakistan

and having ascertained that the arrangements for the return of
the Afghan refugees are satisfactory to them,

Have agreed as follows:

Article I

All Afghan refugees temporarily present in the territory of
the Islamic Republic of Pakistan shall be given the opportunity
to return voluntarily to their homeland in acccordance with
the arrangements and conditions set out in the present
agreement.

Article II

The Government of the Republic of Afghanistan shall take
all necessary measures to ensure the following conditions for
the voluntary return of Afghan refugees to their homeland:

    (a) All refugees shall be allowed to return in freedom to
their homeland;

    (b) All returnees shall enjoy the free choice of domicile
and freedom of movement within the Republic of
Afghanistan;

    (c) All returnees shall enjoy the right to work, to adequate
living conditions and to share in the welfare of the
state;

    (d) All returnees shall enjoy the right to participate on an
equal basis in the civic affairs of the Republic of
Afghanistan. They shall be ensured equal benefits
from the solution of the land question on the basis of
the land and water reform;

    (e) All returnees shall enjoy the same rights and privi-
leges, including freedom of religion, and have the
same obligations and responsibilities as any other
citizens of the Republic of Afghanistan without
discrimination.

The Government of the Republic of Afghanistan undertakes
to implement these measures and to provide, within its
possibilities, all necessary assistance in the process of repatriation.

Article III

The Government of the Islamic Republic of Pakistan shall
facilitate the voluntary, orderly and peaceful repatriation of all

Afghan refugees staying within its territory and undertakes to provide, within its possibilities, all necessary assistance in the process of repatriation.

Article IV

For the purpose of organising, coordinating and supervising the operations which should effect the voluntary, orderly and peaceful repatriation of Afghan refugees, there shall be set up mixed commissions in accordance with the established international practice. For the performance of their functions the members of the Commissions and their staff shall be accorded the necessary facilities, and have access to the relevant areas within the territories of the high contracting parties.

Article V

With a view to the orderly movement of the returnees, the commissioners shall determine frontier crossing points and establish necessary transit centres. They shall also establish all other modalities for the phased return of refugees, including registration and communication to the country of return of the names of refugees who express the wish to return.

Article VI

At the request of the governments concerned, the United Nations High Commissioner for Refugees will cooperate and provide assistance in the process of voluntary repatriation of refugees in accordance with the present agreement. Special agreements may be concluded for this purpose between UNHCR and the high contracting parties.

Article VII

The present agreement shall enter into force on 15 May 1988. At that time the mixed commissions provided in Article IV shall be established and the operations for the voluntary return of refugees under this agreement shall commence.

The arrangements set out in Articles IV and V above shall remain in effect for a period of eighteen months. After that period the high contracting parties shall review the results of the repatriation and, if necessary, consider any further arrangements that may be called for.

Article VIII

This agreement is drawn up in the English, Pashtu, and Urdu languages, all texts being equally authentic. In case of any divergence of interpretation, the English text shall prevail.

Done in five original copies at Geneva this fourteenth day of April 1988.

(Signed by Afghanistan and Pakistan)

ANNEX

MEMORANDUM OF UNDERSTANDING

I. *Basic Requirements*

(a) The parties will provide full support and co-operation to the representative of the Secretary-General and to all the personnel assigned to assist him.

(b) The representative of the Secretary-General and his personnel will be accorded every facility as well as prompt and effective assistance, including freedom of movement and communications, accommodation, transportation and other facilities that may be necessary for the performance of their tasks. Afghanistan and Pakistan undertake to grant to the representative and his staff all the relevant privileges and immunities provided for by the convention on the privileges and immunities of the United Nations.

(c) Afghanistan and Pakistan will be responsible for the safety of the representative of the Secretary-General and his personnel while operating in their respective countries.

(d) In performing their functions, the representative of the Secretary-General and his staff will act with complete impartiality. The representative of the Secretary-General and his personnel must not interfere in the internal affairs of Afghanistan and Pakistan and, in this context, cannot be used to secure advantages for any of the parties concerned.

II. *Mandate*

The Mandate for the implementation-assistance arrangements envisaged in paragraph 7 derives from the instruments comprising the settlement. All the staff assigned to the representative of the Secretary-General will accordingly be carefully briefed on the relevant provisions of the instruments and on the procedures that will be used to ascertain violations thereof.

III. *Modus operandi and Personnel Organization*

The Secretary-General will appoint a senior military officer as deputy to the Representative, who will be stationed in the area, as head of two small headquarters units, one in Kabul and the other in Islamabad, each comprising five military officers, drawn from existing United Nations operations, and a small civilian auxiliary staff.

The deputy to the representative of the Secretary-General will act on behalf of the representative and be in contact with the parties through the liaison officer each party will designate for this purpose.

The two headquarters units will be organized into two inspection teams to ascertain on the ground any violation of the instruments comprising the settlement. Whenever considered necessary by the representative of the Secretary-General or his deputy, up to 40 additional military officers (some 10 additional inspection teams) will be redeployed from existing operations within the shortest possible time (normally around 48 hours).

The nationalities of all the officers will be determined in consultation with the parties.

Whenever necessary the representative of the Secretary-General, who will periodically visit the area for consultations with the parties and to review the work of his personnel, will also assign to the area members of his own office and other civilian personnel from the United Nations Secretariat as may be needed. His deputy will alternate between the two headquarters units and will remain at all times in close communication with him.

IV. *Procedure*

(A) Inspections Conducted at the Request of the Parties

(I) A complaint regarding a violation of the instruments of the settlement lodged by any of the parties should be submitted in writing. In the English language, to the respective headquarters units and should indicate all relevant information and details.

(II) Upon receipt of a complaint the deputy to the representative of the Secretary-General will immediately inform the other party of the complaint and undertake an investigation by making on-site inspections, gathering testimony and using any other procedure which he may deem necessary for the investigation of the alleged violation. Such inspection will be conducted using headquarters staff as referred to above, unless the deputy representative of the Secretary-General considers that additional teams are needed. In that case, the parties will, under the principle of freedom of movement, allow immediate access of the additional personnel to their respective territories.

(III) Reports on investigations will be prepared in English and submitted by the deputy representative of the Secretary-General to the two governments, on a confidential basis. (A third copy of the report will be simultaneously transmitted, on a confidential basis, to United Nations headquarters in New York, exclusively for the information of the Secretary-General and his representative). In accordance with paragraph 7 a report on an investigation should be considered in a meeting of the parties not later than 48 hours after it has been submitted. The Deputy Representative of the Secretary-General will, in the absence of the representative, lend his good offices to the parties and in that context he will assist in the organization of the meetings and participate in them. In the context of those meetings the deputy representative of the Secretary-General may submit to the parties for their coinsideration and approval suggestions and recommendations for the prompt, faithful and complete observance of the provisions of the instruments. (Such suggestions and recommendations will be, as a matter of course, consulted with, and cleared by, the representative of the Secretary-General).

(B) Inspections conducted on the initiative of the deputy representative of the Secretary-General

In addition to inspections requested by the parties, the deputy representative of the Secretary-General may carry out on his own initiative and in consultation with the representative inspections he deems appropriate for the purpose of the implementation of paragraph 7. If it is considered that the conclusions reached in an inspection justify a report to the parties, the same procedure used in submitting reports in connection with inspections carried out at the request of the parties will be followed.

Level of Participation in Meetings

As indicated above, the deputy representative of the Secretary-General will participate at meetings of the parties convened for the purpose of considering reports on violations. Should the parties decide to meet for the purpose outlined in paragraph 7 at the high political level, the representative of the Secretary-General will personally attend such meetings.

V. *Duration*

The deputy to the representative of the Secretary-General and the other personnel will be established in the area not later than twenty days before the entry into force of the instruments. The arrangements will cease to exist two months after the completion of all timeframes envisaged for the implementation of the instruments.

VI. *Financing*

The cost of all facilities and services to be provided by the parties will be borne by the respective governments. The salaries and travel expenses of the personnel to and from the area, as well as the costs of the local personnel assigned to the headquarters units, will be defrayed by the United Nations.

Source: TASS 14 April 1988. Kindly supplied by the USSR Embassy.

DY—L

Studies in Geopolitics and Military Science: Postgraduate Courses

EDWARD FOSTER

The author is a researcher on the RUSI Western European Security Programme

THE FOLLOWING is a list of postgraduate options which may be of interest to those wishing to further their academic studies into areas within the scope of this yearbook. The normal requirement for eligibility on a Master's postgraduate course is a Bachelor's degree, class upper second or higher. However, candidates may still be considered favourably by some departments if they can demonstrate educational qualifications and competence appropriate to their proposed course of study. Where Diplomas are awarded, the work involved is substantially the same as for the MA or equivalent, but the Master's degree is conferred on candidates reaching a higher mark and may require a further stage of examination.

POSTGRADUATE COURSES IN STRATEGIC STUDIES

Department of International Relations
Edward Wright Building
University of Aberdeen
Dunbar Street
ABERDEEN AB9 2TY
Tel. (0224) 272725

In association with the Centre for Defence Studies, Aberdeen offers a one-year MLitt course in Strategic Studies for graduates in International Relations, Politics, Economics, Sociology or History, or those with appropriate career experience. The three terms are given over to Strategic Theory, Defence Economics, and Western European Security since 1945 respectively, and normally include visits to NATO (Brussels) and SHAPE. Enquiries should be directed to the course director, J H Wylie.

Graduate School of International Studies
University of Birmingham
P.O. Box 363
BIRMINGHAM B15 2TT
Tel. (021) 4143344

The Master's Degree in International Studies is a one-year taught course involving core studies of the student's choice, one option being Contemporary Military Science. This is combined with two other papers chosen by the student, drawn from a range of mostly economic subjects. MPhil or PhD research is also undertaken at Birmingham for a minimum of two years, and all the graduates may be followed part-time. Details are available from the School's Director, Dr J D Armstrong.

Department of Politics
University of Bristol
BRISTOL BS8 1TU
Tel. (0272) 303030

The course leading to an MSc in International Relations offers programmes in Competing Political Theories, Politics in the Developed World, and those in the Developing World. Strategic Studies and Political Implications of Modern Warfare are two fields available for coverage. Research studies are also undertaken; details are available from the Department.

School of International Studies
University of Cambridge
CAMBRIDGE CB2
Tel. (0223) 335564

A paper on Strategic Studies is one of the options for the one-year course of instruction leading to an MPhil in International Relations. Research of a more extensive nature results in an MLitt or DPhil. Course Director is R T B Langhorn.

Department of International Relations
Keele University
STAFFORDSHIRE ST5 5BG
Tel. (0782) 621111

Although Keele's postgraduate courses do not include any specific Strategic Studies element, there is a paper devoted to Soviet Policy Studies included in the MA in Diplomatic Studies.

This may be completed full-time or part-time; schedules for a research MA may also be negotiated with the Department. Research is also conducted at MPhil or PhD level. Further details can be obtained from the Director, Dr H Suganami.

Department of International Relations
University of Kent
CANTERBURY CT2 7NX
Tel. (0227) 66822

The MA course in International Relations includes examination of conflict resolution, arms control, international security and disarmament. PhD studies may also be undertaken and last for three years. Enquiries to Professor A J Groom.

Department of Politics
Fylde College
University of Lancaster
LANCASTER LA1 4YF
Tel. (0524) 65201

MA courses in International Relations and Strategic Studies and in Defence and Security Analysis both offer papers in Strategic Theory, Defence Analysis, Arms Control Problems, and related subjects. In addition, Western European Security, Civil-Military Relations and Arms Control are three of the subjects available as part of the Politics/International Relations Diploma course. Supervised research and facilities also for MPhil and PhD students. Enquiries should be addressed to the Director of Graduate Studies.

School of History
Leeds University
LEEDS LS2 9JT
Tel. (0532) 333612

The School is strong in military subjects of the nineteenth and twentieth centuries, which may be incorporated in the MA in British History. This may take the form of course work or research. More advanced research may also be conducted leading to MPhil or PhD. Details available on request from the University Admissions Office.

Department of War Studies
King's College
University of London
Strand
LONDON WC2R 2LS
Tel. (01) 836 5454

The MA in War Studies is a remarkably comprehensive one-year course combining both general and specific analysis. It includes set papers in military history and theory, contemporary strategic issues, and a special subject paper on anything from technology to the literature of the First World War. Diploma students complete the same studies, but not the additional specialist essay. The Department's head is Professor L D Freedman.

Department of International Relations
London School of Economics
Houghton Street
LONDON WC2A 2AE
Tel. (01) 405 7686

The MSc course in International Relations includes a paper in Strategic Studies. The course is of one year's duration, or may be completed part-time in two years. LSE also offers Diploma courses taking in Strategic Aspects of International Relations. MPhil or PhD research degrees also awarded. Enquiries should be made to the Graduate Administrator's Office.

Department of International Relations and Politics
North Staffordshire Polytechnic
College Road
STOKE-ON-TRENT ST4 2DE
Tel. (0782) 744531

This college has developed working links with the services, NATO, and SHAPE, and offers graduates in suitable disciplines the chance to engage in research studies to MPhil or PhD subject to close supervision. Enquiries to Professor A E Thorndike.

Department of History
University of Oxford
OXFORD
Tel. (0865) 270000

The MPhil is a two-year instructional course comprised of general, specialist, and research studies. No papers are given over to specific military or strategic subjects, but a privately chosen area of research could easily be combined with study into, say, British Foreign Policy. MLitt research is generally conducted over three years. DPhil over four years. Details are available from the Graduate Admissions Office.

School for European and International Studies
University of Reading
Whiteknights
PO Box 218
READING RG6 2AA
Tel. (0734) 875123

Papers on Strategy and Security in Europe, Defence Economics, Terrorism, and French Foreign/Defence Policy are options open to those on MA courses in European, International, or Defence and Internal Security Studies. These may be completed full-time or part-time. MPhil or PhD courses may also be followed as research. Applications to the Secretary of the Graduate School.

Department of Politics
University of Southampton
SOUTHAMPTON SO9 5NH
Tel. (0703) 559122

The MSc or Diploma in International Studies is a one-year instructional course offering options in Arms Control and International Order, Soviet Foreign Policy, and Revolution and International Order. It may also be taken part-time over two years. MPhil and PhD research may also be conducted at the Department. Details are available from the Postgraduate Studies Office.

Department of International Relations
University of Sussex
Falmer
BRIGHTON BN1 9RH
Tel. (0273) 606755

Suitably qualified graduates may apply for the MA course in International Relations, which includes second-term options such as Defence Policy and Soviet Policy in Eastern Europe. Part-time students take two years over their MA rather than one, while MPhil and DPhil researchers spend their first two terms on taught work. Enquiries should be addressed to the University's Admissions Office.

COURSES IN SPECIALISED MILITARY FIELDS

Department of Military Studies
University of Manchester
MANCHESTER
Tel. (061) 275 2000

Postgraduate studies take the form of MPhil or PhD research into chosen areas of military theory or history carried out under close supervision. Graduates without a suitable academic background (generally a degree in history or politics) may be advised to follow supervised studies into the organisational theory, psychology, philosophy and History of War drawn from the department's course of undergraduate study. Enquiries should be addressed to Professor R Elliott-Bateman.

Royal College of Defence Studies
Seaford House
37 Belgrave Square
LONDON SW1X 8NS
Tel. (01) 235 1091

The RCDS is a military institution with an annual professional course for officers (of staff rank) of the armed forces of the United Kingdom, those of other friendly states, and government officials. Studies are made of world economics, strategic regions, and Allied defence policies.

DEDICATED STUDIES IN PEACE AND ARMS CONTROL

School of Peace Studies
University of Bradford
BRADFORD BD7 1DP
Tel. (0274) 733466

Studies for the MA or Diploma may be followed full-time or part-time. The precise subjects covered vary, but have included nuclear arms, defence economics, conflict resolution, and Northern Ireland. Recent MPhil and PhD research has concentrated on SDI, UK defence policy, and Soviet disarmament policy. The Professor of Peace Studies is Professor J O'Connell.

Richardson Institute for Conflict and Peace Studies
Department of Politics
University of Lancaster
LANCASTER LA1 4YF
Tel. (0524) 65201

The Institute's MA in Peace Studies was the first of its kind to become available at a British university, and is open to qualified graduates or those with experience of "Peace" work. Its papers on the Arms Race supplement those available for Lancaster's MA in International Relations (see above). For details, apply to the Administrator at the Richardson Institute.

Arms Control Chronology, August 1987—July 1988

DR JOHN WALKER

Dr Walker is a Senior Research Officer in the Arms Control and Disarmament Research Unit, Foreign and Commonwealth Office. This contribution is a research study only and does not necessarily reflect the views of Her Majesty's Government.

GLOSSARY

ALCM	—	Air launched cruise missile
BCA	—	Basing Country Agreement
BWC	—	Biological Weapons Convention
BWCRC	—	Biological Weapons Convention Review Conference
CD	—	Conference on Disarmament
CDE	—	Conference on Disarmament in Europe
CORRTEX	—	Continuous Reflectometry for Radius versus Time Experiments
CSCE	—	Conference on Security and Co-operation in Europe
CST	—	Conventional Stability Talks
CWC	—	Chemical Weapons Convention
GLCM	—	Ground launched cruise missile
JDT	—	Joint Draft Text
JVE	—	Joint Verification Experiment
LRINF	—	Longer Range Intermediate Nuclear Forces
MBFR	—	Mutual Balanced Force Reduction
MoU	—	Memorandum of Understanding
NST	—	Nuclear Space Talks
NTB	—	Nuclear Test Ban
NTM	—	National Technical Means
PNET	—	Peaceful Nuclear Explosions Treaty
SLCM	—	Sea launched cruise missile
SRINF	—	Shorter Range Intermediate Nuclear Forces
START	—	Strategic Arms Reduction Talks
SVE	—	Special Verification Commission
TTBT	—	Threshold Test Ban Treaty

CHEMICAL AND BIOLOGICAL WEAPONS

1987

6 August: Soviet Foreign Minister Shevardnadze announces in his speech to the CD that the

Soviet Union would proceed from the need to make legally binding the principle of mandatory challenge inspections without the right of refusal. He also invites all CD Delegations to visit the Soviet Military facility at Shikhany to see standard items of Soviet chemical weapons; and chemical destruction technology at a mobile facility.

25 August: Finland tables CD/785 at the CD entitled "Air Monitoring as a Means for the Verification of Chemical Disarmament". It concludes that a global network of monitoring stations could be used to reveal possible use, field tests of agents, and transport accidents. Clandestine production and stockpiles of CW would be revealed only in the event of leaks of agents due to faults in handling or ageing of the munitions. If such a monitoring network were established, the governments of the States Parties to the Convention could monitor the compliance in their own countries and the possible atmospheric transport of agents from other countries.

27 August: Ad Hoc Committee on Chemical Weapons, under the Chairmanship of Rolf Ekeus of Sweden produces its report (CD/782) in which the latest draft text of the Chemical Weapons Convention is contained.

3–4 October: CD Delegations visit the Soviet CW Research facility at Shikhany during which standard munitions of Soviet CW stockpile are displayed, as well as a mobile destruction facility. Western delegations doubt whether they were shown full picture of Soviet capabilities.

7–8 October: US and Soviet Union hold bilateral talks on chemical weapons proliferation in Berne.

15–18 November: Soviet officials visit West German CW destruction plant near Münster.

19–20 November: Soviet military experts visit the US CW depot at Tooele.

30 November– 16 December:	Informal consultations on Chemical Weapons in Geneva among some members of CD.
30 November– 17 December:	Seventh Round of Soviet-American bilateral talks on Chemical Weapons takes place in Geneva.
10 December:	The Final Communiqué Washington Summit between President Reagan and General-Secretary Gorbachev includes the following on chemical weapons:

> The leaders expressed their commitment to the negotiation of a verifiable, comprehensive and effective international convention on the prohibition and destruction of chemical weapons.

16 December:	Soviet Union tables CD/789 in the CD containing "Information on the Presentation at the Shikhany Military Facility of Standard Chemical Munitions and of Technology for the Destruction of Chemical Weapons at a mobile unit".
26 December:	Soviet Foreign Ministry announces that, "the stocks of chemical weapons in the USSR do not exceed 50,000 tons in terms of poisonous substances". The statement claims further that this corresponds to the chemical weapons stocks of the United States; moreover, all Soviet chemical weapons are located on Soviet territory.

1988

12–29 January:	Intersessional Meeting of CD Ad Hoc Committee on Chemical Weapons held in Geneva.
25 January:	West Germany tables a paper (CD/792) at the Ad Hoc Committee entitled "Super Toxic Lethal Chemicals (STLCs)". It argues that the CD should accept the task of drawing up a list of relevant STLCs on the basis of research work already likely to have been undertaken. This is the only way that parties' obligations can be defined with the necessary position. So that the most compre-

hensive possible security can be obtained against any future STLC production, this list should focus on toxicity alone. The crucial point is that such a list would create clarity regarding the scope of the reporting obligations and of the further controls which should be accepted.

The West Germans also table (CD/791) entitled "Verification of non-production: the case for ad hoc checks". The paper argues that the international authority should be empowered to carry out on its own initiative *Ad Hoc* checks at short notice in production facilities of the chemical industry. These checks should serve solely to ascertain whether, at the time of the check, substances listed in the Annexes to Article VI and not reported for the facility in question were being produced there.

2 February: Ad Hoc Committee on Chemical Weapons reconvenes for the spring session of the CD. Chairman Rolf Ekeus submits revised rolling text to take account of progress made during two informal sessions held during the winter.

5 February: United States tables a working paper CD/802 entitled "Thresholds for monitoring chemical activities not prohibited by a convention". The paper proposes that the threshold quantities for each of the three schedules of chemicals be subject to various degrees of stringent verification procedures according to the risk posed to the Convention. For example, in Schedule I production of only 10 grams/year of ultratoxic substances would be allowed for permitted purposes. The aggregate threshold for Schedule I chemicals for all permitted purposes would be 10 kg/year–1,000 kg/year. These quantities would be produced at the single small-scale facility subject to systematic international on-site verification, including possible continuous monitoring with instruments.

19 February:	Soviet Union tables CD/808—"Memorandum on multilateral data exchange in connection with the elaboration of a convention on the complete and general prohibition of chemical weapons". The memo invites states participating in the negotiations to provide details on whether they possess CW stocks, CW production facilities on past transfers to other states of CW or of technology or equipment for their production.
26 February:	Argentina tables CD/809 in the CD—"Assistance in relation to protection against chemical weapons". The paper suggests and specifies certain criteria which should govern the provision of assistance in relation to protection against chemical weapons. The paper concludes that the suggested guidelines could serve as a basis for the preparation of article X of the draft CWC, so that a detailed and effective provision on assistance may be formulated in order to strengthen the security of the States parties.
4 March:	GDR tables paper (CD/812) entitled "The Executive Council: Composition, size, decision making and other procedural matters". It proposes a membership of 21 to be divided along the lines of the CD, geographic regions, and those states with developed chemical industries; all members to be elected for two-year term.
8–25 March:	Eighth round of Bilateral Soviet–US CW talks are held in Geneva.
29 March:	The FRG tables Paper (CD/822) entitled "The Order of Destruction of Chemical Weapons" suggesting a procedure in which the security of all states would be maintained during the ten year destruction phase. After the levelling out of the large stocks at the end of the fifth year after entry into force, a review of the results achieved so far would be carried out. During the second phase destruction would be carried out in a linear fashion ie the existing stockpile for each CW-

possessor state would be subdivided into the equal reduction amounts to be destroyed during the remaining five years of the destruction period.

31 March: Canada tables a paper (CD/823) entitled "Chemical Weapons Convention: Article VIII Factors involved in determining verification inspectorate personnel and resource requirement". The paper aims to bring into focus the activities that the verification provisions, which are currently set out in CD/782, will require of the International Inspectorate. These provisions necessitate *inter alia* that it inspect and monitor stockpiles, CW destruction facilities, chemical weapons production facilities, and relevant segments of the civilian chemical industry in order to ensure that States Parties are fulfilling their obligations. From these various verification activities, the paper devises an outline sketch of related resource requirements, particularly the skills and types of personnel needed.

11 April: Iran tables paper (CD/827) entitled "Letter dated 11 April 1988 from the Permanent Representative of the Islamic Republic of Iran addressed to the President of the Conference on Disarmament, containing the list of occasions of use of chemical weapons by Iraq against Iran from January 1981 to March 1988".

12 April: FRG tables a paper "Provision of data relevant to the CWC" (CD/828). It notes that provision multilaterally of essential data prior to the signing of the Convention is required. The paper sets out the types of data which need to be submitted in this regard on a voluntary basis by all states.

15 April: Annual exchange of information and data due from State Parties to BWC as agreed at 1986 BWCRC.

19 April: US tables paper (CD/830) entitled "Letter dated 18 April 1988 from the Representative

of the United States of America addressed to the President of the Conference on Disarmament transmitting the text of a document entitled 'Information presented to the visiting Soviet delegation at the Tooele Army Report, 18–21 November 1987'".

25 April: The mission of the medical specialist dispatched by the UN Secretary-General to Iran and Iraq to investigate allegations lodged by both Governments on the use of CW, concludes that the use of such weapons may have intensified.

24–26 May: Thirteen Soviet military civilian and diplomatic officials visit the Chemical Weapons Defence Establishment at Porton Down as part of a bilateral UK–Soviet reciprocal exchange programme. The purpose of these visits is to build greater confidence in support of the Geneva negotiations.

2 June: The President Reagan–General-Secretary Gorbachev Moscow Summit Communiqué observes in relation to chemical weapons, that both sides reaffirm the importance of resolving outstanding problems at the CD negotiations, and stresses that progress would be furthered by concrete solutions and greater openness. Both leaders condemn the proliferation and illegal use of CW in violation of the 1925 Geneva Protocol and express support for international investigations of suspected violations. They call for wider implementation of export controls to inhibit CW proliferation.

7 June: In his speech to the Third UN Special Session on Disarmament the Foreign Secretary Sir Geoffrey Howe made three proposals to address the growing CW proliferation problem. First, all members of the UN—some 50 in all—who have not yet acceded to the Geneva Protocol should immediately do so. Second, procedures should be agreed without delay for investigating automatically allegations of chemical weapons use. Third,

wherever the use of chemical weapons is clearly established, the international community must take effective and speedy action to cut off the supply of key precursors.

29 June–4 July: British Delegation visits Soviet Military CW Facility at Shikhany: the second leg of the exchange programme aimed at building confidence in support of the CWC negotiations in the CD. British subsequently express disappointment at the attitude to secrecy in the Soviet Union.

11–29 July: Ninth US–Soviet bilateral round of talks on chemical weapons is held.

20 July: Report of the Mission [1–5 July] dispatched by the UN Secretary-General to Investigate Allegations of Use of Chemical Weapons in the conflict between the Islamic Republic of Iran and Iraq concludes *inter alia* that such use has become more intense and frequent.

25 July: Report of the Mission [10–11 July–to Iraq] dispatched by the UN Secretary-General to investigate Allegations of the Use of Chemical Weapons in the conflict between the Islamic Republic of Iran and Iraq concludes *inter alia* that on the basis of clinical examinations of nine Iraqi soldiers, the Mission was able to determine conclusively that their injuries have been produced by mustard gas; and on the basis of this investigation the number of casualties and the extent of their injuries seemed less extensive than in previous investigations.

28 July: US tables paper in CD entitled "Destruction of Chemical Weapons Production Facilities" (CD/849). The paper's purpose was to present for discussion some general concepts concerning destruction methods. These were placed under the following headings: general procedures; destruction of supertoxic chemical facilities; destruction of non-supertoxic chemical facilities; demolition of buildings; demolition of non-chemical facilities and equipment; time and manpower require-

ments; environmental requirements; and monitoring of destruction. Also gave details of five US CW production facilities and their functions.

CONVENTIONAL ARMS CONTROL

1987

28 August:	US inspects Soviet military exercise near Minsk under terms of Stockholm Document.
10–12 September:	UK inspects Soviet/GDR military exercise in Cottbus–Juterborg under terms of Stockholm Document.
22 September:	CSCE Review Conference re-opens in Vienna. Yugoslavia submits proposal for a further stage of CDE. Cyprus also tables a proposal.
24 Sepbember:	43rd Round of MBFR begins in Vienna.
26 September–6 October:	Czechoslovak observers inspect US "Reforger 87" exercise in the FRG.
28 September:	Group of 23 (NATO and WTO) talks on the mandate for the Conventional Stability Talks resume (CST).
5 October:	Warsaw Pact tables language at CST expanding on earlier working document; NATO recirculated "objectives" section of an earlier paper.
5–7 October:	Soviet Union inspects Turkish/US exercise under terms of Stockholm Document.
19 October:	Warsaw Pact tables new approach to inclusion of nuclear weapons in the CST Mandate.
28–30 October:	Soviet Union inspects US military exercise near Nuremberg under terms of Stockholm Document.
11–13 November:	GDR inspects West German military exercise near Kassel.
3 December:	43rd Round of MBFR ends.
10 December:	Reagan–Gorbachev Summit Communiqué section on conventional arms control notes that:

The President and the General Secretary discussed the importance of the task of reducing the level of military confrontation in Europe in the area of armed forces and conventional armaments. The two leaders spoke in favor of early completion of the work in Vienna on the mandate for negotiations on this issue, so that substantive negotiations may be started at the earliest time with a view to elaboration concrete measures. They also noted that the implementation of the provisions of the Stockholm Conference on Confidence- and Security Building Measures and Disarmament in Europe is an important factor in strengthening mutual understanding and enhancing stability, and spoke in favor of continuing and consolidating this process. The President and the General Secretary agreed to instruct their appropriate representatives to intensify efforts to achieve solutions to outstanding issues.

They also discussed the Vienna (Mutual and Balanced Force Reduction) negotiations.

11 December:	NATO/WTO agree a statement of objectives and methods for proposed CST.
14 December:	Group of 23 adjourns. Agreement reached on several items for CST mandate; for example, the objective would be establishment of a stable and secure balance of "conventional arms and equipment at a lower level." The potential for surprise attack and large-scale offensives would be diminished.
18 December:	CSCE adjourns.

1988

22 January:	CSCE Review Conference re-convenes.
25 January:	Group of 23 resumes in Vienna.
27 January–17 March:	44th Round of MBFR.
4–5 February:	US inspects Hungarian/Soviet military exercise.
22 February:	WTO proposes in G23 forum that future negotiations should be based on principles of equal rights.

2–3 March: NATO Leaders Summit meeting held in Brussels. Summit statement in relation to conventional arms control emphasises the need for conventional arms control in Europe to strengthen stability. Conventional arms control in Europe is not merely a technical corrective to a self-contained problem; it should be seen in a coherent political and security framework. NATO's objectives in conventional stability negotiations will be: establishment of a secure and stable balance of conventional forces at lower levels; elimination of disparities prejudicial to stability and security; and as a matter of high priority, the elimination of the capability for launching surprise attack and for initiating large-scale offensive action.

8 March: Warsaw Pact tables a draft military section for the CSCE Final Document.

13–15 March: Soviet Union inspects NATO exercises in North Norway.

21 March: Preamble for Mandate for conventional stability talks agreed by Group of 23 (G23).

23 March: G23 reach agreements on verification and data language for Mandate.

24 March: Neutral and Non-Aligned (NNA) states table a draft military security text for the CSCE Final Document.

25 March: Fifth Session of CSCE Review Conference ends. G23 round also closes.

30 March: Warsaw Pact Foreign Ministers call for exchange of data with NATO, and other measures designed to improve security in Europe during their meeting in Sofia. The communiqué also noted that the subject for the CST should be "the armed forces and conventional armaments and military equipment, including dual-capable means without their nuclear component. As to the nuclear component itself, it could become the subject of separate negotiations which should not be put off indefinitely".

10–12 April: US inspects military exercise in GDR.

15 April:	CSCE Review Reconvenes.
20 April:	G23 Mandate talks reconvene in Vienna.
25 April:	G23 agrees that future CST would be concluded within the framework of the CSCE process, but confined to the 23 NATO and WTO states.
13 May:	NNA produce draft text for a compromise Draft First Document of the CSCE Review Conference.
19 May:	45th Round of MBFR begins.
29 May–2 June:	During Summit meeting in Moscow, Gorbachev suggests a proposal for data exchange, verification of the exchange, and then cuts of 300,000 personnel on each side.
15 June:	Poland circulates revised version of "Jaruzelski Plan" to CSCE States' Embassies in Warsaw. The plan includes *inter alia* a separate negotiation on nuclear weapons not covered by other agreements; no increase of existing levels of nuclear weaponry; "establishment of a zone of dispersed armaments, in which a mutually agreed number of units of equipment combat force . . . roughly comparable quantities of weapons and state of readiness could be deployed at an appropriate distance from the line of contact".
30 June:	45th Round of MBFR ends.

NUCLEAR AND SPACE TALKS (NST): START/ABM–BMD

1987

15–18 September:	Shultz and Shevardnadze meet in Washington. In discussion on *START* issues the Russians state that a 50 per cent cut in SS-18 warheads would be permanent; a 60 per cent ceiling on ICBM warheads was acceptable; bomber counting rules should be maintained, but they would not accept 80–85 per cent limit on ballistic missile warheads; they would make a unilateral statement that they

would cut the throw weight of its ballistic missile force; proposed a limit of 400 on SLCMs.

SPACE: Russians offer a detailed list of objects which should not be launched into space, including limits on size of mirrors and speed of interceptors. They also offer to affirm traditional interpretation of ABM Treaty.

Shultz and Shevardnadze also sign agreement to establish Nuclear Risk Reduction Centres in Moscow and Washington. It provides for notifications of ICBM launchers and transmission of other information.

22–23 October: During Shultz's visit to Moscow the Russians propose a new formula on sub-limits for warheads within the agreed total of 6,000 (3,000–3,300 ICBM warheads; a limit of 1,800–2,000 on SLBM warheads; and no more than 800–900 ALCMs).

7–10 December: In a joint statement issued at the end of their Summit meeting in Washington, President Reagan and Mr Gorbachev announce that considerable progress was made towards an agreement in 50 per cent reductions in strategic nuclear weapons. The negotiators at Geneva were instructed to speed up their work on a treaty text, preferably in time for signature at their next Summit meeting. They should build on the joint draft treaty text being developed in Geneva, which included agreement on ceilings of no more than 1,600 strategic offensive delivery systems, 6,000 warheads, 1,540 warheads on 154 heavy missiles; 4,900 warheads on ballistic missiles; agreed counting rules for heavy bombers and their nuclear weapons arms; and agreement that the aggregate throw weight of the Soviet Union's ICBM/SLBMs will be reduced and should not exceed 50 per cent of the existing level.

SPACE: The relevant section in the communiqué observed that:

Taking into account the preparation of the Treaty on Strategic Offensive Arms, the leaders of the two countries also instructed their delegations in Geneva to work out an agreement that would commit the sides to observe the ABM Treaty, as signed in 1972, while conducting their research, development, and testing as required, which are permitted by the ABM Treaty, and not to withdraw from the ABM Treaty for a specified period of time. Intensive discussions of strategic stability shall begin no later than three years before the end of the specified period, after which, in the event the sides have not agreed otherwise, each side will be free to decide its course of action. Such an agreement must have the same legal status as the Treaty on Strategic Offensive Arms, the ABM Treaty, and other similar, legally binding agreements. This agreement will be recorded in a mutually satisfactory manner. Therefore, they direct their delegations to address these issues on a priority basis.

The sides shall discuss ways to ensure predictability in the development of the US–Soviet strategic relationship under conditions of strategic stability, to reduce the risk of nuclear war.

1988

14 January: Ninth Round of US–Soviet NST begins in Geneva.

15 January: Soviet Union tables draft protocol to the proposed START Treaty which would tie the reductions in nuclear forces directly to constraints on strategic defences and impose restrictions on the SDI programme.

22 January: US tables Draft Treaty on defence and space issues which would be quite separate from the START agreement.

12 February: US tables a draft protocol on inspection at a plenary meeting of START talks.

15 February: Soviet Union tables new proposals for compliance monitoring in START.

22 February: Joint statement issued after Shevardnadze and Shultz meeting in Moscow notes that:

The Ministers reviewed the entire complex of issues associated with the Treaty, with a particular focus on finding mutually acceptable solutions to differences which remain. Emphasising the importance of verification, they directed their negotiators to develop, by the time of the March Foreign Ministers' Meeting: a joint draft Protocol on Inspection, a joint draft Protocol on Conversion or Elimination of strategic offensive arms, and a joint draft memorandum of understanding which will be integral to the Treaty on Reduction and Limitation of Strategic Offensive Arms.

2 March:	US tables a draft Memorandum of Understanding on data exchange.
7 March:	Soviet Union tables drafts for procedures on weapon elimination and inspection of weapon sites.
11 March:	Soviet Union tables proposal on data exchanges.
21–23 March:	Shultz and Shevardnadze meet in Washington, the second such meeting in 1988. Their communiqué noted that:

They reaffirmed their strong commitment made in the Washington Summit joint statement to make an intensive effort to complete a treaty on the reduction and limitation of strategic offensive arms and all integral documents at the earliest possible date, preferably in time for signature of the Treaty during the next meeting of the two leaders. The Ministers reviewed the joint draft texts of a Protocol on Inspection, a Protocol on Conversion or elimination of strategic offensive arms, and a Memorandum of Understanding developed in accordance with their directive at the February Ministerial in Moscow. Re-emphasising their commitment to effective verification measures, they agreed that the Soviet and US negotiators in Geneva will seek to resolve the remaining differences in these documents and report on progress at the next Ministerial.

SPACE: The Americans propose new measures for testing in space. Two Ministers

	direct negotiators to expedite preparation of JDT of a separate agreement building on the 10 December joint statement on space and ABM Treaty.
30 March:	United States proposes package of "predictability measures" in the Space Group of the NST.
31 March:	The Soviet Union proposes measures on inspection of space launchers. These included *inter alia* creation of an International Space Inspectorate; provision of advance information on launches; and permanent presence of inspector groups at all launch sites.
8 April:	Soviet Union proposes measures for constraining SLCMs and mobile ICBMs at the START talks. US proposes measures to eliminate discrepancies in the JDT.
21–22 April:	Third Ministerial meeting, 1988 between Shultz and Shevardnadze, this time in Moscow. The two sides discussed *inter alia* the remaining differences on such questions as verification and counting of long-range air-launched cruise missiles, limitation and verification of long-range sea-based cruise missiles fitted with nuclear warheads, sublimits and mobile ICBMs.
	SPACE: Russians present a space weapons draft treaty in the arms control working group. There is also detailed discussion of the preparation of the JDT of a separate agreement on the basis of the formula contained in the December Summit joint statement.
11–12 May:	Fourth Shultz–Shevardnadze meeting is held in Geneva. Little attention on START, main focus rests on INF. Soviet side proposes 1,600 warheads and 800 launchers for mobile missiles.
29 May–2 June:	Gorbachev–Reagan Summit in Moscow.
31 May:	Shultz and Shevardnadze sign a ballistic missile test notification agreement in which each side would provide, for all launchings of

land-based and sea-based ballistic missiles, at least 24 hours notice of a date of launching, the launching area and impact area.

12 July: Tenth Round of START talks begin in Geneva. Soviet side tables draft protocol on the summary throw-weight of ICBMs and SLBMs, which takes into account the considerations expressed by the American delegation.

INTERMEDIATE/SHORTER RANGE NUCLEAR FORCES

1987

25 August: The United States tables revised and simplified verification proposal in INF negotiations. This was occasioned by Soviet acceptance of the "double zero" option (ie no LRINF and SRINF globally). The proposals were based upon the elimination of all shorter-range INF missile systems within one year and the elimination of all longer-range INF systems within three years; and, a ban on modernisation, production and operational test flights of these missiles.

26 August: Chancellor Kohl announces that provided an INF agreement was concluded, ratified and the contracting parties kept to the agreed time table for the elimination of their systems, he was prepared to state that the Pershing 1A Missiles owned by West Germany would not be modernised, but scaled down.

2 September: The US State Department announces that if the conditions set out in Chancellor Kohl's August statement were met, the US would withdraw the nuclear warheads for the 72 Pershing 1A missiles.

10 September: The US tables revised Treaty language on verification, pace of reductions, and scope of the Treaty in the INF negotiations. These included a ban on conventionally armed

GLCMs; missile destruction to begin immediately Agreement enters into force. The verification proposal called for exchange of baseline data on missile numbers, launchers and support facilities; an initial inspection of facilities to check declared data. Missile destruction would be subject to mandatory on-site inspection.

15–18
September:

US Secretary of State George Shultz and Soviet Foreign Minister Eduard Shevardnadze met in Washington. In their joint statement, issued on Friday 18 September, it was announced that both sides had instructed their delegations to "work intensively to resolve remaining technical issues and promptly to complete a draft treaty". A further meeting between the two Ministers would take place in Moscow in late October to review the results of the negotiations.

22–3
October:

Shultz and Shevardnadze meet in Moscow. Agreement reached on the phasing of missile reductions over 12–18 months for SRINF and over three years for LRINF. Some of the detail of the complex verification regime remained unresolved.

30 October:

In a joint statement the US and Soviet Union announce that General Secretary Gorbachev would visit the US beginning 7 December for two or three days and President Reagan would visit Moscow in the first half of 1988. It was expected, that if the INF negotiations were completed, and the verification details satisfactorily resolved the two leaders would sign the treaty.

15–17
November:

Ambassador Voronstov and Ambassador Kampelman, Heads of the Soviet and US delegations meet in Geneva to work out the outstanding difficulties facing the INF treaty.

18 November:

Soviet Union provides further data on its INF/SRINF Missiles necessary for the Treaty.

23–4 November:	Further Shultz–Shevardnadze meeting held to prepare for the Summit meeting in December.
25 November:	NATO Foreign Ministers agree to halt GLCM deployment as soon as the INF Treaty is signed.
8 December:	At their Summit Meeting in Washington, President Reagan and Mr Gorbachev signed the INF Treaty eliminating all US and Soviet land-based INF missiles worldwide. The Treaty's provisions are set out in Seventeen Articles which set out *linter alia* the destruction process phases, elimination procedures (how the missiles were to be physically destroyed), the inspection regime including base-line, and close-out inspections; and 20 challenge inspections for the first three years. The Treaty also creates a Special Verification Commission to resolve questions on implementation and agree on further measures. Accompanying the main treaty were two Protocols: one on Eliminations, and one on Inspections. Furthermore, a Memorandum of Understanding regarding the establishment of the data base was also signed and is integral to the Treaty. Finally, the details of the inspection provisions as they relate to the basing countries are contained in a document signed between the US, the UK, FRG, Belgium, the Netherlands and Italy to establish the practical procedures and provisions. There was an exchange of notes between the UK and the Soviet Union granting the Soviet Government the right to conduct inspections on British territory. The Soviet government undertakes in return to comply with British laws and procedures. A similar procedure took place for all basing countries, including the GDR and Czechoslovakia who exchanged notes with the US.
11 December:	Signature in Brussels of Basing Country

Agreement between US, UK, FRG, Belgium, Italy and Netherlands. BCA governs conduct of INF Treaty inspections in European Basing countries.

1988

22 January: The Soviet Council of Ministers approves the Treaty.

25 January: The US Senate, Foreign Relations Committee and the Armed Services Committee begin their discussions on treaty ratification. Secretary of State Shultz provides an hour long prepared testimony to the Foreign Relations Committee, and Defense Secretary Carlucci testifies before the Armed Services Committee. Both support the Treaty, and reject criticism levelled at the Agreement by its opponents. The Supreme Soviet's Presidium adopts a resolution on the Treaty, praising it, and passing it to the Foreign Affairs Commissions of the Soviet of the Union and the Soviet of Nationalities.

9 February: A Joint Session of the Supreme Soviet's Foreign Affairs Commissions hears testimonies in support of the Treaty from Foreign Minister Shevardnadze and Defence Minister Yazov. The Two Foreign Relations Commissions form a Preparatory Commission as their working organ, with 10 deputies to draw up proposals and recommendations.

25 February: The Soviet Union begins withdrawal of its SS-12 SRINF launchers from the GDR.

9–12 March: US and Soviet officials meet in Moscow to discuss the implementation details of the Treaty verification provisions.

12–13 March: The first Soviet SRINF missiles removed from Eastern Europe at Sary Ozek, north of Alma Alta in Soviet Kazakhstan, where they will be destroyed.

28 March: Senate Armed Services Committee votes 18–2 in favour of the INF Treaty.

30 March: Senate Foreign Relations Committee sends the Treaty to the floor of the Senate on a vote of 17–2 after 21 days of hearings in

which twenty-nine witnesses provided testimonies.

12 April: US and Soviet officials meet in Washington in
a further effort to finalise the practicalities of
verification. These included the logistics of
moving personnel and equipment. The US
side was led by Brigadier-General Roland
Lajoie, the Soviet team by General Vladimir
Medvedev.

14 April: Secretary of State George Shultz requests
from Shevardnadze a definitive clarification
from the Soviet Union that the Treaty bans
"futuristic weapons".

15 April: Shevardnadze confirms in a letter to the
State Department that "the definitive view of
the Soviet side is that the Treaty on the
elimination of intermediate range and shorter-range missiles bans those two classes of
missiles, however equipped, nuclear or non-
nuclear".

26 April: The House of Commons passes the Arms
Control and Disarmament (Privileges and
Immunities) Act 1988. The Act is principally
concerned with the granting of diplomatic
privileges and immunities to observers and
inspectors carrying out functions in the UK
under the terms of the Stockholm agreement. Its provisions will also cover Soviet
inspectors under the INF Treaty.

28 April: The Senate informs the Reagan Administration that a US–Soviet agreement on "exotic-
weapons" was required before the Senate
would go forward to vote on ratification.

29 April: Senator Byrd states that the floor debate on
the Treaty would not begin until 11 May.
There were three areas of concern in the
Treaty's provisions: weapons definition;
futuristic weapons, inspection procedures;
and separately the need to improve US
indigenous NTM capabilities.

9 May: Senator Byrd says Treaty would not be
brought to the floor until the inspection
differences were resolved.

11–12 May: Shultz and Shevardnadze meet in Geneva. In the course of their discussions and those amongst their experts, the outstanding issues are resolved. These covered four main verification issues: SS-20 stages, the US right to look inside structures large enough to hide a SS-20 stage; access to structures inside missile sites and launcher installations; and inspection of missile canisters at the Soviet production facility at Votkinsk; the Russians agree to abandon their demand for a veto over the use of photographic equipment.

20 May: Third meeting on inspection practicalities is held.

21–22 May: The US and USSR exchanged notes with respect to site diagrams in the Treaty's MoU.

23 May: The Foreign Affairs Commissions of the Chambers of the Supreme Soviet unanimously recommend ratification of the Treaty after receiving the report of the Preparatory Commission.

28 May: The Presidium of the Supreme Soviet ratifies INF Treaty.

1 June: General-Secretary Gorbachev and President Reagan sign the Protocols on INF Treaty Ratification in Moscow.

6 June: The Special Verification Commission created by the INF Treaty holds its first session in Geneva.

8 June: In his speech at the Third UN Special Session on Disarmament, Shevardnadze invites international representatives to witness Soviet destruction of some of their SS-20, 12 and 22 missiles.

13 June: The US and USSR exchange final data on their INF/SRINF forces now the Treaty has entered into force.

1 July: Reciprocal inspection in accordance with the INF Treaty begins. First US baseline on-site inspection team arrives in Moscow.

2 July: Seventy Soviet baseline inspectors arrive at Travis Air Force Base in California. Twenty-

two would go to the Magna Plan in Utah with the remaining 48 divided into groups to carry out inspections at all, or some, of the five facilities in the Western USA where GLCMs/PIIs are to be destroyed.

4 July: SS-12 missiles in the Soviet Far-East are being dismantled at a base near Novosysoyevka before being sent to sites for destruction.

5–6 July: Soviet inspectors arrive in West Germany to inspect Pershing II and GLCM bases at Mutlangen and Wueschheim.

19 July: First visit of Soviet inspectors to the UK under the INF Treaty began with the arrival of a team of 20 at Greenham Common. Inspectors also visit Molesworth.

22 July: The first explosion of a SS-20 missile held at Kapustin Yar in the presence of US inspectors.

27 July: SVC completes its first meeting in Geneva. The meeting achieved major progress in elaborating measures to facilitate the effective implementation of INF Treaty provisions.

NUCLEAR TESTING

1987

28 August: The CD Summer Session ends with no Ad Hoc Committee on the NTB being convened owing to inability of CD to agree mandate.

16 September: During Shultz–Shevardnadze meeting in Washington, Russians announce in a Press Briefing that they were ready to accept four 1 kiloton tests a year.

17 September: The US and Soviet Union in the course of the Washington Foreign Ministers meeting announce an agreed mandate for nuclear testing talks:

> The sides, as the first step, will agree upon effective verification measures, and proceed to negotiating further intermediate limitations on

nuclear testing, leading to the ultimate objective of the complete cessation of nuclear testing, as part of an effective disarmament process.

The talks were to begin before 1 December.

9 November: US–Soviet Nuclear Testing talks begin in Geneva. US Delegation leader Robert Barker, Soviet leader Igor Palenykh.

20 November: First round of testing talks end with agreement to exchange of visits at their respective test sites in January to familiarise each side with the conditions and operations at the other's test site. Joint experiments would be held in about six months time in which the US would use the CORRTEX system for evaluating the yield of a Soviet nuclear test, and the Soviet Union would use teleseismic methods to monitor a US test.

10 December: At the Reagan–Gorbachev Summit in Washington the text of the agreement on exchange visits and the Joint Verification Experiments (JVEs) is issued. It notes that "The experiment will ... provide the basis for agreeing on those verification measures which could be used by either side to verify compliance by the other side with the provisions of the 1974 (TTBT) and 1976 (PNET) treaties."

1988

11–15
January: US experts visit Soviet test site at Semipalatinsk.

25–30
January: Soviet experts visit the US Department of Energy's Nuclear Test Site at Nevada.

15 February: Second round of the Nuclear Testing Talks begins.

16 February: Australia suggests that the CD agree to a mandate for an Ad Hoc Committee on Testing which should be based on the terms of UN First Committee Resolution A/RES/42/27.

7–18 March: The Ad Hoc Group of Seismic Experts holds its first meeting of the year in Geneva. The Group was designing an international data exchange system and planning the conduct of a large scale experiment.

9 March: US tables draft Protocol for the TTBT containing revised verification procedures. It included the right to on-site observation of all tests over 50 kilos. In the event of no tests being conducted above that level, the US would have the right to inspect the two largest tests below 50 kt.

18 March: The Soviet Union tables its own supplementary Protocol to the TTBT at the nuclear testing talks.

23 March: In the Communiqué following a Shultz–Shevardnadze meeting, the two ministers instruct their delegations to design and conduct as soon as possible the JVE; complete a detailed plan and schedule for the JVE by the April Ministerial meeting; prepare a joint draft of the TTBT Protocol by the time of the JVE to be finalised through the conduct and analysis of the experiment; and to accelerate work on verification issues for the PNET.

31 March: US tables draft PNET Protocol at the Nuclear Testing Talks in Geneva.

13 April: Soviet Union tables its draft PNET Protocol.

19 April: The Group of 21 table mandate for Ad Hoc Committee on Nuclear Test Ban at the CD (same as July 1987 and as such unacceptable to West).

21–22 April: Shultz–Shevardnadze Ministerial in Moscow. Communiqué states in relation to the Testing Talks:

> Having discussed the range of questions related to bilateral full-scale talks on problems of nuclear tests the ministers approved the text of the agreement between the United States of America and the Union of Soviet Socialist Republics, initialled by their representatives at the Geneva talks, on the holding of the joint verification experiment. They instructed their delegations at the talks to complete as speedily as possible the work on the supplement to the

agreement so that the agreement together with the supplement could be signed within the shortest possible period of time.

The ministers also approved a rough schedule of preparations for and the holding of the joint verification experiment at the testing site in Nevada and the testing site in Semipalatinsk.

31 May: Shultz and Shevardnadze sign the JVE Agreement at the Moscow Summit. The test in Nevada would take place on 17 August and in Semipalatinsk on 14 September. The Summit Joint Statement notes inter alia that "they confirmed their understanding that verification measures for the TTBT will, to the extent appropriate, be used in further nuclear test limitation agreements which may subsequently be reached."

28 June: The Second Round of Nuclear Testing Talks end. The sides are now close to agreement on the verification protocol for the PNET. Talks to resume after second JVE in late September.

17 July: American officials arrive in Soviet Union with CORRTEX equipment for the JVE.

BIBLIOGRAPHY AND FURTHER READING

The Arms Control Reporter. A Chronicle of Treaties, Negotiations, Proposals, Weapons and Policy 1987–88. Institute for Defence and Disarmament Studies. Editor Chalmers Hardenbergh (Brookline, Mass.)

Arms Control Today. A Publication of the Arms Control Association, Dupont Circle Washington DC. Published monthly except for two bimonthly issues appearing July/August and January/February.

ADIU Report. Armament and Disarmament Information Unit, Science Policy Research Unit, University of Sussex. Published bimonthly.

Arms Control and Disarmament Quarterly Review. Arms Control and Disarmament Research Unit, FCO. Numbers 7 (October 1987)–10 (July 1988).

Notes on Arms Control. Arms Control and Disarmament Research Unit, FCO. Numbers 1 (January/February)–6 (July) 1988.

Special Report of the Ad Hoc Committee on Chemical Weapons to the Conference on Disarmament CD/831, 20 April 1988.

Summary of World Broadcasts Part 1 The USSR. Printed and published by the monitoring service of the BBC (August 1987–July 1988).

Strategic Survey 1987–1988. International Institute for Strategic Studies, London.

Soviet News. Published by the Press Department of the Soviet Embassy in London.

Official Texts. United States Information Service, US Embassy,

Advertising Office

For information regarding availability and cost of advertising space in this and other Brassey's publications please contact:

Richard A Ewin or David Harrison
Overseas Publicity Ltd
46 Keyes House
Dolphin Square
London SW1V 3NA

Tel: 01–834–5566
Telex: 24924
Fax: 01–630–5878